THE MATHEMATICAL
THEORY OF
TONE SYSTEMS

PURE AND APPLIED MATHEMATICS

A Program of Monographs, Textbooks, and Lecture Notes

MONOGRAPHS AND TEXTBOOKS IN
PURE AND APPLIED MATHEMATICS

116. *H. Strade and R. Farnsteiner*, Modular Lie Algebras and Their Representations (1988)
117. *J. A. Huckaba*, Commutative Rings with Zero Divisors (1988)
118. *W. D. Wallis*, Combinatorial Designs (1988)
119. *W. Więslaw*. Topological Fields (1988)
120. *G. Karpilovsky*, Field Theory (1988)
121. *S. Caenepeel and F. Van Oystaeyen*, Brauer Groups and the Cohomology of Graded Rings (1989)
122. *W. Kozlowski*, Modular Function Spaces (1988)
123. *E. Lowen-Colebunders*, Function Classes of Cauchy Continuous Maps (1989)
124. *M. Pavel*, Fundamentals of Pattern Recognition (1989)
125. *V. Lakshmikantham et al.*, Stability Analysis of Nonlinear Systems (1989)
126. *R. Sivaramakrishnan*, The Classical Theory of Arithmetic Functions (1989)
127. *N. A. Watson*, Parabolic Equations on an Infinite Strip (1989)
128. *K. J. Hastings*, Introduction to the Mathematics of Operations Research (1989)
129. *B. Fine*, Algebraic Theory of the Bianchi Groups (1989)
130. *D. N. Dikranjan et al.*, Topological Groups (1989)
131. *J. C. Morgan II*, Point Set Theory (1990)
132. *P. Biler and A. Witkowski*, Problems in Mathematical Analysis (1990)
133. *H. J. Sussmann*, Nonlinear Controllability and Optimal Control (1990)
134. *J.-P. Florens et al.*, Elements of Bayesian Statistics (1990)
135. *N. Shell*, Topological Fields and Near Valuations (1990)
136. *B. F. Doolin and C. F. Martin*, Introduction to Differential Geometry for Engineers (1990)
137. *S. S. Holland, Jr.*, Applied Analysis by the Hilbert Space Method (1990)
138. *J. Okniński*, Semigroup Algebras (1990)
139. *K. Zhu*, Operator Theory in Function Spaces (1990)
140. *G. B. Price*, An Introduction to Multicomplex Spaces and Functions (1991)
141. *R. B. Darst*, Introduction to Linear Programming (1991)
142. *P. L. Sachdev*, Nonlinear Ordinary Differential Equations and Their Applications (1991)
143. *T. Husain*, Orthogonal Schauder Bases (1991)
144. *J. Foran*, Fundamentals of Real Analysis (1991)
145. *W. C. Brown*, Matrices and Vector Spaces (1991)
146. *M. M. Rao and Z. D. Ren*, Theory of Orlicz Spaces (1991)
147. *J. S. Golan and T. Head*, Modules and the Structures of Rings (1991)
148. *C. Small*, Arithmetic of Finite Fields (1991)
149. *K. Yang*, Complex Algebraic Geometry (1991)
150. *D. G. Hoffman et al.*, Coding Theory (1991)
151. *M. O. González*, Classical Complex Analysis (1992)
152. *M. O. González*, Complex Analysis (1992)
153. *L. W. Baggett*, Functional Analysis (1992)
154. *M. Sniedovich*, Dynamic Programming (1992)
155. *R. P. Agarwal*, Difference Equations and Inequalities (1992)
156. *C. Brezinski*, Biorthogonality and Its Applications to Numerical Analysis (1992)
157. *C. Swartz*, An Introduction to Functional Analysis (1992)
158. *S. B. Nadler, Jr.*, Continuum Theory (1992)
159. *M. A. Al-Gwaiz*, Theory of Distributions (1992)
160. *E. Perry*, Geometry: Axiomatic Developments with Problem Solving (1992)
161. *E. Castillo and M R. Ruiz-Cobo*, Functional Equations and Modelling in Science and Engineering (1992)
162. *A. J. Jerri*, Integral and Discrete Transforms with Applications and Error Analysis (1992)
163. *A. Charlier et al.*, Tensors and the Clifford Algebra (1992)
164. *P. Biler and T. Nadzieja*, Problems and Examples in Differential Equations (1992)
165. *E. Hansen*, Global Optimization Using Interval Analysis (1992)
166. *S. Guerre-Delabrière*, Classical Sequences in Banach Spaces (1992)
167. *Y. C. Wong*, Introductory Theory of Topological Vector Spaces (1992)
168. *S. H. Kulkarni and B. V. Limaye*, Real Function Algebras (1992)
169. *W. C. Brown*, Matrices Over Commutative Rings (1993)
170. *J. Loustau and M. Dillon*, Linear Geometry with Computer Graphics (1993)
171. *W. V. Petryshyn*, Approximation-Solvability of Nonlinear Functional and Differential Equations (1993)
172. *E. C. Young*, Vector and Tenser Analysis: Second Edition (1993)
173. *T. A. Bick*, Elementary Boundary Value Problems (1993)
174. *M. Pavel*, Fundamentals of Pattern Recognition: Second Edition (1993)
175. *S. A. Albeverio et al.*, Noncommutative Distributions (1993)
176. *W. Fulks*, Complex Variables (1993)

177. *M. M. Rao*, Conditional Measures and Applications (1993)
178. *A. Janicki and A. Weron*, Simulation and Chaotic Behavior of α-Stable Stochastic Processes (1994)
179. *P. Neittaanmäki and D. Tiba*, Optimal Control of Nonlinear Parabolic Systems (1994)
180. *J. Cronin*, Differential Equations: Introduction and Qualitative Theory, Second Edition (1994)
181. *S. Heikkilä and V. Lahshmikantham*, Monotone Iterative Techniques for Discontinuous Nonlinear Differential Equations (1994)
182. *X. Mao*, Exponential Stability of Stochastic Differential Equations (1994)
183. *B. S. Thomson*, Symmetric Properties of Real Functions (1994)
184. *J. E. Rubio*, Optimization and Nonstandard Analysis (1994)
185. *J. L. Bueso et al.*, Compatibility, Stability, and Sheaves (1995)
186. *A. N. Michel and K. Wang*, Qualitative Theory of Dynamical Systems (1995)
187. *M. R. Darnel*, Theory of Lattice-Ordered Groups (1995)
188. *Z. Naniewicz and F. D. Panagiotopoulos*, Mathematical Theory of Hemivariational Inequalities and Applications (1995)
189. *L. J. Corwin and R. H. Szczarba*, Calculus in Vector Spaces: Second Edition (1995)
190. *L. H. Erbe et al.*, Oscillation Theory for Functional Differential Equations (1995)
191. *S. Agaian et al.*, Binary Polynomial Transforms and Nonlinear Digital Filters (1995)
192. *M. I. Gil'*, Norm Estimations for Operation-Valued Functions and Applications (1995)
193. *P. A. Grillet*, Semigroups: An Introduction to the Structure Theory (1995)
194. *S. Kichenassamy*, Nonlinear Wave Equations (1996)
195. *V. F. Krotov*, Global Methods in Optimal Control Theory (1996)
196. *K. I. Beidar et al.*, Rings with Generalized Identities (1996)
197. *V. I. Amautov et al.*, Introduction to the Theory of Topological Rings and Modules (1996)
198. *G. Sierksma*, Linear and Integer Programming (1996)
199. *R. Lasser*, Introduction to Fourier Series (1996)
200. *V. Sima*, Algorithms for Linear-Quadratic Optimization (1996)
201. *D. Redmond*, Number Theory (1996)
202. *J. K. Beem et al.*, Global Lorentzian Geometry: Second Edition (1996)
203. *M. Fontana et al.*, Prüfer Domains (1997)
204. *H. Tanabe*, Functional Analytic Methods for Partial Differential Equations (1997)
205. *C. Q. Zhang*, Integer Flows and Cycle Covers of Graphs (1997)
206. *E. Spiegel and C. J. O'Donnell*, Incidence Algebras (1997)
207. *B. Jakubczyk and W. Respondek*, Geometry of Feedbade and Optimal Control (1998)
208. *T. W. Haynes et al.*, Fundamentals of Domination in Graphs (1998)
209. *T. W. Haynes et al.*, eds., Domination in Graphs: Advanced Topics (1998)
210. *L. A. D'Alotto et al.*, A Unified Signal Algebra Approach to Two-Dimensional Parallel Digital Signal Processing (1998)
211. *F. Halter-Koch*, Ideal Systems (1998)
212. *N. K. Govil et al.*, eds., Approximation Theory (1998)
213. *R. Cross*, Multivalued Linear Operators (1998)
214. *A. A. Martynyuk*, Stability by Liapunov's Matrix Function Method with Applications (1998)
215. *A. Favini and A. Yagi*, Degenerate Differential Equations in Banach Spaces (1999)
216. *A. Illanes and S. Nadler, Jr.*, Hyperspaces: Fundamentals and Recent Advances (1999)
217. *G. Kato and D. Struppa*, Fundamentals of Algebraic Microlocal Analysis (1999)
218. *G. X.-Z. Yuan*, KKM Theory and Applications in Nonlinear Analysis (1999)
219. *D. Motreanu and N. H. Pavel*, Tangency, Flow Invariance for Differential Equations, and Optimization Problems (1999)
220. *K. Hrbacek and T. Jech*, Introduction to Set Theory, Third Edition (1999)
221. *G. E. Kolosov*, Optimal Design of Control Systems (1999)
222. *N. L. Johnson*, Subplane Covered Nets (2000)
223. *B. Fine and G. Rosenberger*, Algebraic Generalizations of Discrete Groups (1999)
224. *M. Väth*, Volterra and Integral Equations of Vector Functions (2000)
225. *S. S. Miller and P. T. Mocanu*, Differential Subordinations (2000)
226. *R. Li et al.*, Generalized Difference Methods for Differential Equations: Numerical Analysis of Finite Volume Methods (2000)
227. *H. Li and F. Van Oystaeyen*, A Primer of Algebraic Geometry (2000)
228. *R. P. Agarwal*, Difference Equations and Inequalities: Theory, Methods, and Applications, Second Edition (2000)
229. *A. B. Kharazishvili*, Strange Functions in Real Analysis (2000)
230. *J. M. Appell et al.*, Partial Integral Operators and Integro-Differential Equations (2000)
231. *A. I. Prilepko et al.*, Methods for Solving Inverse Problems in Mathematical Physics (2000)
232. *F. Van Oystaeyen*, Algebraic Geometry for Associative Algebras (2000)
233. *D. L. Jagerman*, Difference Equations with Applications to Queues (2000)

234. *D. R. Hankerson et al.*, Coding Theory and Cryptography: The Essentials, Second Edition, Revised and Expanded (2000)
235. *S. Dăstălescu et al.*, Hopf Algebras: An introduction (2001)
236. *R. Hagen et al.*, C*-Algebras and Numerical Analysis (2001)
237. *Y. Talpaert*, Differential Geometry: With Applications to Mechanics and Physics (2001)
238. *R. H. Villarreal*, Monomial Algebras (2001)
239. *A. N. Michel et al.*, Qualitative Theory of Dynamical Systems: Second Edition (2001)
240. *A. A. Samarskii*, The Theory of Difference Schemes (2001)
241. *J. Knopfmacher and W-B. Zhang*, Number Theory Arising from Finite Fields (2001)
242. *S. Leader*, The Kurzweil–Henstock Integral and Its Differentials (2001)
243. *M. Biliotti et al.*, Foundations of Translation Planes (2001)
244. *A. N. Kochubei*, Pseudo-Differential Equations and Stochastics over Non-Archimedean Fields (2001)
245. *G. Sierksma*, Linear and Integer Programming: Second Edition (2002)
246. *A. A. Martynyuk*, Qualitative Methods in Nonlinear Dynamics: Novel Approaches to Liapunov's Matrix Functions (2002)
247. *B. G. Pachpatte*, Inequalities for Finite Difference Equations (2002)
248. *A. N. Michel and D. Liu*, Qualitative Analysis and Synthesis of Recurrent Neural Networks (2002)
249. *J. R. Weeks*, The Shape of Space: Second Edition (2002)
250. *M. M. Rao and Z. D. Ren*, Applications of Orlicz Spaces (2002)
251. *V. Lakshmikantham and D. Trigiante*, Theory of Difference Equations: Numerical Methods and Applications, Second Edition (2002)
252. *T. Albu*, Cogalois Theory (2003)
253. *A. Bezdek*, Discrete Geometry (2003)
254. *M. J. Corless and A. E. Frazho*, Linear Systems and Control: An Operator Perspective (2003)
255. *I. Graham and G. Kohr*, Geometric Function Theory in One and Higher Dimensions (2003)
256. *G. V. Demidenko and S. V. Uspenskii*, Partial Differential Equations and Systems Not Solvable with Respect to the Highest-Order Derivative (2003)
257. *A. Kelarev*, Graph Algebras and Automata (2003)
258. *A. H. Siddiqi*, Applied Functional Analysis (2004)
259. *F. W. Steutel and K. van Harn*, Infinite Divisibility of Probability Distributions on the Real Line (2004)
260. *G. S. Ladde and M. Sambandham*, Stochastic Versus Deterministic Systems of Differential Equations (2004)
281. *B. J. Gardner and R. Wiegandt*, Radical Theory of Rings (2004)
262. *J. Haluška*, The Mathematical Theory of Tone Systems (2004)

Additional Volumes in Preparation

E. Hansen and G. W. Waister, Global Optimization Using Interval Analysis: Second Edition, Revised and Expanded (2004)

THE MATHEMATICAL THEORY OF TONE SYSTEMS

Ján Haluška

Mathematical Institute
Slovak Academy of Sciences
Bratislava, Slovakia

Žilina University
Žilina, Slovakia

CRC Press
Taylor & Francis Group
Boca Raton London New York

CRC Press is an imprint of the
Taylor & Francis Group, an **informa** business

CRC Press
Taylor & Francis Group
6000 Broken Sound Parkway NW, Suite 300
Boca Raton, FL 33487-2742

First issued in paperback 2019

ISBN-13: 978-0-8247-4714-5 (hbk)
ISBN-13: 978-0-367-39470-7 (pbk)
ISBN-13: 978-80-88683-28-5 (Ister Science)

Visit the Taylor & Francis Web site at
http://www.taylorandfrancis.com

and the CRC Press Web site at
http://www.crcpress.com

Library of Congress Cataloging-in-Publication Data
A catalog record for this book is available from the Library of Congress.

Preface

Intervals in music are rather to be judged intellectually through numbers than sensibly through the ear.

Pythagoras

Definition of tone system The sculptor working in marble has set his limit by the choice of this material to the exclusion of all other materials. Analogously, the musician has to select the tone system he wants to use. This strong emphasis on the necessity of limitation reflects not a subjective prejudice but it is a fundamental artistic law. There is no art without limitation.

There are four important and mutually interacting attributes that we can manipulate to create or describe any sound. And we can work with these attributes in two different ways: we can measure them and we can hear them. If we measure them, they are physical attributes; if we hear them, they are perceptual attributes. The four physical attributes are: frequency, amplitude, waveform, and duration. Their perceptual counterparts are: pitch, loudness, timbre, and (psychological) time. There is similarity between hearing and measuring these attributes; however, it is a complex correlation. The two are not exactly parallel.

There are more or less suitable mathematical structures that reflect the nature of the various classes $\mathcal{T}$ of tones and distinguish them from other hearable sound. Dealing with tones as physical sound, elements of $\mathcal{T}$ can be modelled as Fourier series (Chapter 2) or wavelets (Chapter 3). Considering tones as perceptual objects, classes $\mathcal{T}$ are usually mentioned as the sets of numbers, e.g., the set of all integer numbers $\mathbb{Z}$ (Equal Temperaments, c.f. Chapter 5), the

iii

set of all fuzzy numbers $\mathbb{F}$ (Well Temperaments, c.f. Chapter 7), the set of all numbers of the form 2^p3^q, where $p, q \in \mathbb{Z}$ (Pythagorean System, c.f. Chapter 6) or $p, q \in \mathbb{Q}$ (Euler music space, c.f. Chapter 4), or, they are mentioned as a finite sets which are "acoustically dense" (53-tones per octave in Turkish music, c.f. Chapter 5).

What is tone system from the mathematical viewpoint? Having defined the *class of tones* $\mathcal{T}$ (what tones are), we can start with the following very simplified definition.

Definition 1 Let $\mathcal{T}$ be a class of tones. *Tone system in a broader sense* is a couple $(\mathbb{T}, \Omega)$ where $\mathbb{T}$ is a subset of $\mathcal{T}$ and $\Omega : \mathcal{T} \to \mathbb{R}$ is a real function called the *pitch function*. The set $S = \Omega(\mathbb{T}) = \{\Omega(T) \in \mathbb{R}; \ T \in \mathbb{T}\}$ is called the *tone system in a narrower sense*.

There are many examples in this book showing the merits of Definition 1. However, it is clear that tone system in Definition 1 is such an abstract notion, so that we are not able to distinguish exactly which structures in nature and art are music-inspired or related to music and which are not. After some years of considerations of systems of tones used in music, our present opinion is that the "usual" tone system should involve the following four additional concepts which, in general, we formulate only verbally:

(S1) construction or selection algorithm for $\mathbb{T} \subset \mathcal{T}$;

(S2) notion of symmetry;

(S3) characteristic or typical relation, fundamental equation;

(S4) uncertainty measure.

On the one hand, Example 4, c.f. Chapter 1, is a collection of number sets such that no concepts (S1)–(S2) but trivial was found in these African tone systems. On the other hand, if the concepts (S1)–(S4) are nontrivial, then they are mathematically very heterogeneous and also qualitatively very different for different tone systems. An exemplary explicit formulation of (S1)–(S4) can be found in Chapter 7, Definition 24.

There are three main mutually interacting sources that lead to useful tone systems in music (in this book we do not consider applications of the tone system notion to other disciplines of science and art): *acoustics* (e.g., physical properties of sound, the building acoustics, spectral analysis, timbre, turbulence of air); *psychology* (e.g., the Weber–Fechner law, linear part of the pitch perception curve, dissonance curves, reference tone, individual perception); and *human spiritual culture* (e.g., language, art—including music and dance, philosophy, science, instrument building).

Unifying the theory The set T may be equipped with more unary (e.g., scalar multiplication), binary (e.g., addition, join, meet), ternary, etc., n-ary operations (e.g., this aggregation operation is produced by an n-member orchestra), and also with different orders (e.g., the linear order according to the frequency of tones, the "spiral of fifths", c.f. also List of intervals), topological, and other mathematical structures.

The sense of Definition 1 consists of the idea that the set $\Omega(\mathbb{T})$ of numbers (quantity) should reflect extracted mathematical structures of the set $\mathbb{T}$ of tones (quality). The pitch function Ω is an element of the "first dual" T' of T (should be defined exactly; to have a vector structure, tone pitches can be mentioned as logarithms of relative frequencies of tones). If the first dual is bounded with hearing, then the second dual T'' is a manifestation of measuring. Or vice-versa, c.f. the psychological model in Chapter 4, Figure 4.1.

While the first extremal approach to create a unifying tone system theory (*the concentration on mere quantitative elements*—psychological statistics, physics of sound) isolates us from the inner world of music (and therefore, it is not unifying), the second approach (*the concentration on mere qualitative elements*—spiritual, musical, religious, metaphysical, philosophical) affects on few concrete tone systems that may be used practically. This second approach yields a number of particular, often contentious, theories about tone systems which are concentrated on one side of the topic: national music, production of musical instruments, philosophy, one music style.

The truth should be somewhere in between. This book is an attempt to describe elements of a unifying mathematical theory of tone systems on the basis of the *uncertainty-knowledge-based information theory*. On the other hand, the study of tone systems provides an excellent mathematical laboratory to study all types of uncertainty.

Moreover, the tone system notion mediates not only a special mathematical duality. Mathematics is very apt to reflect the *spiritual duality "natural science ↔ art"* which has its manifestation in the notion of tone system, [39]. This duality was one of three main motivations for writing this book. Maybe the tone system notion is unique in this way. This duality has two directions ("natural science → art" and "art → natural science") which are mutually symbiotic and reflective.

The second motivation was finding out that the *interdisciplinarity is the inner property of tone systems* and that the topic of tone systems is a very mathematical subject. There are three general arguments for the unambiguous and resolute quoted claim of Pythagoras that we should care much more for tone systems: Theory of tone systems is a *very interdisciplinary topic* within mathematics itself and is only inspired by and applied to music. There are papers concerning the tone systems in set theory, harmonic analysis, number theory, fuzzy theory, genetic algorithms, statistics, graph theory, algebra, theoretic arithmetic, differential equations, Diophantine equations, dynamical systems, geometry, logic, discrete mathematics, functional analysis. The theory of tone systems *can be applied not only to music* but also to medicine, acoustics, linguistics, quantum mechanics, information science, and psychology. Problems about tone systems were studied at the *origin of mathematics as a science.* They are their own legacy of mathematics. Many results and even whole branches of mathematics (e.g., partial differential equations, Fourier analysis) arose from problems about tone systems during its history.

The third main motivation for writing this book was my opinion that *the most valuable technical advance for music in general is the development of tone systems.*

Part I: Fundamentals We explain that the essential approach for the study of tone systems is use of uncertainty-based information theory which gives us an integrated view on the subject and, at the same time, forms the theory of tone systems as an open system. From the contents it is immediately clear that there are connections with harmonic analysis, mathematical psychology, and fuzzy set and systems theory.

The notion of the *geometric net* (of numbers, operators) is a generalization of the elementary notion of geometric progression to more quotients. It was developed for describing tone systems. This technique is systematically used through the special systems part of the book. But we do not study systematically the theory of geometric net. The concept of geometric net is related to the notion of *analytic algebra* which is a generalization of power series.

The mathematical terminology is standard. It is worth noting two items. First, in the multiplicative group $(0, \infty)$, the length of an interval $[a, b]$, where $a \leq b$, is b/a. Borrowing the usual musical terminology, we will simply say that b/a is an interval. The second terminological peculiarity is the use of special historical names of some lengths of intervals or numbers, c.f. List of intervals. For the compilation of this list, various sources were used such as [26]. Intervals listed here were ordered by M. Op de Coul. For further names of intervals (also outside octave), c.f. e.g. [26] or the Bohlen–Pierce site

http://members.aol.com/bpsite/index.html.

A consistent system of nomenclature is described in

http://uq.net.au/~ zzdkeena/Music/IntervalNaming.htm

(D. Keenan: A note on the naming of musical intervals).

Part II: Special systems The rest of the chapters describes classes of tone systems. It should be underlined that we classified (or covered) all *practical* and also theoretical tone systems (in the narrower sense at least). The families of tone systems described in this part are not disjoint. On the contrary, there are many set and also idea intersections among the tone systems families.

At the end of the chapters there are "Remarks and ideas for exploring". Here are noted also some suggestions to prove or calculate some additional or complementary facts to the basic text. Some of the exercises are unsolved problems and are challenges to make new discoveries.

There is a question of unambiguity and systemizing of tone systems in the literature. The names of tone systems in this book are built according to the following pattern:

$NAME1/.../NAMEp(COM1/.../COMq)\text{-}N,$

$\omega_1, \omega_2, \ldots, \omega_N$

where *NAMEp* are tone system names in the literature (often the creators' names); *COMq* are comment acronyms (not necessarily present), c.f. List of abbreviations and acronyms, making the scale unambiguously identified; N denotes the cardinality of the discrete tone system or the cardinality of its period if the tone system is periodical. If N is not discrete (e.g., the case of gamelan tone systems), then it is the number of continuous zones or continuous zones within a period. After the tone system name, there is a short text comment and then the sequence $\omega_1, \omega_2, \ldots, \omega_N$ of pitch values (or zone means) follows in the footnote size. The pitch values are either in cents or expressed as ratios (or both, mixed); they are not in Hz. If S is not periodical, then $S = \{\omega_1, \omega_2, \ldots, \omega_N\}$ and $\omega_1 = 1$ (0 cents). *The absence of value 1 (the unison) means that S is periodical.* In this case, the last file value is this period interval (mostly $2 = 1\,200$ cents, the octave) with the exception that for tetrachordal scales, two periods are printed.

We illuminated the text with more than 200 tone systems used in practice (their names are alphabetically listed in the Index) and more than 800 systems are listed in Appendix A. An excellent collection of examples is a permanently updated archive of used tone systems created by the M. Op de Coul (together with other composers), the Fokker Foundation, c.f. [192], which contains more than 3000 items.

Literature   The key figures from the past are: Archytas, Avicena, Euler, Fibonacci, Fourier, Helmholtz, Kepler, Mersenne, Petzval, Pólya, Ptolemy, Pythagoras, and Stevin. We collected *literature about tone systems reviewed in Mathematical Reviews* since its beginning in 1940. Most of these publications are listed in the bibliography. We omitted the popularized works and works about the history of tone systems, c.f. [193].

Although there are some short passages which aim to illuminate the correlation between time and the conceptual development of tone systems, the *book is not about the history of tone systems.* Rather, the most complete present special and permanently updated literature list about tone systems and their history is collected by M.

Op de Coul, B. McLaren, F. Jedrzejewski, [192]. This bibliography contains more than 4 000 items and also internet links to specialized journals and other internet pages.

There are many book sources concerning the *musicological terminology used in tone systems*. This book is self-contained. For those who are curious, a musicological dictionary with definitions that lead easily to mathematically formalized definitions of terms was created and is continually improved on the Web by the composer J. L. Monzo, c.f. [195].

Acknowledgements A. Romanowska, the mathematician and violinist, wrote Section 5.1. I. Ortgies, the organ scholar, organ builder and theorist, wrote Subsection 6.3.3 and Appendix B. M. Pfliegel is the author of the photographs on the cover page. There are two pictures from the collection of paintings on wood on the balustrade of the largest wooden church (more than 6 000 seats) all over the world, Svätý Kríž, Slovakia, 1693.

The author thanks composers of contemporary music and computer music theorists for their information-sharing and feedback concerning currently used tone systems. My thanks go to D. Benson, D. Canright, P. Erlich, K. Grady, M. Haluška, Y. Hellegouarch, B. Hero, D. Keenan, V. A. Lefebvre, H. L. Mittendorf, J. L. Monzo, G. Morrison, G. Mazzola, E. Neuwirth, M. Op de Coul, R. Ružička, M. Schulter, J. Starrett, Ch. Stoddard, and W. Sethares.

The author wishes to thank to all members of the seminar Mathematics and Music, Bratislava, since 1984, and especially to R. Berger and B. Riečan, for encouraging his work and stimulating his mathematical interest in the topic of tone systems.

In May 1999, the author was a member of a team that received the scientific award *Prize of the Slovak Academy of Sciences*. The prize collection of papers *Fuzzy sets and their applications* included a series of the author's articles about tone systems.

The author would like to express his gratitude to G. J. Klir for the gift of the book [33] and to D. Stanzial for the gift of the proceedings [6].

The author is also grateful for the fruitful personal discussions with and hospitality from H. G. Feichtinger in Vienna, G. Di Maio in Caserta, T. Noll in Berlin, and A. Romanowska in Warsaw.

Most of the work presented in this monograph was done by the author at the Mathematical Institute of the Slovak Academy of Sciences, branch Košice, in the frame of the project Tone systems, 1998–2000. The work over the book was partially supported by Grant VEGA 1141 and by the international grant SAS (Bratislava)—CNR (Rome) (Integration in vector spaces equipped with additional structures), 2000–2003.

I hope that the book will be interesting and useful for mathematicians, musicians, teachers, tuners, producers of musical instruments, and university students. The book is specially recommended to those mathematicians who play some musical instrument.

Ján Haluška

Contents

List of Tables

List of Figures

List of symbols

$\emptyset$	empty set
$\mathbb{P}$	the set of all prime numbers
$\mathbb{C}$	the set of all complex numbers
$\mathbb{E}^n$	n-dimensional Euler music space
$\mathbb{L}$	$((0, \infty), \cdot, 1, \leq)$, multiplicative group on $\mathbb{R}^+$
$\mathbb{N}$	set of all natural numbers
$\mathbb{N}^+$	set of all nonnegative integer numbers
$\mathbb{Q}$	set of all rational numbers
$\mathbb{R}$	set of all real numbers
$\mathbb{R}^+$	positive cone of $\mathbb{R}$
$\mathbb{Z}$	set of all integer numbers
$\mathbb{F}$	set of all fuzzy numbers
$[\alpha]$	integer part of α, $\alpha \in \mathbb{R}$
$\Re A$, $\Re(A)$	real part of A
$\Im A$, $\Im(A)$	imaginary part of A
$\ldots, G_\flat, D_\flat, A_\flat, E_\flat, B_\flat, F, C, G, D, A, E, B, F_\sharp, C_\sharp, G_\sharp, D_\sharp, A_\sharp, \ldots$	
	"the spiral of fifths"
I	directed partially ordered set
E_N	N-tone equal tempered system, $\{ \sqrt[N]{2^z}; \ z \in \mathbb{Z}\}$
ϕ	Euler ϕ-function
$\ln$	logarithm to the base e (natural logarithm)
$\mathcal{K}$	Pythagorean comma, $531\,441/524\,288$
$\mathcal{D}$	Comma of Didymus, syntonic comma, $81/80$
K	comma (in general)
ω	relative frequency
τ_ω	temperature of ω
μ_ω	mistuning of ω

Π	Pythagorean System, $\mathcal{T} = \{2^p 3^q;\ p, q \in \mathbb{Z}\}$
Θ_ω	tempered value (approximation) of $\omega \in \mathbb{R}$
W	whole tone, tonos
$X, Y, \ldots, Z$	semitones
N	cardinality of one scale periode
$\Omega : \mathcal{T} \to \mathbb{R}$	pitch function
T	tone (sound signal or perceptual object)
$\mathcal{T}$	class of all tones of a certain type
$\mathbb{T}$	set of tones in the tone system, $\mathbb{T} \subset \mathcal{T}$
$(\mathbb{T}, \Omega)$	tone system in the broader sense
$\Omega(\mathbb{T}),\ S$	tone system in the narrower sense
$\mathfrak{S}_\mathbb{F}$	class of all fuzzy tone systems
$\mathfrak{S}_E$	class of all equal temperaments
$\Delta,\ \delta$	dissonance function

List of abbreviations and acronyms

$APPR$ = approximation, $ARCH$ = archaic, $ARCHY$ = Archytas, $ARISTO$ = Aristoxenos, AUG = augmented, CET = non-octave equal temperament, $CHRO$ = chromatic, $CSOUND$ = Computer sound, $DIAT$ = diatonic, $DIDY$ = Didymus, ENH = enharmonic, $HARM$ = harmonic, HEM = hemiolitic, HYP = Hyperenharmonic, $HYPOD$ = hypodorian, $HYPOL$ = hypolydian, $HYPOP$ = hypophrygian, INV = inverted, JI = Just Intonation, LIM = limit, MAJ = major, MIN = minor, $MIDI$ = Musical Instrument Digital Interface, MIX = mixed, $MIXOL$ = mixolydian, MOS = the most optimal system, $NEUTR$ = neutral, PE = Pelog, $PHRYG$ = Phrygian, PIS = Perfect Immutable System, PTO = Ptolemy, $PYTH$ = Pythagorean, RAT = rationalized, SLE = Slendro, $SOFT$ = soft, $SUPER$ = superparticular, $SUB/HARM$ = subharmonic, TET = Equal Temperament with the octave equivalence, $WERC$ =Werckmeister, WIL = Wilson

List of intervals

Ratio	Name
1/1	unison, perfect prime, tonic
2/1	octave
3/2	perfect fifth
4/3	perfect fourth
5/3	major sixth
5/4	major third
6/5	minor third
7/3	minimal tenth
7/4	harmonic seventh
7/5	septimal or Huygens' tritone, Bohlen–Pierce fourth
7/6	septimal minor third
8/5	minor sixth
8/7	septimal whole tone
9/4	major ninth
9/5	just minor seventh
9/7	septimal major third
9/8	major whole tone
10/7	Euler's tritone
10/9	minor whole tone
11/5	neutral ninth
11/6	21/4-tone, undecimal neutral seventh
11/7	undecimal augmented fifth
11/8	undecimal semi-augmented fourth
11/9	undecimal neutral third
11/10	4/5-tone, Ptolemy's second
12/7	septimal major sixth
12/11	3/4-tone, undecimal neutral second
13/7	16/3-tone
13/8	tridecimal neutral sixth
13/12	tridecimal 2/3-tone
14/9	septimal minor sixth
14/11	undecimal diminished fourth

14/13	2/3-tone
15/7	septimal minor ninth, Bohlen–Pierce ninth
15/8	classic major seventh
15/14	major diatonic semitone, Cowell just semitone
16/7	septimal major ninth
16/9	Pythagorean minor seventh
16/11	undecimal semi-diminished fifth
16/13	tridecimal neutral third
16/15	minor diatonic semitone
17/8	septendecimal minor ninth
17/12	2nd septendecimal tritone
17/14	supraminor third
17/16	17th harmonic, overtone semitone
18/11	undecimal neutral sixth
18/17	Arabic lute index finger
19/15	undevicesimal ditone
19/16	19th harmonic, overtone minor third
19/18	undevicesimal semitone
20/9	small ninth
20/11	large minor seventh
21/16	narrow fourth
21/17	submajor third
21/20	minor semitone
22/21	hard semitone
23/16	23rd harmonic
24/17	1st septendecimal tritone
25/9	classic augmented eleventh, Bohlen–Pierce twelfth
25/12	classic augmented octave
25/14	middle minor seventh
25/16	classic augmented fifth
25/18	classic augmented fourth
25/21	Bohlen–Pierce second, quasi-tempered minor third
25/24	classic chromatic semitone, minor chroma
26/25	Avicenna 1/3-tone
27/14	septimal major seventh
27/16	Pythagorean major sixth
27/20	acute fourth
27/22	neutral third, Zalzal's wosta
27/25	large limma, alternate Renaissance semitone
27/26	tridecimal comma
28/15	grave major seventh
28/17	submajor sixth
28/25	middle second
28/27	1/3-tone, Archytas inferior 1/4-tone
30/19	undevicesimal minor sixth

31/16	31st harmonic
31/30	31st-partial chroma, Didymus superior 1/4-tone
32/15	minor ninth
32/17	17th subharmonic
32/19	19th subharmonic
32/21	wide fifth
32/25	classic diminished fourth
32/27	Pythagorean minor third
32/31	Greek enharmonic 1/4-tone, Didymus inferior 1/4-tone
33/25	2 pentatones
33/32	undecimal comma, 33rd harmonic
34/21	supraminor sixth
35/18	septimal semi-diminished octave
35/24	septimal semi-diminished fifth
35/27	9/4-tone, septimal semi-diminished fourth
35/32	septimal neutral second
36/25	classic diminished fifth
36/35	1/4-tone, septimal diesis
40/21	acute major seventh
40/27	grave fifth
42/25	quasi-tempered major sixth
44/27	neutral sixth
45/32	tritone
45/44	1/5-tone
46/45	23rd-partial chroma, Ptolemy inferior 1/4-tone
48/25	classic diminished octave
48/35	septimal semi-augmented fourth
49/25	Bohlen–Pierce eighth
49/30	larger approximation to neutral sixth
49/36	Arabic lute acute fourth
49/40	larger approximation to neutral third
49/45	Bohlen–Pierce minor semitone
49/48	1/6-tone, 2nd septimal diesis
50/27	octave-large limma
50/33	3 pentatones
50/49	Erlich's decatonic comma, tritonic diesis
51/50	17th-partial chroma
54/35	septimal semi-augmented fifth
54/49	Zalzal's mujannab
55/54	Ptolemy enharmonic
60/49	smaller approximation to neutral third
63/25	quasi-equal major tenth, Bohlen–Pierce eleventh
63/32	octave- septimal comma
63/40	narrow minor sixth
63/50	quasi-equal major third

64/35	septimal neutral seventh
64/45	2nd tritone
64/63	septimal comma
65/64	13th-partial chroma
68/35	23/4-tone
72/49	Arabic lute grave fifth
75/49	Bohlen–Pierce fifth
75/64	classic augmented second
80/49	smaller approximation to neutral sixth
80/63	wide major third
81/44	2nd undecimal neutral seventh
81/50	acute minor sixth
81/64	Pythagorean major third
81/68	Persian wosta
81/70	Al-Hwarizmi's lute middle finger
81/80	syntonic comma, Didymus comma
88/81	2nd undecimal neutral second
89/84	approximation to equal semitone
91/59	15/4-tone
96/95	19th-partial chroma
99/70	2nd quasi-equal tritone
99/98	small undecimal comma
100/63	quasi-equal minor sixth
100/81	grave major third
105/64	septimal neutral sixth
125/64	classic augmented seventh, octave- minor diesis
125/72	classic augmented sixth
125/96	classic augmented third
125/108	grave augmented whole tone
125/112	classic augmented semitone
126/125	small septimal comma
128/75	diminished seventh
128/81	Pythagorean minor sixth
128/105	septimal neutral third
128/125	minor diesis, diminished second
131/90	13/4-tone
135/128	major limma, major chroma, limma ascendant
140/99	quasi-equal tritone
144/125	classic diminished third
145/144	29th-partial chroma
153/125	7/4-tone
160/81	octave-syntonic comma
161/93	19/4-tone
162/149	Persian neutral second
192/125	classic diminished sixth

216/125	acute diminished seventh
225/128	augmented sixth
225/224	septimal kleisma
231/200	5/4-tone
241/221	Meshaqah's 3/4-tone
243/125	octave-maximal diesis
243/128	Pythagorean major seventh
243/160	acute fifth
243/200	acute minor third
243/242	neutral third comma
245/243	minor Bohlen–Pierce diesis
246/239	Meshaqah's 1/4-tone
248/243	tricesoprimal comma
250/153	17/4-tone
250/243	maximal diesis
256/135	octave-major chroma
256/225	diminished third
256/243	limma, Pythagorean minor second
261/256	vicesimononal comma
320/243	grave fourth
375/256	double augmented fourth
375/343	Bohlen–Pierce major semitone
400/243	grave major sixth
405/256	wide augmented fifth
512/375	double diminished fifth
512/405	narrow diminished fourth
513/512	undevicesimal comma
625/324	octave-major diesis
525/512	Avicenna enharmonic diesis
625/567	Bohlen–Pierce great semitone
648/625	major diesis
675/512	wide augmented third
687/500	11/4-tone
729/400	acute minor seventh
729/512	Pythagorean tritone
729/640	acute major second
736/729	vicesimotertial comma
749/500	ancient Chinese quasi-equal fifth
750/749	ancient Chinese tempering
800/729	grave whole tone
1 024/675	narrow diminished sixth
1 024/729	Pythagorean diminished fifth
1 029/1 024	gamelan residue
1 125/1 024	double augmented prime
1 215/1 024	wide augmented second

1 280/729	grave minor seventh
1 732/1 731	approximation to 1 cent
1 875/1 024	double augmented sixth
2 025/1 024	2 tritones
2 048/1 125	double diminished octave
2 048/1 215	narrow diminished seventh
2 048/1 875	double diminished third
2 048/2 025	diaschisma
2 058/2 057	xenisma
2 187/2 048	apotome, Pythagorean major semitone
2 187/2 176	septendecimal comma
2 401/2 400	Breedsma
3 125/3 072	small diesis
3 125/3 087	major Bohlen–Pierce diesis
3 375/2 048	double augmented fifth
4 096/2 187	Pythagorean diminished octave
4 096/3 375	double diminished fourth
4 375/4 374	ragisma
4 608/4 235	Arabic neutral second
5 120/5 103	Beta 5
5 625/4 096	double augmented third
6 144/3 125	octave-small diesis
6 561/4 096	Pythagorean augmented fifth
6 561/6 125	Bohlen–Pierce major link
6 561/6 400	Mathieu superdiesis
8 192/5 625	double diminished sixth
8 192/6 561	Pythagorean diminished fourth
9 801/9 800	kalisma
10 125/8 192	double augmented second
10 935/8 192	fourth + schisma,
15 625/15 309	great Bohlen–Pierce diesis
15 625/15 552	kleisma, semicomma majeur
16 384/10 125	double iminished seventh
16 384/10 935	fifth-schisma,
16 875/16 807	small Bohlen–Pierce diesis
19 683/10 000	octave-minimal diesis
19 683/16 384	Pythagorean augmented second
20 000/19 683	minimal diesis
32 768/19 683	Pythagorean diminished seventh
59 049/32 768	Pythagorean augmented sixth
59 049/57 344	Harrison's comma
65 536/32 805	octave-schisma
65 536/59 049	Pythagorean diminished third
83 349/78 125	Bohlen–Pierce minor link
177 147/131 072	Pythagorean augmented third

262 144/177 147	Pythagorean diminished sixth
393 216/390 625	Wuerschmidt's comma
413 343/390 625	Bohlen–Pierce small link
531 441/262 144	Pythagorean augmented seventh
531 441/524 288	Pythagorean comma, ditonic comma
1 048 576/531 441	Pythagorean diminished ninth
1 594 323/1 048 576	Pythagorean double augmented fourth
2 097 152/1 594 323	Pythagorean double diminished fifth
2 109 375/2 097 152	semicomma
4 782 969/4 194 304	Pythagorean double augmented prime
8 388 608/4 782 969	Pythagorean double diminished octave
14 348 907/8 388 608	Pythagorean double augmented fifth
16 777 216/14 348 907	Pythagorean double diminished fourth
33 554 432/33 480 783	Beta 2, septimal schisma
34 171 875/33 554 432	Ampersand's comma
43 046 721/33 554 432	Pythagorean double augmented second
67 108 864/43 046 721	Pythagorean double diminished seventh
129 140 163/67 108 864	Pythagorean double augmented sixth
134 217 728/129 140 163	Pythagorean double diminished third
387 420 489/268 435 456	Pythagorean double augmented third
536 870 912/387 420 489	Pythagorean double diminished sixth
1 162 261 467/536 870 912	Pythagorean double augmented seventh

19 073 486 328 125/19 042 491 875 328
'19-tone' comma
36 893 488 147 419 103 232/36 472 996 377 170 786 403
'41-tone' comma
19 383 245 667 680 019 896 796 723/19 342 813 113 834 066 795 298 816
Mercator's comma

Part I

Fundamentals

Chapter 1

Tone systems and uncertainty theory

Music creates order out of chaos; for rhythm imposes unanimity upon the divergent, melody imposes continuity upon the disjointed, and harmony imposes compatibility upon the incongruous.

Yehudi Menuhin

1.1 Introduction

1.1.1 A piece of metaphysics

The idea that universe is created out of music is a very ancient one. Remember the Pythagorean school, Kepler's "Harmony of spheres," and many elder assertions of philosophical schools from India, Summer, and China. The reformer Martin Luther also claimed that music is a gift of God to man. We are not able to judge metaphysical problems of this sort, about relation of music and the material world. Or, in particular, to judge what is music or what is not. The obvious fact is that music (or, what is commonly understood as music) is composed or ordered somehow. In other words, the basic material of music, tones, can be structured. These structures are tone systems.

According to comments to Definition 1 of tone system in the Preface, it is clear that tone systems can be of various degrees of complexity—from a simple finite list of rational numbers to classes with cardinality greater than continuum (functions acting in functional spaces). There are wonderful clear relations in the frame of natural numbers understood for pupils of basic schools and also very ingenious constructions from functional and harmonic analysis, general algebra, and other parts of mathematics which need the university level of knowledge of mathematics.

One interesting aspect is that when considering tone systems, we obtain a tangible feeling that mathematics is an experimental science. The golden treasures are shallow under ground and, finding any, you will ask whether it is possible that people have searched around a thousand years and did not find or observe it. You can experiment using the computer. However, as a rule, easily formulated hypotheses cannot be proved easily.

The present-day musicians, who seriously think about suitable tone systems for their compositions, use various names to distinguish themselves from the huge mass of musicians who suppose tone systems as something a priori given, mostly—the 12-tone Equal Temperament in the western cultural zone. In Europe, we find the following adjectives for denotations of such non-rigid musicians or music: experimental, digital, new, fractal, contemporary, electronic, computer. In America, the most frequently used adjectives are rather the following: microtonal or xenharmonic. Underline that this concerns Europe and America since music of other parts of Earth, although relatively mutually isolated, was never bounded with any unambiguous tone system. From the view of the present world globalization, it is a qualitative and very positive moment.

1.1.2 Practical manners of the tone system description

Terms such as musical interval, relative frequency, and positive number are used as synonyms depending on the context. The situation is similar when using terms point, vector, and n-tuple in analytical geometry. This became quite evident when we realize that there is a bijection between the set of all positive numbers with the group operation of multiplication and the set of all musical intervals with the vector space structure. Simply put, the Weber–Fechner's psychological law says that we hear (locally) the class of tones T "as a vector space;" we hear not frequencies of tones but their logarithms. Thus we can (locally) apply all techniques and the theory of vector spaces to the theory of tone systems.

So, we can evaluate pitches measuring either frequencies and their ratios or logarithms of these frequencies and their differences. This important isomorphism between the multiplicative and additive structures on $\mathbb{R}^+$ and $\mathbb{R}$, respectively, is given with the *logarithm* function and its inverse, the *exponential* function. In the theory of tone systems, the special logarithm bases are considered. If ω is the relative frequency of an musical interval (the ratio of frequencies expressed, e.g., in Hz), then the size c of this interval in units called *cents* we obtain with the formula:

$$c = \log_{2^{1/1200}} \omega = 1\,200 \log_2 \omega, \quad \omega = 2^{c/1\,200} = \left(2^{1/1\,200}\right)^c.$$

Thus, one octave $2/1 = 1\,200$ cents, one equally tempered semitone $\sqrt[12]{2} = 100$ cents, the pure fifth is $3/2 \approx 702$ cents. For imagination, the sensibility of the human ear is approximately 5 cents.

The system of cents was introduced by J. Ellis in the 1930s. Besides $2^{1/1\,200}$, there are also other logarithm bases used, e.g., $2^{1/1\,000}$, $2^{1/1\,060}$.

For the purpose of tuning musical instruments, the set of pitches $S = \Omega(\mathbb{T})$ is described in practice as a set of: (1) relative frequencies [in ratios or in cents] of tones + the reference frequency [in Hz]; (2) absolute frequencies [in Hz] of tones; or (3) technological measurements of generators (e.g., the pipe lengths in feet and inches, etc.) of tones. Each of these manners has its advantages and disadvantages. We will use the first one.

The *pitch standard* (or also—the *reference frequency*) is a frequency of the pitch a' which serves as the origin for instrument tuning. If we have the reference frequency and the deterministic construction or selection algorithm (S1), c.f. Preface, then the set S can be derived. For instance, the set S of theoretical frequencies of all strings of the classical piano tuned to 12-tone Equal Temperament we obtain so that we multiply the reference frequency $a' = 440\,\text{Hz}$ with $\sqrt[12]{2^i}$ where $i = -48, -47, \ldots, 36$, c.f. also Section 7.1.2.

If the algorithm (S1) is deterministic, then the source of uncertainty of the tone system S is reduced to the uncertainty of the reference frequency which is a result of compromises among different tasks. From the viewpoint of building acoustics, the ideal pitch standard is individual for each concert place (hall, church, etc.). Psychologically, it should be in the middle of the zone where the human ear hears the pitch linearly. The third set of conditions is implied by brilliance of timbres of various instruments in various pitch registers.

The pitch standard has varied considerably over the years. In fact, before the 15th century, the pitch standard had ranged a' from 504.2 Hz to 377 Hz. The Mean Pitch, proposed by M. Praetorius in 1619 set the reference at 424.2 Hz. This standard more or less lasted for over two centuries and agreed closely with Handel's own fork (422.5 Hz in 1751) and that of the London Philharmonic (423.3 in 1820). In 1859 a French Government Commission was set up to establish a new standard as earlier in the century, with the development of brass instruments, pitch standards had been steadily climbing because of the increased brilliance of tone these instruments displayed at higher tunings – in 1858 the standard at the Paris Opera was 428 Hz and in Vienna 456.1 Hz. 435.4 Hz was embodied by Lissajous in a standard fork 'diapason normal'. An International Conference in 1939 finally nailed the lid shut on the debate and set a reference of a' (440 Hz and 20°C). For the historical review, c.f. [26], the Ellis's appendix, pp. 493–511.

For most people frequencies must be between 20–20 000 Hz to be heard as pitch, and the upper half of that range is more important to our perception of brilliance than to musically useful pitch. Even the highest tone of picollo is only about 3 000 Hz—far short of 20 000.

1.2   Current trends

We observe these nondisjoint qualitative trends in the present tone system experience and theory, c.f. [153]: *adaptive systems* and *segmentation*. The sense of the adjective "adaptive" is intuitively clear and in practice it means: computer-controlled. Under segmentation one can understand how to find the basic units of speech and music.

1.2.1   America

An overview of the American microtonal scene up to 1991 was published by D. Kaisler, [141]. Pick up some concrete fact about the
situation of using tone systems in the so called computer music after
1991.

B. Denckla, [97], controls the Just Intonation in real time via
MIDI. That is, he wrote the software for Machover's hyperinstru-
ments in the "Brain Opera." Other Boston composers work on the
borderline between noise and pitched sounds. For example, D. Miller
used his "granular synthesis." Pitch tracking, live synthesized sound
generation, and algorithmic composition are sometimes used.

The member of the Dartmouth faculty L. Polansky developed the
interactive real-time computer composition program, called HMSL,
and he has worked with live electronics and acoustic ensembles in
17-limit just intonation since the mid-1970s. He is interesting also
in measuring of morphological similarity in music, c.f. [162].

Example 1

DUDON/HARRISON/POLANSKY(SEPT/SLE/NO.1)-5;
From HMSL Manual of L. Polansky;
8/7, 64/49, 3/2, 12/7, 2

G. L. Nelson at Oberlin college has worked with algorithmic composition techniques and *MIDI* wind controllers in 96 Tone Equal
Temperament.

H. Waage, Pennsylvania, built a set of logic-circuit-controlled
keyboards (and more recently *MIDI* software) which automatically
adjusts vertical harmonies in just intonation to produce just harmonies (with just 7th chords) that stay in tune as the music changes
key.

W. A. Sethares, [60], Winconsin University, represents a break-
through in the effort to quantify the relationship between timbre
and tuning. Sethares has also proposed a digitally-controlled dyna-
mically-retuned keyboard to perform complex microtonal music in
real time, a proposal which duplicates in digital terms I. Darreg's
analog vacuum-tube-based Elastic Tuning Organ. On Darreg's in-

struments, the chords "pulled" toward the nearest just ratios over time, regardless of the microtonal tuning used.

Example 2

DARREG(NO.1)-9; I. Darreg's Mixed JI Genus 1 (Archytas Enhar-monic, Ptolemy Soft Chromatic, Didymus Chromatic Genera);
28/27, 16/15, 10/9, 4/3, 3/2, 14/9, 8/5, 5/3, 2

DARREG(NO.2)-9; I. Darreg's Mixed JI Genus 2 (Archytas Enhar-monic and Chromatic Genera);
28/27, 16/15, 9/8, 4/3, 3/2, 14/9, 8/5, 27/16, 2

Stanford University is home to C. Chafe, J. Chowning, L. Rush, and W. Schottstaedt, all of whom have produced microtonal compositions for both algorithmic-controlled or entirely computer-resident instruments. Schottstaedt and Chowning work in microtonal equal temperaments and non-just non-equal-tempered tone systems rather than just intonation.

Example 3

CHOWNING(STRIA)-9; Tuning used in J. Chowning's Stria, $\sqrt[9]{\pi}$;
92.566, 185.131, 277.697, 370.262, 462.828, 555.394, 647.959, 740.525, 833.090

At Berkeley's Center for New Music and Technology, D. Wessel explores new synthesis methods and real-time technology by which computers listen and respond musically to human performers (similar to work at the M.I.T. Media Laboratory). In 1986 Wessel reportedly convinced Yamaha to include user retunability in its DX7II synthesizers, leading to the current generation of commercially available off-the-shelf retunable digital synthesizers.

The most recent generation of microtonal American composers appears to favor a panintonational approach (non-just, non-equal temperaments). These composers and theorists, who have grown up with post-1983 *MIDI* synthesizers and digital technology, tend to use a wide variety of different just, equal-tempered and non-just non-equal-tempered tone systems. This new generation appears less attached to traditional 19th-century paradigms of music-making, and typically make some use of algorithmic or other computer-based compositional elements in their performances.

E. Moreno, Palo Alto, has composed in and theorized extensively about the psychoacoustics of non-octave tone systems.

The Bay area is also the spark point of the American Gamelan movement, and Mills College gives regular performances with its own gamelan Udan Kyai Mas.

B. Hopkin edits the journal Experimental Musical Instruments. This magazine often features microtonal articles and Hopkin himself has explored microtonality as well as noise composition and acoustic music based more on timbre than specific tunings.

J. Gibbon lives between San Diego and Los Angeles, and has for many years explored the implications of the harmonic series and the possibilities of microtonal percussion instruments, most notably with his microtonal "bell garden."

The Southern California Microtonal group (members: B. Wesley, J. Glasier, J. Stayton, W. Parsons and B. McLarren), San Diego, employs both home-built acoustic instruments and digital synthesizers in combination and singly. Concerts in Southern California typically take place in storefronts, bookstores and art galleries. Whereas the main concentration in New York is on divisions of the whole-tone, and in Boston on 72-tone and 19-tone, and in the St. Louis area on 31-tone, and in the Bay area of Northern California on just intonation, composers in Southern California tend to work in a wide array of different tunings. The Southern California Microtonal Group has performed and composed in every equal temperament from 5 through 53 tones per octave, many different forms of just intonation, the harmonic series in many forms, and various non-just non-equal-tempered tunings including Pelog, Slendro, the vibrational modes of the free-free and free-clamped metal bar, the vibrational modes of a free drumhead, and various mathematical series.

1.2.2 Europe

The situation is much more rich and complicated than in North America since there are plenty of national schools of experimental music (Czech, Dutch, French, German, Italian, British, Swiss, Polish, Portuguese, Slovak, Spanish, etc.) and, in fact, the whole both theoretical and practical tradition of the tone systems consideration (e.g., A. Hába, H. Helmholtz, D. Fokker, J. M. Petzval, etc., etc., going back to ancient Greece). These national schools are often grouped at state (public) radio studios or musical universities. This is a historical consequence of the fact that the speed of development of computers in America went faster than in Europe. However, with the appearance of the PC, this advantage of America was lost.

There are also some long-years seminars called commonly as "Mathematics and music" (Bratislava, Dortmund, Lisbon, Paris, Rome, Vienna, Zurich). Besides these activities, let us mention specially the following projects: the G. Mazzola's project Rubato and R. Wille's project and instrument Mutabor using special logic adapting harmony during the play. Further, there is I.R.C.A.M. in Paris, Fokker foundation activities in Amsterdam, international conference "Understanding and Creating Music" in Caserta, special issues "Harmonic analysis and tone systems" of the Tatra Mountains Mathematical Publications journal published by the Mathematical Institute of the Slovak Academy of Sciences, proceedings from conferences, etc. In general, while musical enthusiasts are ready to study mathematics "give the tone" in America and mathematicians are not very interested in this, the situation in Europe is vice versa. European professional composers and interpreters are much more conservative and the initiative in tone system development in Europe is often coming from the side of theorists which are at the same time also musicians and therefore they see many connections. We can observe also common efforts of linguists, musicologists, mathematicians and also other natural scientists (name for instance the international group: M. Boroda–musicologist, G. Altmann and R. Köhler–linguists, G. Wimmer–statistician) to find the basic units of the human speech and, specially, music.

Say some words about segmentation on the pitch level in general. The dominated power of the present western music can be denoted concisely with the word "clavier." Grubby spoken, music scores are still pressed into the frame of N different music granules (qualitative degrees) within the octave, the discrete choice of pitch frequencies, and the octave equivalence. Usually, $N = 12$ for the western cultural zone. The similar general picture holds also for rhythm. The exception proves the rule.

In fact, musical intervals are doubtless classical kinds of segmentation of western music. Tone intervals are the smallest tone groupings and they chain the tones in the composition into one entity. The idea is not new: to consider as segments all relatively frequency intervals within a tone system. *Every interval (unit) should be derived from a few basic intervals (units) and the minimal set of all basic intervals (units) must carry the whole information about the tone system.* In general, the basic intervals (units) of a tone system need not be the smallest ones and the tone system can be composed from "atoms" (like DNK in genetics, it has only 4 "bricks").

Using only the 12-tone Equal Temperament, the segmentation of the music on the pitch level is trivial and not interesting: there is only one semitone, $\sqrt[12]{2}$. What will happen when we deal, e.g., with the Pythagorean System which has also a relatively simple structure (comparing with other historical tone systems)? *Is there also a finite number of basic intervals, or the set of all basic intervals is infinite?* This question was not solved in the literature and the answer seems not to be trivial. For instance, for Pythagorean System, it is known that the class of all tones $T = \Pi = \{2^p 3^q;\ p, q \in \mathbb{Z}\}$ is a dense subset of the real halfline. So, there are infinitely many different intervals in this tone system. There are sequences of intervals tending to 1, to ∞, or to any number you wish (say, to the perfect fifth, 3/2).

We solved the problem for 12 granules within the octave (for $M \in \mathbb{N}$, $M \neq 12$, the concrete answer can be given using our techniques). *In the case $M = 12$, there are 23 semitone couples which can form the octave and perfect fifth in real numbers,* c.f. Section 6.5. One couple is rational (the well-known semitones limma and apotome) and 22 irrational. There are 29 different semitones creating these couples. The other result, interesting from the philosophical viewpoint but not surprising one, consists of the fact that there exists no semitone couple in transfinite numbers which yields any Pythagorean System.

Look for some analogies about segmentations described in the natural sciences with the purpose to apply then these reflections to music.

Firstly, there are *segmentations on various levels which are ordered hierarchically.* Some examples. Molecules are segments of matter on the chemical level, they are the smallest parts chemically. However, on the atomic level we have a more tiny segmentation of the matter than molecules. Each atom has also its segments (electron, positron, neutron, etc.) which are composed into structures. But these subatomic units are segmented, too! A higher segmentation level than molecule is based on cells. Cells are organized into an biological organism. A social community (not necessarily the human one) has its segmentation given by individuals. There is also another segmentation sequence over the molecule level: the Earth, the solar system, the galaxy, etc. So, we see that the segmentation hierarchies have not a linear order.

The second reflection is that *time, respectively an evolution, is rather not used for segmentations*. A segmentation takes into consideration only the final state.

The third observation is a claim that we need a microscope or a telescope, a specially developed and powerful tool. When observing a cell, we need a optical microscope; for depicting of atoms, we need an electronic microscope; for describing of the quantum world, we need appropriate instruments—they interact with the observed object (it is very close to the psychological interaction: interpreter ↔ listener). What is a common and important quality of all "microscopes and telescopes"? All they *transform the images*. We see and hear only images of objects which are zoomed, colored, get larger, get smaller, turned, flipped, viewed from inside, upside down, etc. No atom looks like we see it prepared in an electronic microscope. The question is not to obtain a true image (what is it?). We tend to obtain such an image which shows explicitly the segmentation and structure of the entity. Finally, our senses are also only special transform tools (which differs from senses of insects, for instance).

The fourth observation is that there are only a *finite number of all basic segments and also of all segments* (atoms, molecules, cells, men, stars, etc.) together on each segmentation level. Some segments we identify as equivalent ones and therefore there is a question to find the "Mendelejev table" of basic segments for a given segmentation level. Elements of this table should generate every segment and the table should contain the minimal number of elements. As we will see in the case of 12 qualitative granules, the role of basic segments in the Pythagorean System play 29 semitones (which can be grouped specially into 23 couples).

The fifth reflection. Music abstracts both the material world and soul. Therefore, the very *music segmentations may be as complex as segmentations of the world and soul.*

1.2.3 Africa

Under the history of tone systems we understand a time period which ends with the appearance of Helmholtz's book [26]. The early history can be described as an existence of relatively independent, qualitatively different, original, and non-comparable old cultural zones over the Earth: Africa, Asia has its subzones: China, India, Middle Asia, South-East Asia. Then there is the Pacific Ocean region. There are also other, smaller regions, not mentioned here.

At present, African tone systems are not known in detail. It is assumed that some these tone systems certainly go back to the origin of existence of mankind. We do not know whether African systems are mere relic events of tone systems or they represent the fundamental trend to which the tone system development will tend.

Of course, there are also many interactions with the further development and the neighbor influences mainly from Asia and, in the present time, also from Europe and America. We can observe tone systems which are not octave-based. Pentatonic is a very frequent structure. Since in African music hit instruments prevail, there is a sense to study besides discrete pitch spectra also continuous ones (like in the Pacific Region).

P. F. Berliner describes a number of tunings used for the mbira, which is a kalimba-like African instrument common in Zimbabwe (the Shona people). Apparently each region has its own tuning, and different instrument-makers tune their instruments differently. The prevailing theory of Shona mbira tuning is that "mbira makers and players use a distinctive, well defined scale, with only slight variation in different parts of the country .... It can be described as a seven-tone scale...." Cf. [4], p. 66. Berliner found in a sample of tunings that the variation between adjacent scale values was very large, varying between 37 to 286 cents, and not equal at all. The various mbira players select their instruments based on a variety of factors, including tuning, which they refer to collectively as the "chuning" of the instrument. We cite again [4]:

"...I asked several musicians who owned these mbira ... to select from a set of fifty-four forks tuned 4 cents apart (from 212 cents to 424 cents) the individual forks which each thought matched the tuning of the keys on his respective instrument. The fact that they sometimes said that the pitch of an mbira key fell between two tuning forks demonstrated that the musicians could discern fine variations in tuning."

BANDAMBIRA(NO.1)-8 is one of the tone systems Berliner gives, c.f. Example 4. Note that the octave is not a factor of two in this tone system. Apparently the "octaves" are highly variable in mbira tone systems, c.f. *BANDAMBIRA(NO.2)-22*.

Most of the tone systems in Example 4 are results of empirical measurements. These are examples of nonperiodical tone systems. Usually, each tone in the tone system determines a new musical interval. There are a lot of African musical tone systems such that we do not know any but seemingly trivial (irregular, but neither random nor ambiguous!) mathematical structure. The acoustical, psychological, or musical reasons for existence of these tone systems stay hidden for us. We bring a set of examples of such tone systems.

Example 4

BADUMA-8; African Baduma Sanza (idiophone, set of lamellas, thumb-plucked);
0, 694, 997, 1 200, 1 404, 1 611, 1 711, 1 898

BAKWESE-18; Marimba of the Bakwese, SW Zaire, $a' = 140.5$ Hz;
0, 145, 346, 468, 609, 785, 966, 1 123, 1 279, 1 474, 1 580, 1 772, 1 952, 2 146, 2 344, 2 438, 2 674, 2 780

BANDAMBIRA(NO.1)-8; Mubayiwa Bandambira's tuning, c.f. [4];
0, 185, 389, 593, 756, 914, 1 051, 1 302

BANDAMBIRA(NO.2)-22;
Mubayiwa Bandambira's Mbira DzaVadzimu tuning, $a' = 114$ Hz;
0, 355, 554, 650, 829, 982, 1 400, 1 169, 1 850, 1 732, 2 038, 2 207, 2 400, 1 531, 2 415, 2 600, 2 804, 3 008, 3 171, 3 329, 3 466, 3 717

BAPARE-11; African Bapare Xylophone, idiophone, loose-log;
0, 128, 317, 502, 699, 888, 1 141, 1 345, 1 431, 1 604, 1 744

BAPERE-6; African, Bapere Horns Aerophone, made of reed, one tone each;
0, 599, 813, 1 011, 1 217, 1 510

GAMBIA-8; Mandinka balafon scale;
0, 151, 345, 526, 660, 861, 1 025, 1 141

GONDO-22; J. Gondo's Mbira DzaVadzimu tuning, a' = 122 Hz;
0, 323, 480, 644, 830, 981, 1 330, 1 179, 1 888, 1 697, 2 025, 2 189, 2 371, 1 517, 2 390, 2 569, 2 787, 2 923, 3 105, 3 256, 3 417, 3 609

KUKUYA-5;
African Kukuya Horns (aerophone, ivory, one tone only);
0, 279, 562, 789, 980

KUNAKA(NO.1)-8; J. Kunaka's mbira tuning;
0, 196, 377, 506, 676, 877, 1 050, 1 148

KUNAKA(NO.2)-22; J. Kunaka's Mbira Dza Vadzimu, a' = 113 Hz;
0, 455, 547, 757, 935, 1 089, 1 501, 1 260, 1 972, 1 763, 2 153, 2 317, 2 478, 1 638, 2 464, 2 660, 2 841, 2 970, 3 140, 3 341, 3 514, 3 612

MAMBUTI-9; African Mambuti Flutes (aerophone, vertical wooden, one note each);
0, 204, 411, 710, 1 000, 1 206, 1 409, 1 918, 2 321

MUDE-22; Hakurotwi Mude's Mbira Dza Vadzimu, a' = 132 Hz;
0, 174, 289, 575, 612, 770, 976, 1 146, 1 326, 1 467, 1 678, 1 848, 1 987, 2 115, 2 117, 2 348, 2 528, 2 646, 2 860, 3 032, 3 205, 3 465

MUJURU-22;
Ephat Mujuru's Mbira Dza Vadzimu tuning, a' = 106 Hz;
0, 126, 243, 399, 713, 818, 1 232, 1 082, 1 706, 1 443, 1 858, 1 955, 2 219, 1 371, 2 210, 2 400, 2 556, 2 699, 2 918, 3 069, 3 197, 3 437

MBOKO(ZITHER)-8;
African Mboko Zither (chordophone; idiochordic palm fibre, plucked);
0, 206, 345, 528, 720, 814, 1 024, 1 166

SANZA-9; African N'Gundi Sanza (idiophone; thumb-plucked, set of lamellas);
0, 197, 288, 485, 692, 986, 1 204, 1 423, 1 494

SHONA-8; Mbira scale from Zimbabwe (Shona people);
0, 98, 271, 472, 642, 771, 952, 1 148

YASWA-11; African Yaswa Xylophones
(idiophone; calbash resonators with membrane);
0, 209, 416, 686, 926, 1 144, 1 213, 1 377, 1 530, 1 826, 2 025

Example 5 *There are three African scales with the octave period:*
AFRICAN-7; Xylophone from West Africa;
152, 287, 533, 724, 890, 1039, 1 200

BANYORO-5; African Banyoro Xylophone (idiophone; loose log);
193, 494, 3/2, 909, 1 200

VOLANS-7; African scale according to Volans;
171, 360, 514, 685, 860, 1060, 1 200

1.3 Uncertainty-based information theory

1.3.1 Classical harmonic analysis theory

According to classical harmonic analysis theory, the class of tones $\mathcal{T}$ is modelled as follows:

$$\mathcal{T} = \left\{ T : \mathbb{R} \to \mathbb{R}; \quad T(t) = \sum_{k=1}^{\infty} \alpha_k \sin(2\pi\omega_k t + \varphi_k), t \in \mathbb{R} \right\}, \quad (1.1)$$

where t is time, ω_k is the frequency of the k-th harmonic of the tone associated with its *pitch*, α_k is the amplitude of the k-th harmonic (corresponding to its *loudness*), φ_k is the phase of the k-th harmonic (conventionally interpreted as the *entry delay* of the given partial). The values ω_k, α_k, φ_k are supposed to be constants with respect to time t. The pitch function $\Omega : T \to \mathbb{R}$ is defined simply:

$$\Omega(T) = \omega_1 \quad \text{or} \quad \Omega(T) = \log_b \omega_1,$$

where, e.g., $b = 2, 2^{1/1\,200}, 2^{1\,000}, 2^{1\,060}$. According to Fourier's theorem, any tone with a periodic waveform is a sum of *harmonics*, i.e. harmonics with frequencies ω_k satisfying the *harmonic frequency ratio*

$$\omega_1 : \omega_2 : \cdots : \omega_k : \cdots = 1 : 2 : \cdots : k : \ldots.$$

The modelling of physical tones via the class $\mathcal{T}$ of all Fourier series has its psychological counterpart in the Ptolemy System (and vice-versa, c.f. Chapter 2).

Definition 2 We say that a class of tones $\tilde{\mathcal{T}}$ is called the *Ptolemy System* if

$$\tilde{\mathcal{T}} = \{p_1^{\alpha_1} p_2^{\alpha_2} \dots, p_n^{\alpha_n}; \ p_1, p_2, \dots p_n \in \mathbb{P}, \ \alpha_1, \alpha_2, \dots, \alpha_n \in \mathbb{Z}, n \in \mathbb{N}\}.$$

We say also, that a tone system $(\mathbb{T}, \Omega)$, $\mathbb{T} \subset \tilde{\mathcal{T}}$, is *based on prime numbers* $p_1, p_2, \dots, p_n \in \mathbb{P}$ if

$$\mathbb{T} = \{p_1^{\alpha_1} p_2^{\alpha_2} \dots, p_n^{\alpha_n}; \ p_1, p_2, \dots p_n \in \mathbb{P}, \ \alpha_1, \alpha_2, \dots, \alpha_n \in \mathbb{Z}_1\},$$

where $\Omega(T) = \log(T)$ (or often simply $\Omega(T) = T$), $T \in \mathbb{T}$, $\mathbb{Z}_1 \subset \mathbb{Z}$ and $n \in \mathbb{N}$. If the tone system $(\mathbb{T}, \Omega)$ is based on the set of prime numbers $\{(1), 2, 3, 5, \dots, p\}$, then we will say that it is a *p-limit* tone system.

Example 6
5-limit tone systems:

MALCOLM(NO.1)-12; Malcolm's Monochord I;
16/15, 9/8, 6/5, 5/4, 4/3, 45/32, 3/2, 8/5, 5/3, 16/9, 15/8, 2

RING(K/NO.1)-7; H. Partch's double-tie circular mirroring of 4:5:6 and 5-limit tonality Diamond;
6/5, 5/4, 4/3, 3/2, 8/5, 5/3, 2

TURKISH-7; Turkish tone system, 5-limit From Palmer on an album of Turkish music;
16/15, 5/4, 4/3, 3/2, 5/3, 16/9, 2
7-limit tone systems:

FOKKER-12; Fokker's 7-limit 12-tone just scale;
15/14, 9/8, 7/6, 5/4, 4/3, 45/32, 3/2, 45/28, 5/3, 7/4, 15/8, 2

FORTUNA-12; 11-limit tone system from [17];
21/20, 8/7, 7/6, 14/11, 21/16, 10/7, 32/21, 11/7, 12/7, 7/4, 40/21, 2

OCTONY(7-LIM)-8; 7-limit Octony, c.f. L. Euler: Genus Musicum, Chapter 6, p. 118, on white keys $+ B_\flat$;
35/32, 5/4, 21/16, 3/2, 105/64 7/4, 15/8, 2

OTHER-12; Other Music, 7-limit black keys;
15/14, 9/8, 7/6, 5/4, 4/3, 7/5, 3/2, 14/9, 5/3, 7/4, 15/8, 2
    13-limit tone system:

PARTCH-41; 13-limit tone system after H. Partch: Genesis of a Music, p. 454, 2nd edition;
14/13, 13/12, 12/11, 11/10, 10/9, 9/8, 8/7, 7/6, 13/11, 6/5, 11/9, 16/13, 5/4,
14/11, 9/7, 13/10, 4/3, 11/8, 18/13, 7/5, 10/7, 13/9, 16/11, 3/2, 20/13, 14/9,
11/7, 8/5, 13/8, 18/11, 5/3, 22/13, 12/7, 7/4, 16/9, 9/5, 20/11, 11/6, 24/13,
13/7, 2
    17-limit tone systems:

MALCOLM(NO.2)-12; Malcolm's Monochord II;
17/16, 9/8, 19/16, 5/4, 4/3, 17/12, 3/2, 19/12, 5/3, 85/48, 15/8, 2

Besides harmonics, there are sounds with no salient pitch, e.g., c.f. [133], which are of various types. For instance, the ratio of their partial frequencies ω_k is not harmonic but *inharmonic*. For instance, the ratio $1 : \sqrt[12]{2}$ is used in the 12-tone Equal Temperament. There are some tone systems with inharmonic intervals:

Example 7

ARABIC(NO.1)-12; From [17]. *Try C or G major;*
100, 200, 300, 350, 500, 600, 700, 800, 900, 1 000, 1 050, 1 200

ARISTO(SYN/DIAT)-7;
Aristoxenos Diatonon Syntonon, Dorian Mode;
100, 300, 500, 700, 800, 1 000, 1 200

ARISTO(CHRO/UNM)-7; ;
Unmelodic Chromatic, genus of Aristoxenos, Dorian Mode, 4.5 + 3.5 + 22 parts;
75, 400/3, 500, 700, 775, 2 500/3, 1 200

UR-TEMES-5; Ur-Temes's 5-tone ϕ scale;
273.000, 366.910, 466.181, 560.090, 833.090

DIAT/TET-7; Equal Diatonic, Islamic form;
500/3, 500·(2/3), 500, 700, 700+500/3, 700 + 500·(2/3), 1 200

An example of sound with no salient pitch is *noise*. There are several types of noises. Another types of musical sounds will be discussed in Chapter 3.

The standard musicological definition is that a musical interval is consonant if it sounds pleasant or restful; a consonant interval has little or no musical tension or tendency to change. Dissonance, on the other hand, is the degree to which an interval sounds unpleasant or rough; dissonant intervals generally feel tense and unresolved.

In [26], Helmholtz offers a classical, in the present-day physiologically-based, explanation for consonance that is based on the phenomenon of *beats*. If two tones are sounded at almost the same frequency, then a distinct beating occurs that is due to interference between the two tones (piano tuners use this effect regularly). The beating becomes slower as the two tones move closer together, and completely disappears when the frequencies are identical. Typically, slow beats are perceived as a pleasant vibrato while fast beats tend to be rough and annoying. Recalling that any timbre can be decomposed into sine wave components, Helmholtz theorized that dissonance between two tones is caused by the rapid beating of various sine wave components. Consonance, according to Helmholtz, is the absence of such dissonant beats.

Example 8

HELM-7; Helmholtz's Chromatic scale;
16/15, 5/4, 4/3, 3/2, 8/5, 15/8, 2

HELM(PURE/NO.1)-24;
Helmholtz's two-keyboard harmonium scale, untempered;
135/128, 16/15, 10/9, 9/8, 75/64, 32/27, 5/4, 81/64, 675/512, 4/3, 45/32, 729/512, 6 075/4 096, 3/2, 25/16, 405/256, 5/3, 27/16, 225/128, 3 645/2 048, 15/8, 243/128, 2 025/1 024, 2

HELM(PURE/NO.2)-24; Simplified HELM(PURE/NO.1)-24;
135/128, 16/15, 10/9, 9/8, 75/64, 32/27, 5/4, 512/405, 675/512, 4/3, 45/32, 64/45, 40/27, 3/2, 25/16, 128/81, 5/3, 27/16, 225/128, 16/9, 15/8, 256/135, 160/81, 2

HELM(TEMP)-24;
Helmholtz's two-keyboard harmonium tempered scale;
91.446, 111.976, 182.892, 203.422, 274.338, 294.868, 5/4, 406.843, 477.760, 498.289, 589.735, 610.265, 681.181, 701.711, 25/16, 793.157, 884.603, 905.132, 976.049, 996.578, 1 088.025, 1 108.554, 1 179.471, 2

Besides the beat theory, there are also other theories of consonance. It would be naive to suggest that truly musical properties can be measured as a simple tonal consonance or psychological ability. Even in the realm of harmony, consonance is not the whole story. Indeed, a harmonic progression that was uniformly consonant would likely be boring. Harmonic interest arises from a complex interplay of dissonance (restlessness) and consonance (rest). We cannot ignore musically essential aspects such as melody and rhythm. So, each music deals with a set of tones equipped with a special structure. The theoretical definition of tone as a stationary air/liquid vibration representable via Fourier series is not satisfactory for the present time.

The theory of tone systems in this book is a presentation of a new approach which is able to integrate the many various and different mathematical, psychological, psychoacoustical, musically-spiritual concepts of tone systems, including nonstationary signals and noises. It is based on the notion of uncertainty-based information which can be three types and their mixtures.

Roughly speaking, to have the simplified view, mathematical and acoustical motives in tone system building lead to uncertainty called *nonspecificity*, c.f. Chapter 3. Psychological and psychoacoustical reasons lead to uncertainty called *fuzziness*, c.f. Chapter 2. Cultural-musical and spiritual–reasons lead to uncertainty called *strife*, c.f. Section 4.1. In fact, each tone system has these three uncertainty sides which are inseparable.

On this very general (philosophical) level we adopted the viewpoint of G. J. Klir and M. J. Wierman, which they presented in the book [33].

1.3.2   Significance of uncertainty

When dealing with real-world problems (and, in particular, problems of tone systems), we can rarely avoid uncertainty. At the empirical (psychoacoustical) level, uncertainty is an inseparable companion of almost any measurement, resulting from a combination of inevitable measurement errors and resolution limits of measuring instruments. At the cognitive level, it emerges from the vagueness and ambiguity inherent in musical language. At the psychological level, uncertainty has even uses and it is often created and maintained by people for different purposes (individual dissonance curves, education, etc.). As a result of the famous Gödel theorem, we are now aware that even mathematics is not immune from uncertainty.

When encountered with uncertainty regarding our purposeful actions, we are forced to make decisions. This is a subtle connection between uncertainty and decision making: in a predestinate world, decision would be *illusory*; in a world of a perfect foreknowledge—*empty*; in a world without natural order—*powerless*. Our intuitive attitude to life implies non-illusory, non-empty, non-powerless decision ... Since decision in this sense excludes both perfect foresight and anarchy, it must be defined as choice in the face of bounded uncertainty.

Western intellectual culture has been preoccupied with the pursuit of absolutely certain knowledge or, barring that, the nearest possible approximation of it. This preoccupation appears to be responsible not only for the neglect of uncertainty by Western art and science, but also, prior to the 1960s, the absence of adequate conceptual frameworks for seriously studying it.

The change of attitude towards uncertainty since the 1960s was caused by a sequence of events. The emergence of computer technology in the 1950s opened qualitative new possibilities. These, in turn, aroused interest of some researchers to study certain problems that, due to their enormous complexities, were previously beyond the scope of scientific inquiry (problems of *organized complexity*). These problems involve nonlinear systems with large numbers of components and rich interactions among the components, which are usually nondeterministic, but not as a result of randomness that could yield meaningful statistical averages. They are typical in life, behavioral, social, and environmental sciences, as well as in applied fields such as modern technology or medicine. In particular—in tone systems.

Shortly after the emergence of computer technology, it was the common belief that the level of complexity we can handle is basically a matter of the level of the computational power at our disposal. In the early 1960s, this naive belief was replaced with a more realistic outlook. There are definite limits in dealing with complexity, which neither our human capabilities nor any computer technology can overcome. One such limit was determined by H. Bremermann in 1962 by simple considerations based on quantum theory. Bremermann calculates the total number of bits processed by a hypothetical computer the size of the Earth within a time period equal to the estimated age of the Earth. This imaginary computer would not be able to process more than 10^{93} bits (Bremermann's limit). Problems that require processing more than 10^{93} bits of information are called *transcomputational problems.*

Many problems dealing with systems of even modest size (smaller than 10^{93}) exceed the limit in their information-processing demands. The nature of these problems has been extensively studied within an area referred to as the *theory of computational complexity*, which emerged in the 1960s as a branch of the *general theory of algorithms*. The main challenge in pursuing the problems that possess the characteristics of organized complexity narrows down fundamentally to one question: *how to deal with systems and associated problems whose complexities are beyond our information-processing limits? That is, how can we deal with these systems and problems if no computational power alone is sufficient? The only possible answer is that we must adequately simplify them to make them computationally tractable.* In every simplification, unfortunately, we lose something.

In general, we deal with problems in tone systems that are constructed either as models of some aspects of an existing musical reality (e.g., folk songs, sounds from nature, acoustics of instrument material, and building, etc.) or as models of some desirable man-made objects (e.g., musical compositions, instrument production, education, etc.). The purpose of constructing models of the former type is to understand some phenomena of reality, be it natural or man made, making adequate predictions or retrodictions, learning how to control the phenomena in any desirable way, and utilizing all these capabilities for various ends; models of the latter type are constructed for the purpose of prescribing operations by which a conceived artificial object can be constructed in such a way that desirable objective criteria are satisfied within given constraints. In constructing a model, we always attempt to maximize its usefulness in a predefined sense. This aim is closely connected with the relationship among three key characteristics of every systems model: *complexity, credibility,* and *uncertainty.* This relationship, c.f. [33], is not as yet fully understood. We only know that uncertainty has a pivotal role in any efforts to maximize the usefulness of systems models. Although usually undesirable when considered alone, uncertainty becomes very valuable when considered in connection to the other characteristics of systems models: a slight increase in uncertainty may often significantly reduce complexity and; at the same time, increase credibility of the model.

Together with the emotional side, every music has also an informational content. The concept of information is intimately connected with the concept of uncertainty. The most fundamental aspect of this connection is that the uncertainty involved in any problem-solving situation is a result of some information deficiency. Information (pertaining to the model within which the situation is conceptualized) may be incomplete, imprecise, fragmentary, not fully reliable, vague, contradictory, or deficient in some other way. These various information deficiencies may result in different types of uncertainty.

Assume that we can measure the amount of uncertainty involved in a problem solving situation conceptualized in the mathematical theory of tone systems. Assume further that the amount of uncertainty can be reduced by obtaining relevant information as a result of some action (finding a relevant new fact, designing a relevant experiment and observing the experimental outcome, receiving a requested message, discovering a relevant historical record). Then, the amount of information obtained by the action may be measured by the reduction of uncertainty that results from the action. Let us call it *uncertainty-based information.*

The nature of uncertainty-based information depends on the theory within which uncertainty pertaining to various problem-solving situations is formalized. Each formalization of uncertainty in a problem-solving situation is a mathematical model of the situation. When we commit ourselves to the mathematical theory of tone systems, our modelling becomes necessarily limited by the constraints of the theory. A more general theory is capable of capturing uncertainties of some problem situations more faithfully than its less general competitors. As a rule, however, it involves greater computational demands.

Uncertainty-based information was first conceived in terms of classical set theory by Hartley in 1928 and, later, in terms of probability theory by Shannon in 1948. Research on a broader conception of uncertainty-based information, liberated from the confines of classical set theory and probability theory, began in the early 1980s (Higashi, Klir, Höhle, Yager, etc.). The name *generalized information theory* was coined for a theory based upon this broader conception by Klir. The ultimate goal of generalized information theory is to capture the properties of uncertainty-based information formalized within any feasible mathematical framework. Although this goal has not been fully achieved as yet, substantial progress has been made in this direction since the early 1980s.

1.3.3   Uncertainty forms and principles

When the seemingly unique connection between uncertainty and probability theory was broken, and uncertainty began to be conceived in terms of the much broader frameworks of fuzzy set theory and fuzzy measure theory, it soon became clear that uncertainty can manifest itself in different forms. These forms represent distinct types of uncertainty. Three types of uncertainty are now recognized in the following five theories, which are currently the only theories in which measurement of uncertainty is well established: classical set theory, fuzzy set theory, probability theory, possibility theory, and evidence theory.

The three uncertainty types are:
• *fuzziness* (lack of definite or sharp distinction, vagueness, haziness, cloudiness, uncleanness, indistinctness, sharplessness), which results from the imprecise boundaries of sets; the "psychological phenomenon" in tone systems theory; gamelan orchestras;
• *nonspecificity* (disagreement in choosing among several possible alternatives, principal imprecision); it may be connected with sizes (cardinalities) of relevant sets of alternatives; dissonance, incongruence, discrepancy, discord, imprecision); the "physical phenomenon" in the tone systems theory; using different frequency values for the

same note depending on the harmonic context (tritone values, different whole tones, minor thirds, etc.);

- *strife* (conflicts among the various sets of alternatives, variety, discord, generality, diversity, equivocation) results from the choosing one from two or more alternatives; Garbuzov theory; the "musically-cultural phenomenon" in tone systems theory; Indian ragas.

Two of the uncertainty types, nonspecificity and strife, are viewed as species of a higher uncertainty type, which seems well captured by the term *ambiguity*. The latter is associated with any situation in which it remains unclear which of several alternatives should be accepted as the genuine one. In general, ambiguity results from the lack of certain distinctions characterizing an object (nonspecificity), from conflicting distinctions, or from both of these. The third uncertainty type, fuzziness, is different from ambiguity; it results from the lack of sharpness of relevant distinctions. It is conceivable that other types of uncertainty will be discovered when the investigation of uncertainty extends beyond the boundaries of the above mentioned five theories of uncertainty. We will discuss about such a candidate type in Section 4.2.3 (a join uncertainty, *harmony-melody* uncertainty).

Rather than speculating about this issue, we restrict our explanations to the three currently recognized types of uncertainty (and the associated information). The measures of the three types of uncertainty in classical set theory, fuzzy set theory, probability theory, possibility theory, and evidence theory are summarized in [33].

The use of uncertainty measures for dealing with tone systems problems is well captured by three general principles of uncertainty. These principles are: a *principle of minimum uncertainty*, a *principle of maximum uncertainty*, and a *principle of uncertainty invariance*. Due to the connection between uncertainty and uncertainty-based information, these principles may also be interpreted as *principles of information*.

Let us summarize now basic ideas of the three uncertainty principles. Given an experimental frame within which we choose to operate, and some evidence regarding entities of this frame, we express information based upon the evidence in terms of some theory of uncertainty. This expression may be viewed as our information domain within which we can deal with a variety of associated problems.

Note that problems that involve neither a change in information domain nor a change of the uncertainty theory employed can be solved by appropriate chains of logical inference within the information domain. No additional principles are needed. Three classes of problems bounded with the uncertainty principles are readily distinguished:

• *Reduction of the information domain, i.e. information loss.* As a rule, these problems result in multiple solutions. A principle is needed by which we choose meaningful solutions from a given solution set. Since the loss of information is generally undesirable, the *principle of minimum uncertainty* is pertinent. It ensures that we solve the problem with minimum information loss;

• *Extension of the information domain.* Information contained in the information domain is not sufficient to determine unique solutions to these problems. To choose meaningful solutions from all solutions compatible with the information domain, we need to resort to a principle. Since any introduction of information from outside the information domain (i.e., information not supported by evidence) is unwarranted on epistemological grounds, we need to avoid it. This can be accomplished by the *maximum uncertainty principle.* By maximizing uncertainty subject to the constraints expressed by the given information domain, a solution is chosen (or constructed) whose amount of information does not exceed the amount of information associated with the information domain;

• *Information domain is transformed.* Information domain expressed in one theory is transformed into a corresponding expression in another theory. While information is represented differently in the two theories, no information should be unwittingly added or eliminated solely by changing the representation. The *uncertainty invariance principle* ensures that the amount of information expressed by the evidence remains constant no matter in which form it is expressed.

Chapter 2

Fuzziness and sonance

Psychologically, our music perception depends on our cultural environment, education, age, temperament, mood, etc. In Table 2.1, we see various series of consonant musical intervals according to Pythagoras, Helmholtz, Stumpf, Plomp–Levelt, and the Middle Ages period, [26], [31], [161].

This perception can be measured and resulted into our individual *dissonance function* which depends on the timbre. The principle of local consonance describes a relationship between the timbre and a tone system in the narrower sense in which the timbre will sound most consonant. We are able to answer two following complementary questions: Given a timbre, what tone system should it be played in? Given a tone system, how can consonant timbres be chosen? In this chapter we explain the basic ideas of the theory of R.Plomp and W. J. M. Reiner about relation tone system ↔ timbre from the view of W. Sethares, c.f. [60].

TABLE 2.1: Series of consonant intervals

	Pythagoras	Middle Ages	Helmholtz	Stumpf	Plomp–Levelt
1	octave	octave	octave	octave	octave
2	fifth	fifth	fifth	fifth	fifth
3	fourth	major third	fourth	fourth	major sixth
4	major third	major sixth	major sixth	major third	fourth

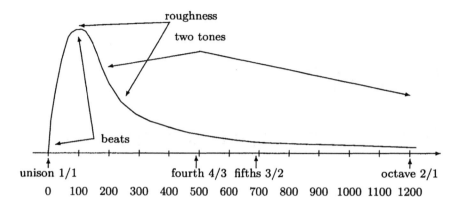

FIGURE 2.1: Sensory dissonance function of two sinusoidal tones

The present theory of dissonance functions is based on three simplifications. We suppose that
• the *dissonance function and timbre do not depend on the reference tone.* This is approximately true only locally (approximately one and a half octaves), c.f. Subsection 1.1.2;
• *tones are represented as a linear combination of sine harmonics* (not wavelets);
• *we deal with only finite many harmonics* (synthesized timbres).

2.1   Dissonance functions

Let a tone T is given as a set of n partial sine tones with frequencies $\omega_i \in \mathbb{R}$ and amplitudes $a_i \in \mathbb{R}^+$, $i = 1, 2, \ldots, n$, c.f. Section 1.3.1, (1.1).

Consider a psychological interaction function of a *couple of sine tones with equal amplitudes* (the tuning-fork tones) when one of them is stable and the second one is moving in pitch. Let the horizontal axis be logarithm scaled with pitch values ω in cents. Let the vertical axis be a measure scale of the subjective feeling of dissonance. Figure 2.1 shows an "averaged" shape of the dissonance function (one half, the function is symmetrical with respect to the axis y; the "consonance curve" is simply flipped upside-down). We can see that dissonance begins at zero (the relative interval of tones: a unison) increases rapidly to a maximum (the relative interval of tones: a minor second, about 100 cents), and then falls back towards zero. Note that the musically consonant intervals are undistinguished— there is no dip in the curve at the fourth, 4/3, fifth, 3/2, or even the octave, 2/1.

Here are two examples of analytically expressed dissonance functions (fuzzy sets) $\delta_{a_*,\omega_*}(a,\omega)$ of two sine tones, where we fix (a_*,ω_*) and move (a,ω), where $a, a_* \in \mathbb{R}^+$ and $\omega, \omega_* \in \mathbb{R}$.

Example 9

(1)

$$\delta_{a_*,\omega_*}(a,\omega) = a_* \cdot a \quad 2^{-\frac{1}{(\omega-\omega_*)^2}+(\omega-\omega_*)^2}$$

(2) W. Sethares, c.f. [60], considered the following dissonance function:

$$\delta_{a_*,\omega_*}(a,\omega) = a_* \cdot a \left[e^{-\frac{0.8(\omega-\omega_*)}{s(\omega)}} - e^{\frac{-1.38(\omega-\omega_*)}{s(\omega)}} \right],$$

where $\omega \leq \omega_*$ and $s(\omega) = 0.021\omega + 19$.

If we know the dissonance function δ_{a_*,ω_*} of two sine tones, we may construct the dissonance function Δ_T (which is psychologically subjective, of course) of the tone $T(a,\omega)$ (timbre) composed of n partial tones. Consider first the dissonance function in the form:

$$\Delta_T(a,\omega) = \sum_{n=1}^{N} \delta_{a_n,\omega_n}(a,\omega),$$

where $\omega \in \mathbb{R}$, $\omega_n \in \mathbb{R}$ are frequencies of tones and $a\mathbb{R}^+$, $a_n \in \mathbb{R}^+$ are their amplitudes, respectively, $n = 1, 2, \ldots, N$.

Example 10 Let the dissonance function of sine tones

$$\delta_{a_*,\omega_*}(a,\omega) = \left(\frac{2}{1+a-a_*} \right)^{-\frac{1}{(\omega-\omega_*)^2}+(\omega-\omega_*)^2}, \tag{2.1}$$

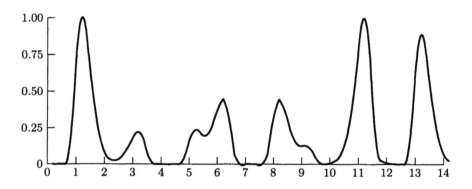

FIGURE 2.2: A dissonance function (one half)

where $\omega, \omega_* \in \mathbb{R}$, $a, a_* \in [0,1]$. In Figure 2.2, there is a (normalized, one-half) graph of the function

$$\begin{aligned}\Delta_T(1,\omega) = \quad &\delta_{1,0}(1,\omega) + \delta_{0.88,1\,200}(1,\omega) + \delta_{0.88,-1\,200}(1,\omega) + \\ &\delta_{0.44,700}(1,\omega) + \delta_{0.44,-700}(1,\omega) + \delta_{0.22,400}(1,\omega) + \\ &\delta_{0.22,-400}(1,\omega) + \delta_{0.11,1\,000}(1,\omega) + \delta_{0.11,-1\,000}(1,\omega),\end{aligned}$$

where values of frequencies of partial tones $\omega_i = 0$, $1\,200$, 700, 400, $1\,000$ are in cents (in the 100 cent scale in Fig 2.2).

It can be easily computationally verified, that the shape of this and also of similar functions Δ_T are very sensitive on small changes in the parameters a_n, ω_n of functions δ_{a_n,ω_n}, $n = 1, 2, \ldots, n$. Reversely, choosing these parameters a_n, ω_n and function δ_{a_*,ω_*}, we can model any dissonance function Δ_T, in fact, very freely. Consequently, the psychological factor plays a very important role when studying tonal sonance much more than e.g. psychoacoustical factors. It is clear also that the notion of dissonance function is of a great degree of complexity and we need to use the computer help when obtaining computational results about these functions.

Deal now with the process of the sequential enriching of a sine tone with additional one, two, and more partial sine tones. For the sake of simplicity, use the aggregation function Δ_T in the form (2.1) and the following dissonance function of sine tones $\delta_{a_*,\omega_*}(a,\omega) =$

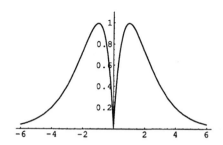

FIGURE 2.3: Dissonance function $|\omega|e^{-|\omega|}$

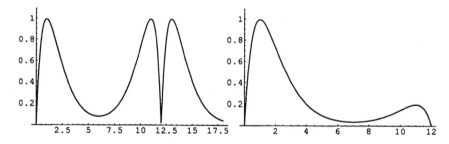

FIGURE 2.4: Dissonance functions, two harmonics

$a \cdot a_* |\omega - \omega_*| e^{-|\omega - \omega_*|} \in [0, 1]$, where $\omega, \omega_* \in \mathbb{R}$ are frequencies of tones (in the 100 cent scale in Fig 2.3) and $a, a_* \in \mathbb{R}^+$ are their amplitudes, respectively. For the first two harmonics, the unison $\omega_1 = 0$ and octave $\omega_2 = 1\,200$, the function $\Delta_T(a, \omega) = \delta_{a_1,0}(a, \omega) + \delta_{a_2,1\,200}(a, \omega)$ (its one half) is graphed in the Figure 2.4, where $a_1 = a_2 = a = 1$. For a curiosity, let us find the ratio of amplitudes a_2/a_1 such that the intersection of functions $\delta_{a_1,0}(a, \omega)$ and $\delta_{a_2,1\,200}(a, \omega)$ will occurs exactly in the pure fifth (or, equivalently saying, which ratio a_2/a_1 of amplitudes of the unison and octave produces the local minimum in the pure fifth?). Applying the well-known calculus theorems to the function $\Delta_T(a, \omega)$ it can be easily verified that this ratio is $a_2/a_1 \approx 1/5$, c.f. Figure 2.4.

Analogously, we can consider more harmonics. In Figure 2.5 are graphed two curves of $\Delta_T(a, \omega)$ as examples: (1) $\Delta_T(a, \omega) = \delta_{1,0}(a, \omega) + \delta_{0.66,1\,200}(a, \omega) + \delta_{0.5,700}(a, \omega)$, where $\omega_1 = 0$, $\omega_2 = 1\,200$, $\omega_3 = 700$ cents, respectively. (2) For the unison, octave, fifth, ma-

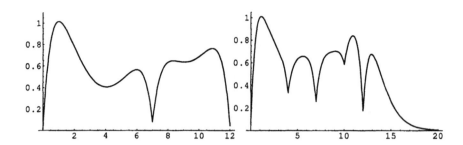

FIGURE 2.5: Dissonance functions, three and five harmonics

jor third, natural seventh, fourth, with $\omega_i = 0,\ 1\,200,\ 700,\ 400,\ 1\,000$ cents, and weights $a_i = 1,\ 0.66,\ 0.4,\ 0.3,\ 0.2$, respectively.

What about the process of *real aggregation of harmonics to one dissonance function*? W. Sethares, [60], suggested the following analytical definition of the dissonance function of tones which are compositions of n partial tones.

Definition 3 Let $\delta_{a_*,\omega_*} : \mathbb{R}^+ \times \mathbb{R} \to \mathbb{R}^+$, $a_* \in \mathbb{R}^+$, $\omega_* \in \mathbb{R}$ be a dissonance function of sine tones. Let $\omega_1 < \omega_2 < \cdots < \omega_n$, $0 \le a_i$, $i = 1, 2, \ldots, n$. We say that $\Delta_T : \mathbb{R}^+ \times \mathbb{R} \to \mathbb{R}^+$ is a *dissonance function of the tone* (timbre) $T(t) = \sum_{i=1}^{n} a_i \sin(2\pi\omega_i t + \varphi_i)$, $t \in \mathbb{R}$, if

(1) $\Delta_T(a, \omega) = \sum_{i=1}^{n} \sum_{j=1}^{i} \delta_{a_i - a_j, \omega_i / \omega_j}(a, \omega)$;

(2) $\Delta_T(a, \omega) \ge \Delta_T(a, 0)$;

(3) $\log \Delta_T(a, \omega) = \log \Delta_T(a, -\omega)$;

(4) $\forall a > 0$, $\exists \lim_{\omega \to \infty} \Delta_T(a, \omega)$; $\lim_{\omega \to \infty} \Delta_T(a, \omega) \le \Delta_T(a, 0)$;

(5) The function $\Delta_T(a, \cdot)$, $a > 0$, has at most $2n(n-1)+1$ minima;

(6) $\Delta_T(a, \omega_i / \omega_j) = \min_{\omega \in (\omega_i / \omega_j - \varepsilon, \omega_i / \omega_j + \varepsilon)} \Delta_T(a, \omega)$,
     where $a \in \mathbb{R}^+$, $\omega \in \mathbb{R}^+$ and $i, j = 1, 2, \ldots, n$.

The first item of Definition 3 says that Δ_T is a fuzzy set (we may normalize the function Δ_T with respect to 1 if we wish) such that the resulting dissonance is the sum of dissonances of all pairs of partial tones. The second item says that in the unison is the global minimum of Δ_T. The third item is a symmetry expression of Δ_T on the logarithm scale. The fourth item says that Δ_T is vanishing under the dissonance of the unison value. The fifth condition says that functions $d(a_i, \omega_i)$ are "sufficiently smooth." Clearly, these minima located on a logarithmic scale are symmetrically situated. The sixth item says that, up to half the local minima occur at intervals $a = \omega_i/\omega_j$, where ω_i and ω_j are frequencies of partial tones, $i, j = 1, 2, \ldots, n$. This condition is known also as the *principle of local consonance*.

2.2 Relatedness tone system $\leftrightarrow$ timbre

2.2.1 Direction: timbre $\rightarrow$ tone system

For an arbitrary timbre (one tone T in the form (1.1) defines timbre; perhaps one whose spectrum does not consist of a standard harmonic series) and using construction of dissonance function given by Definition 3, it is straightforward to draw the dissonance function generated by T. The local minima of this function occur at values which are good candidates for values of a tone system, since they are local points of minimum dissonance.

The optimal tone system for a given timbre is found simply by locating the local minima of the dissonance function.

Definition 4 *A timbre T and a tone system $(\mathbb{T}, \Omega)$ are said to be related* (more precisely, Δ_T-related, where Δ_T is a dissonance function) *if the timbre has a dissonance function Δ_T whose local minima occur at tone system values $\omega \in \Omega\mathbb{T}$).*

Example 11 Given a dissonance function, the points of local consonance for the harmonic timbre with partial tones at $(\omega, 2\omega, \ldots, n\omega)$ are located at integer numbers. Indeed, candidate points of local consonance are at intervals α for which $\omega_i = \alpha\omega_j$. Since the frequencies of partial tones are integer multiples of ω, $\alpha = n/m$ for integers $i, j = 1, 2, \ldots, n$, the principle of local consonance says that the most appropriate tone system for harmonic timbres are located at such α. This provides a psychoacoustic basis for justly intonated temperaments.

2.2.2   Direction: tone system $\rightarrow$ timbre

Given a tone system S (in the narrower sense, a set of N real numbers) and a dissonance function δ of sine tones, to find timbres which will generate dissonance functions Δ with local minima at precisely the tone system frequencies. This problem of finding an optimal timbre for a given tone system is not as simple. In general, there is a class of "best" timbres for a given tone system. However, for certain classes of tone systems (such as equal temperaments) the properties of the dissonance function, c.f. Definition 3, can be exploited to solve the problem efficiently.

The timbre selection is an optimization problem. We have to chose a set of n partial tones with frequencies $(\omega_1, \omega_2, \ldots, \omega_n)$ and amplitudes $(a_1, a_2, \ldots, a_n)$ to minimize the sum of the dissonances over the $N-1$ intervals,

$$C(a,\omega) = w_1 \cdot \left(\sum_{i=1}^{n} \sum_{j=1}^{n} \delta_{a_i-a_j, \omega_i/\omega_j}(a,\omega) \right) + w_2 \cdot N,$$

where N is the cardinality of the tone system, and w_1, w_2 are weighting factors (obtained experimentally). The sum is in fact over the $N-1$ intervals of local minima. Minimizing the cost C tends to place the scale steps at local minima as well as to minimize the value of the dissonance function.

This approach to the problem of timbre selection can lead to following trivial solutions: zero dissonance can be achieved by setting all the amplitudes to zero, or by allowing the a_i to become arbitrarily large, c.f. properties of dissonance functions. To avoid these solutions, the following constraints are necessary: (i) we do not allow the amplitudes to change, $a(\omega) = const(t)$ (i.e., the abbreviation to the stationary tone signals is used essentially [no wavelets!]); and, (ii) force all frequencies to lie in a predetermined range of interest (e.g., within one octave [1,2] using the so-called octave equivalency).

Minimizing the cost C is an n-dimensional optimization problem with a highly complex error surface. Fortunately, such problems can be solved adequately (though not necessarily optimally) using a variety of *random search* methods such as *simulated annealing*. A *genetic algorithm* seems to work well. It requires that the problem be coded in a finite string, the *gene* and that a *fitness* function be defined. Genes for the timbre selection problem are formed by concatenating binary representations of the ω_i. The fitness function of the gene $(\omega_1, \omega_2, \ldots, \omega_n)$ is measured as the value of the cost C above, and timbres are judged "more fit" if the cost C is lower. The genetic algorithm searches n-dimensional space measuring the fitness of timbres. The most fit are combined (via a "mating" procedure) into "child timbres" for the next generation. As generations pass, the algorithm tends to converge, and the most fit timbre is a good candidate for the minimizer of C. Indeed, the genetic algorithm tends to return timbres which are well matched to the desired tone system in the sense that scale steps tend to occur at points of local consonance and the total dissonance at scale steps is low. For example, when the 12 tone equal tempered scale is specified, the genetic algorithm converges near harmonic timbres quite often. This is a good indication that the algorithm is functioning and that the free parameters have been chosen sensibly.

The following examples of synthesizing timbres for given tone systems are results of Slaymaker, Mathews, Pierce, and Sethares, c.f. [60].

Example 12 10-tone Equal Temperament For certain tone systems, e.g., equal tempered tone systems, properties of the dissonance function can be exploited to quickly and easily design timbres, thus bypassing the need to run an optimization program. Recall that the ratio between successive scale steps in N-tone Equal Temperament with the octave period is $X = \sqrt[N]{2}$. Consider timbres for which successive partials are ratios of powers of X. Each partial of such a timbre, when transposed into the same octave as the fundamental, lies on a note of the tone system. Such a timbre is said to be induced by the N-tone Equal Temperament.

Induced timbres are good candidate solutions to the optimization problem. By Definition 3, points of local consonance tend to be located at intervals a for which $\omega_i = a\,\omega_j$ where ω_i and ω_j are partials of the searched timbre T. Since the ratio between any pair of partials in an induced timbre is X raised to the k power for some integer k, the dissonance function will tend to have points of local consonance at such ratios: these ratios occur precisely at steps of the scale. Such timbres tend to minimize the cost C.

This insight can be exploited in two ways. First, it can be used to reduce the search space of the optimization routine. Instead of searching over all frequencies in a bounded region, the search need only be done over induced timbres. More straightforwardly, the timbre selection problem for equal tempered tone systems can be solved by careful choice of induced timbres.

As an example, consider the problem of designing timbres to be played in 10-tone Equal Temperament. This tone system is often considered as one of the worst temperaments for harmonic music, since the steps of the 10-tone Equal Temperament are distinct from the (small) integer ratios, implying that harmonic timbres are very dissonant. The principle of local consonance asserts that these intervals will become more consonant if played with correctly designed timbres. Here are three timbres (given as the sets of partials) induced by the 10-tone Equal Temperament.

$$(\omega, X^{10}\omega, X^{17}\omega, X^{20}\omega, X^{25}\omega, X^{28}\omega, X^{30}\omega),$$
$$(\omega, X^{7}\omega, X^{16}\omega, X^{21}\omega, X^{24}\omega, X^{28}\omega, X^{37}\omega),$$
$$(\omega, X^{7}\omega, X^{13}\omega, X^{17}\omega, X^{23}\omega, X^{28}\omega, X^{30}\omega),$$

where $X = \sqrt[10]{2}$. They really are consonant when played on a 10 tone equal tempered scale. Not surprisingly, the same tones sound quite dissonant when played in a standard 12 tone scale. Analogous arguments suggest that the consonance of 12-tone equal tempered tuning can be maximized by moving the partials away from the harmonic series to a series based on $X = \sqrt[12]{2}$.

Example 13 Brasses, Strings, Bells Any arbitrary timbre (set of frequencies and amplitudes) can be realized with the aid of a computer. Is it always possible to design acoustic instruments that will have a given timbre? How about brasses? Fletcher and Rossing proclaim that "If the flaring part of the horn extends over a reasonable fraction of the total length, for example around one third, then there is still enough geometric flexibility to allow the frequencies of all modes to be adjusted to essentially any value desired." With stringed instruments, the trick is to find a variable thickness string that will vibrate with partials at the desired frequencies. The partials of a drumhead can be tuned by weighting or layering sections of the drumhead. The partials of reed instruments can be manipulated by the contour of the bore as well as the shape and size of the tone holes. Bells can be tuned by changing the shape and thickness of the walls.

Example 14 Stretched and compressed timbres Pierce, Slaymaker, and Mathews have investigated timbres with partials at $\omega_j = \omega \beta^{\log_2(j)}$. When $\beta = 2$, this is simply a harmonic timbre, since $\omega_j = \omega 2^{\log(j)} = j\omega$. When $\beta < 2$, the frequencies of the timbre are compressed, while when $\beta > 2$, the partials are stretched. The most striking aspect of compressed and stretched timbres is the lack of a real octave. The frequency ratio β plays the role of the octave, which Mathews and Pierce call the *pseudo-octave*. Real octaves sound dissonant and unresolved when β is different from 2 while the pseudo-octaves are highly consonant. More importantly, the graph of each function has a similar contour. Points of local consonance occur at (or near) the twelve equal steps of the pseudo-octaves. *Pseudo-fifths*, *pseudo-fourths*, and *pseudo-thirds* are readily discernible. This suggests that much of music theory and practice can be transferred to to compressed and stretched timbres, when played in compressed and stretched scale.

Example 15 Xylophone tone systems It is well known that xylophones, and other instruments which consist of beams with free ends, have partials which are not harmonically related. The principle of local consonance suggests that there is a natural scale, defined by the timbre of the xylophone, in which it will sound most consonant. The first seven frequencies of an ideal beam which is free to vibrate at both ends are given by Fletcher and Rossing as

$$\omega, 2.758\,\omega, 5.406\,\omega, 8.936\,\omega, 13.35\,\omega, 18.645\,\omega, 24.82\,\omega.$$

The dissonant function has numerous minima which are spaced unevenly.

Chapter 3

Wavelets and nonspecificity

In this chapter, we will consider tone systems in the broader sense. We will deal with various mathematical models of sound signals T. Which class of sound signals we wish to claim to be the class T of musical tones, it depends, in fact, on our will. We suppose that tones are represented as real or complex valued functions. The aim of this chapter is also to show how the uncertainty (namely–nonspecificity) of information in tone systems arises from the insufficiency of mathematics.

Specially, we pick up the following literature sources: [95], [96], [14], [89], [66], [41], [46], [158].

3.1   Time-frequency plane

The *time-frequency plane*, c.f. Figure 3.1 serves the mathematicians analogously as the scores, c.f. Figure 3.2 serves the musicians. The difference is in the level of abstraction of information coding. The mathematical time-frequency plane is precise if we consider tones from the physical viewpoint; the musical score uses, in fact, only integer numbers. To continue this metaphor, the tone analysis that we seek to effect should be compared to an exercise called a "musical dictation," which consists in writing down the notes on hearing

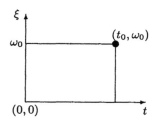

FIGURE 3.1: A point in the time-frequency plane

FIGURE 3.2: Coding of the time-frequency plane: score

a passage of music.

D. Gabor, 1946, and J. Ville, 1947, both addressed the problem of developing a mixed tone representation in terms of a double sequence of *elementary tones*, each of which occupies a certain domain in the time-frequency plane. R. Balian [70] tackled the same problem and expressed the motivation for his work in these terms:

"One is interested, in communication theory, in representing an oscillating tone as a superposition of *elementary wavelets*, each of which has a rather well defined *frequency* and *position in time*. Indeed, useful information is often conveyed by both the emitted frequencies and the tone's temporal structure (music is a typical example). The representation of a tone as a function of time provides a poor indication of the spectrum of frequencies in play, while, on the other hand, its Fourier analysis masks the point of emission and the duration of each of the tone's elements. An appropriate representation ought to combine the advantages of these two complementary descriptions; at the same time, it should be discrete so that it is better adapted to communication theory."

Similar criticism of the usual Fourier analysis, as applied to acoustic tones, is found in the celebrated work of Ville:

"If we consider a passage *of music* containing several measures (which is the least that is needed) and if a note, a' for example, appears once in the passage, harmonic analysis will give us the corresponding frequency with a certain amplitude and a certain phase, without localizing the a' in time. But it is obvious that there are moments during the passage when one does not hear the a'. The *Fourier* representation is nevertheless mathematically correct because the phases of the notes near the a' are arranged so as to destroy this note through interference when it is not heard and to reinforce it, also through interference, when it is heard; but if there is in this idea a cleverness that speaks well for mathematical analysis, one must not ignore the fact that it is also a distortion of reality: indeed when the a' is not heard, the true reason is that the a' is not emitted."

Thus it is desirable to look for a mixed tone construction advocated by Gabor: at each instance, a certain number of frequencies are present, giving volume and timbre to the sound as it is heard; each frequency is associated with a certain partition of time that defines the intervals during which the corresponding note is emitted. One is thus led to define an instantaneous spectrum as a function of time, which describes the structure of the tone at a given instant; the spectrum of the tone (in the usual sense of the term), which gives the frequency structure of the tone based on its total duration, is then obtained by putting together all of the instantaneous spectrums in a precise way by integrating them with respect to time. In a similar way, one is led to a distribution of frequencies with respect to time; by integrating these distributions, one reconstructs the tone ...

Ville thus proposed to unfold the tone in the time-frequency plane in such a way that this development would lead to a mixed representation in time-frequency atoms. The choice of these time-frequency atoms would be guided by an energy distribution of the tone in the time-frequency plane.

3.1.1 Gabor's time-frequency atoms

In the Fourier tone representation, elementary tones are sine functions. In the description of elementary tones proposed by Gabor, he called them *the time-frequency atoms*, are constructed from the

FIGURE 3.3: Time-frequency atom of Gabor

function

$$g(t) = \pi^{\frac{l}{4}} e^{-\frac{t^2}{2}}, \qquad l \in \mathbb{R}^+,$$

and are defined as complex valued function by the formula

$$T(t) = \mathbf{A}_G(t; \omega, t_0, h) = h^{-\frac{1}{2}} e^{i\omega t} g\left(\frac{t - t_0}{h}\right).$$

In Figure 3.3, we see the shape of the real or imaginary part of the function $\mathbf{A}_G(t; \omega, t_0, h)$. The parameters ω and t_0 are arbitrary real numbers, whereas h is positive. The meaning of these three parameters is the following: ω is the average frequency of $\mathbf{A}_G(t; \omega, t_0, h)$ and $t_0 - h$, $t_0 + h$, $h > 0$, are the start and finish of the time-frequency atom $\mathbf{A}_G(t; \omega, t_0, h)$. Naturally, all this depends on the convention used to define the "pass band" of $g(t)$.

The essential problem is *to describe an algorithm that allows a given sound signal (call it the tone) to decompose, in an optimal way, into a linear combination of judiciously chosen time-frequency atoms.* The set of all time-frequency atoms (with ω and t_0 varying arbitrarily in the time-frequency plane and $h > 0$ covering the whole axis) is a collection of elementary tones that is much too large to provide a unique representation of a tone as a linear combination of time-frequency atoms. Each tone admits an infinite number of representations, and this leads us to choose the best among them according to some criterion. This criterion might be the one suggested by Ville: The decomposition of a tone in time-frequency atoms is related to a synthesis, and this synthesis ought logically to be done in accordance with an analysis. The analysis proposed by Ville will be described in the ongoing text. However, Ville did not explain how the results of the analysis would lead to an effective synthesis.

3.1.2 Liénard's time-frequency atoms

A similar program (time-frequency atoms, analysis, synthesis) was proposed by J.-S. Liénard [41]:

We consider the sound signal to be composed into elementary waveforms (windowed sinusoids), each one defined by small number of parameters. A waveform model is a sinusoidal tone multiplied by a windowing function. It is not to be confused with the tone segment, that it is supposed to approximate. Its total duration can be decomposed into attack (before the maximum of the envelope), and decay. In order to minimize spectral ripples, the envelope should present no 1st or 2nd order discontinuity. The initial discontinuity is removed through the use of an attack function (raised sinusoids) such that the total envelope is null at the origin, and maximum after a short time. Although exponential damping is natural in the physical world, we choose to model the decaying part of the waveform with another raised sinusoid. Actually we see the waveform as a perceptual unit, and not necessarily as the response of a format filter to a voicing impulse. Liénard's time-frequency atoms are thus different from those used by Gabor. They are, however, based on analogous principles. We have

$$T(t) = \mathbf{A}_L(t; \omega, \varphi, A) = A(t) \sin(2\pi\omega t + \varphi),$$

where the meaning of the parameters ω, φ, A is the following: ω represents the average frequency of the emitted tone and the envelope $A(t)$ incorporates the attack and decay (the real valued function, a windowing function), $\varphi \in \mathbb{R}$. The principal difference is that, in the atoms of Liénard, the duration of the attack and that of the decay are independent. Thus Liénard's atoms may depend on more independent parameters, c.f. (1.1) where $A(t) = \text{const}(t)$, and the optimal representation of a speech tone as a linear combination of time-frequency atoms is more difficult to obtain. Some empirical methods exist, and they lead to wonderful results for synthesizing the singing voice.

3.2   Wigner–Ville transform

3.2.1   Definition of the transform

We begin by presenting the point of view of Ville. We will then
indicate how to interpret the results in terms of the theory of pseudodifferential operators as expressed in H. Weyl's formalism. This
will bring us back to work done by Eugene P. Wigner in 1932.

Definition 5 (Wigner–Ville transform) Let $T(t)$ be a complex
function, the sound signal with the finite energy, i.e.

$$\int_{-\infty}^{\infty} |T(t)|^2 \, dt < \infty.$$

Denote by

$$W_T(t,\xi) = \int_{-\infty}^{\infty} T\left(t + \frac{\tau}{2}\right) \overline{T}\left(t - \frac{\tau}{2}\right) e^{-i\xi\tau} \, d\tau, \qquad (3.1)$$

where $\overline{T}$ denotes the conjugate to T.

Re mark 1 *Definition 5 deal with arbitrary sound signals, in particular, musical tones but also noise.*

Ville, searching for an "instantaneous spectrum," wanted to display
the energy of a tone $T(t)$in the time frequency plane and to obtain
an *energy density* $W(t,\xi)$ having (at least) the following properties:

$$\int_{-\infty}^{\infty} W(t,\xi) \frac{d\xi}{2\pi} = |T(t)|^2, \qquad (3.2)$$

$$\int_{-\infty}^{\infty} W(t,\xi) \, dt = |\tilde{T}(t)|^2, \qquad (3.3)$$

where $\tilde{T}(\xi)$ denotes the Fourier transform of $T(t)$. These two properties reflect the program that we presented in the introduction: at
each instant t, the function $W(t,\xi)$ gives an instantaneous Fourier
analysis of the tone $T(t)$, and (3.2) is the Plancherel formula. The
same remark holds for (3.3): $|\tilde{T}|^2$ comes from the contributions of

all instants t, and we hope that $|\tilde{T}(\xi)|^2$ is more precisely analyzed by using these "individual contributions."

Properties (3.2) and (3.3) are clearly not sufficient to define $W(t,\xi) = W_T(t,\xi)$.

We impose two other conditions, namely, "Moyal's formula"

$$\iint W_T(t,\xi)W_g(t,\xi)\,dt\frac{d\xi}{2\pi} = \left|\int T(t)\overline{g(t)}\,dt\right|^2, \qquad (3.4)$$

which plays the role of Parseval's identity, and the requirement that if $T(t)$ is a time-frequency atom for a fixed frequency ω,

$$T(t) = \mathbf{A}(t,\omega) = h^{-\frac{1}{2}}e^{i\omega t}g\left(\frac{t-t_0}{h}\right), \qquad (3.5)$$

then

$$W_T(t,\xi) = 2e^{-\frac{(t-t_0)^2}{h^2}}e^{-h^2(\xi-\omega)^2}. \qquad (3.6)$$

Let us stop a moment to examine (3.6). The second member is a function of (t,ξ) that is localized on the rectangle of the time-frequency plane defined by $|t-t_0| \leq h$, $|\xi-\omega| \leq \frac{1}{h}$. This localization corresponds exactly to the frequency content of the time-frequency atom $T(t)$. Up to a normalization factor, the second member of (3.6) is the solution to the localization problem in the time-frequency plane that we want for our time-frequency atom.

The proofs of the following three lemmas are easy and omitted.

Lemma 1 $W_T(t,\xi)$ *is real and continuous in both variables.*

Lemma 2 $W_T(t,\xi)$ *has the properties* (3.2), (3.3), (3.4), (3.6).

Lemma 3 *If* $T(t) = T(-t)$, *then* $W_T(0,0) = 2\int_{-\infty}^{\infty}|T(\tau)|^2\,d\tau > W_T(t,\xi)$ *for all other pairs* (t,ξ).

3.2.2 Computation of Wigner–Ville transforms

We begin by treating the case of sound signals with finite energy.

Theorem 4 *Let* $\int_{-\infty}^{\infty}|T(t)|^2\,dt < \infty$, *i.e.* $T \in L^2(\mathbb{R})$. *If* $W_T(t,\xi)$ *is the Wigner–Ville transform of* $T(t)$, *then*

(i) $W_T(t, \xi - \omega)$ is the transform of $e^{i\omega t}T(t)$;

(ii) $W_T(t - t_0, \xi)$ is the transform of $T(t - t_0)$;

(iii) $W_T\left(\frac{1}{a}, a\xi\right)$ is the transform of $\frac{1}{a}T\left(\frac{t}{a}\right), a > 0$,

(iv) the Wigner–Ville transform of

$$\frac{1}{\sqrt{h}}e^{i\omega t}g\left(\frac{t - t_0}{h}\right)$$

is

$$2e^{-\frac{(t-t_0)^2}{h^2}-h^2(\xi-\omega)^2}.$$

Proof.   We show only (iii): The transform of

$$g(t) = \pi^{-1/4}e^{-\frac{t^2}{2}}$$

is

$$2e^{-t^2-\xi^2}.$$

This implies the assertion.   □

Here are some other useful observations. The Wigner–Ville transformation of a function characterizes the function up to multiplication by a constant of modulus 1. We will prove this in the next section when we establish the connection between the Wigner–Ville transform and the pseudodifferential calculus. The Wigner–Ville transform of $T(-t)$ is $W_T(-t, -\xi)$ when the transform of $T(t)$ is $W_T(t, \xi)$. Multiplying $T(t)$ by a real or complex constant λ results in the transform $W_T(t, \xi)$ being multiplied by λ^2. Thus we need to consider only the case where $\int_{-\infty}^{\infty} |T(t)|^2\, dt = 1$ when we are working with tones of finite energy.

Not all functions $W(t, \xi)$ of the two variables t and ξ are the Wigner–Ville transform of some tone $T(t)$. To see this, we have the following theorem.

Theorem 5 Let

$$Q(t, \xi) = p\xi^2 + 2r\xi t + qt^2,$$

be a positive definite quadratic form, $p > 0$, $q > 0$, $pq > r^2$. Let $T(t)$ be a sound signal such that $T(-t) = T(t)$, $\int_{-\infty}^{\infty} |T(t)|^2\, dt = 1$, and

$$W_T(t, \xi) = 2e^{-Q(t,\xi)}.$$

Then the following assertions are equivalent:
(i) *W_T is the Wigner–Ville transform of T,*   (ii) *$pq - r^2 = 1$.*

Proof.   (i) We consider a positive-definite quadratic form

$$Q(t, \xi) = p\xi^2 + 2r\xi t + qt^2,$$

where $p > 0$, $q > 0$, $pq > r^2$, and ask when $2e^{-Q(t,\xi)}$ is the Wigner–Ville transform of a sound signal $T(t)$.

Since $\int_{-\infty}^{\infty} |T(t)|^2\, dt = 1$, then $\int\int W_T(t,\xi)\, dt\, d\xi = 2\pi$, which implies that $pq - r^2 = 1$.

(ii) We will show that this necessary condition is also sufficient for $2e^{-Q(t,\xi)}$ to be the Wigner–Ville transform of a sound signal.

For this, we observe that if $W_T(t, \xi)$ is the Wigner–Ville transform of $T(t)$, then the transform of $T(t)e^{i\omega t^2/2}$ is $W_T(t, \xi - \omega t)$, where ω is a real constant.

Our quadratic form $p^2 + 2r\xi t + qt^2$ can also be written as $Q(t, \xi) = p(\xi - (r/p)t)^2 + (t^2/p)$ since $pq - r^2 = 1$. For $g(t) = \pi^{-1/4}e^{-t^2/2}$, the Wigner–Ville transform of $p^{-1/4}g(t/\sqrt{p})$ is $2e^{-(t^2/p) - p\xi^2}$, and thus the Wigner–Ville transform of $p^{-\frac{1}{4}}g(t/\sqrt{p})e^{i(r/2p)t^2}$ is $2e^{-Q(t,\xi)}$. □

More generally,

Theorem 6 *For the sound signal (which is called a "chirp")*

$$T(t) = p^{-1/4}g\left(\frac{t - t_0}{\sqrt{p}}\right) e^{i\frac{r}{2p}(t-t_0)^2} e^{i\omega t}, \tag{3.7}$$

the Wigner–Ville transform

$$W_T(t, \xi) = 2e^{-Q(t-t_0, \xi-\omega)}.$$

The quadratic form $Q(t, \xi) = p\xi^2 + 2r\xi t + qt^2$ is subject to the condition $pq - r^2 = l$.

Here is an important identity involving the Wigner–Ville transform. We have

Theorem 7 Let $\hat{T}$ denote the Fourier transform of T. Then

$$\frac{1}{2\pi} \int_{-\infty}^{\infty} \hat{T}(\omega + \xi/2)\overline{\hat{T}}(\omega - \xi/2)e^{i\xi t}\, d\xi \\ = \int_{-\infty}^{\infty} T(t + \tau/2)\overline{T}(t - \tau/2)e^{-i\omega\tau}\, d\tau. \tag{3.8}$$

In other words, if $W_T(t,\omega)$ is the Wigner–Ville transform of $T(t)$, then $W(\omega, -t)$ is that of $\frac{1}{\sqrt{2\pi}}\hat{T}(\xi)$. Another very useful fact is that the Wigner–Ville transform of an arbitrary function is always real, but it is not always positive. This second remark is the source of a great deal of difficulty in the interpretation of $W_T(t,\xi)$. Ville interpreted the Wigner–Ville transform $W_T(t,\xi)$ of a normalized sound signal $T(t)$ as a probability density in the time-frequency plane. If this probability density were concentrated in several well-delimited rectangles in the time-frequency plane, this would lead to a decomposition of the tone in terms of the corresponding time-frequency atoms.

This program has not led to an effective algorithm. The reason for this failure is that if $T(t)$ is the sum of two time-frequency atoms,

$$T(t) = T_1(t) + T_2(t) = e^{i\omega_1 t}g(t - t_1) + e^{i\omega_2 t}g(t - t_2),$$

then

$$W_T(t,\xi) = W_1(t,\xi) + W_2(t,\xi) + W_3(t,\xi) + W_4(t,\xi),$$

where the terms $W_1(t,\xi)$ and $W_2(t,\xi)$ are the "square" terms already calculated and where W_3 and W_4 are two "cross" terms. But these cross terms do not tend to zero if $\omega_2 - \omega_1$ or if $t_2 - t_1$ tends to infinity. In fact,

$$|W_3(t,\xi)| = |W_4(t,\xi)| = 2e^{-(t-t_3)^2-(\xi-\omega_3)^2},$$

where $t_3 = (t_1 + t_2)/2$ and $\omega_3 = (\omega_1 + \omega_2)/2$. These "cross" terms are thus artifacts that are localized in the time-frequency plane midway between the corresponding square terms.

The fact that the Wigner–Ville transform is not, in general, positive and the fact that its localization in the time-frequency plane does not necessarily imply the presence of time-frequency atoms are two independent properties. This can be seen by considering the sound signal $T(t) = e^{-t}$ for $t \geq 0$, and $T(t) = 0$ elsewhere. Then $W_T(t, \xi) = e^{-2t \frac{\sin 2t\xi}{\xi}}$ if $t \geq 0$ and $W_T(t, \xi) = 0$ otherwise.

There exists, however, a simple way to make the Wigner–Ville transform positive. It suffices to smooth it appropriately. Indeed, if T_1 and T_2 are two arbitrary functions (with finite energy), then

$$\iint W_{T_1}(t - u, \xi - v) W_{T_2}(u, v) \, du \, dv$$
$$= 2\pi \left| \int_{-\infty}^{\infty} T_1(s) \overline{T_2(t - s)} e^{-i\xi s} \, ds \right|^2. \tag{3.9}$$

If, in particular, $T_2(t) = \frac{1}{\sqrt{h}} \, g\left(\frac{1}{h}\right)$, where $g(t)$ is the normalized Gaussian function and where $h > 0$ is arbitrary, then we have $W_{T_2}(u, v) = 2e^{-(u^2/h^2) - h^2 v^2}$ and the smoothing function is a Gaussian kernel. The mean value one obtains is the square of the modulus of the scalar product of T_1 and the time-frequency atom

$$\frac{e^{i\xi s}}{\sqrt{h}} g\left(\frac{s - t}{h}\right),$$

centered at t, with width h and average frequency ξ.

But the Wigner–Ville transform $W_T(t, \xi)$ of a sound signal T can also be smoothed by using a kernel of the form $\frac{1}{\pi} e^{-Q(t,\xi)}$, where $Q(t, \xi)$ is one of the quadratic forms previously studied. One obtains a positive contribution that is the square of the modulus of the scalar product of T with a "chirp."

3.3 Decomposition problem

3.3.1 A pseudodifferential calculus

The following considerations allow us to relate the Wigner–Ville transform to quantum mechanics and the work of Wigner. We are going to forget signal-processing problems for the moment and go directly to dimension n. The analogue of the time-frequency plane is the phase space $\mathbb{R}^n \times \mathbb{R}^n$ whose elements are pairs (x, ξ), where x is a position and ξ is a frequency. We start with a "symbol" $\sigma(x, \xi)$

defined on phase space. Certain technical hypotheses have to be made about this symbol to ensure convergence of the following integral when T belongs to a reasonable class of test functions, and we will deal with this in a moment.

Following the formalism of Weyl, we associate with the symbol $\sigma(x, \xi)$ the pseudodifferential operator $\sigma(x, D)$ defined by

$$(2\pi)^n \sigma(x, D)|T|(x) = \int\int \sigma\left(\frac{x+y}{2}, \xi\right) e^{i(x-y)\xi} T(y)\, dy\, d\xi, \quad (3.10)$$

where the integral is over $\mathbb{R}^n \times \mathbb{R}^n$. Define the kernel $\mathrm{Ker}(x, y)$ associated with the symbol $\sigma(x, \xi)$ by

$$\begin{aligned}(2\pi)^n \mathrm{Ker}(x, y) &= \int \sigma\left(\frac{x+y}{2}, \xi\right) e^{i(x-y)\xi}\, d\xi \\ &= (2\pi)^n L\left(\frac{x+y}{2}, x-y\right).\end{aligned} \quad (3.11)$$

This says that the symbol $\sigma(x, \xi)$ is the partial Fourier transform, in the variable u, of the function $L(x, u)$ and that the kernel that interests us

$$\mathrm{Ker}(x, y) = L\left(\frac{x+y}{2}, x-y\right).$$

We can also write, in the inverse sense,

$$L(x, y) = \mathrm{Ker}\left(x + \frac{y}{2}, x - \frac{y}{2}\right),$$

and this allows us to recover the symbol $\sigma(x, \xi)$ by writing

$$\sigma(x, \xi) = \int \mathrm{Ker}\left(x + \frac{y}{2}, x - \frac{y}{2}\right) e^{-iy\xi}\, dy. \quad (3.12)$$

Thus we are led to hypotheses about the symbols that are the reflections, through the partial Fourier transform, of hypotheses that we may wish to make about the kernels. If we admit all the kernel distributions $\mathrm{Ker}(x, y)$ belonging to $S'(\mathbb{R}^n \times \mathbb{R}^n)$, then there will be no restrictions on $\sigma(x, \xi)$ other than the condition that

$$\sigma(x, \xi) \in S'(\mathbb{R}^n \times \mathbb{R}^n)$$

An immediate consequence of (3.12) is this: If $\sigma(x, \xi)$ is the symbol for an operator H, then $\overline{\sigma(x, \xi)}$ is the symbol for the adjoint operator H^*.

Finally, we consider a function T belonging to $L^2(\mathbb{R}^n)$ and satisfying $\|T\|_2 = 1$. Let P_T denote the orthogonal projection operator that maps $L^2(\mathbb{R}^n)$ onto the linear span of T. Then the kernel $\mathrm{Ker}(x, y)$ of P_T is $T(x)\overline{T(y)}$ and the corresponding Weyl symbol is

$$\sigma(x, \xi) = \int T\left(x + \frac{y}{2}\right)\overline{T}\left(x - \frac{y}{2}\right) e^{-iy\xi}\, dy. \qquad (3.13)$$

Returning to dimension one, we have the following result:

The Wigner–Ville transform of the function T is the Weyl symbol of the orthogonal projection operator onto that function T.

From this it is clear that the Wigner–Ville transform of T characterizes T, up to multiplication by a constant of modulus 1.

3.3.2 Instantaneous frequency

In his fundamental work (which has essentially been the source for this chapter), Ville makes a careful distinction between the instantaneous frequency of a sound signal (assumed to be real) and the instantaneous spectrum of frequencies given by the Wigner–Ville transform.

More precisely, let $T(t)$ be a real sound signal with finite energy. Ville writes $T(t) = \Re F(t)$, where $F(t)$ is the corresponding analytic sound signal: $F(t)$ is the restriction to the real axis of a function $F(z)$ that is homomorphic in the upper half-plane $\Im z > 0$ and belongs to the Hardy space $H^2(\mathbb{R})$. Ville then writes $F(t) = A(t)e^{i\varphi(t)}$, where $A(t)$ is the modulus of $F(t)$ and $\varphi(t)$ is its argument. He defines the instantaneous frequency of $T(t)$ by $\frac{d}{dt}\varphi(t)$.

This definition requires the function $T(t)$ to have additional regularity properties. Otherwise $\varphi(t)$ could be as irregular as an arbitrary bounded, measurable function, and the instantaneous frequency would then be a very singular object. This also raises a problem about the continuity of $\varphi(t)$ so as not to introduce Dirac measures in $\frac{d}{dt}\varphi(t)$.

We will not deal with these difficulties, and we assume that Ville's formal definitions make sense. This, of course, clearly limits the class of analyzed tones. Following Ville, we define the instantaneous spectrum of $T(t)$ as the Wigner–Ville transform $W_T(t, \xi)$ of the analytic tone $F(t)$.

An easy calculation shows that, for all real or complex-valued functions $u(t)$, one has

$$\int\int \xi u(t+\tau/2)\overline{u}(t-\tau/2)e^{-i\tau\xi}\,d\tau = -\pi i(u'(t))\overline{u}(t) - u(t)\overline{u}(t).$$

Applying this identity to $u(t) = F(t)$, it becomes

$$\frac{1}{2\pi}\int_\infty^\infty \xi W_T(t,\xi)\,d\xi = \varphi'(t)|F(t)|^2 = \varphi'(t)\left(\frac{1}{2\pi}\int_\infty^\infty W_T(t,\xi)\,d\xi\right).$$

If $W_T(t,\xi)$ is positive or zero, then $\frac{1}{2\pi}W_T(t,\xi)$ will be a probability density (when $\int_\infty^\infty |F(t)|^2\,dt = 1$) and the *instantaneous frequency will be the average of the frequency ξ computed with respect to the instantaneous spectrum.*

Similarly, we can try to compute the analogue of the variance of the variable ξ with respect to the instantaneous spectrum. This is

$$\int_\infty^\infty (\xi - \varphi'(t))^2 W_T(t,\xi)\,d\xi.$$

The calculation is completely general and does not rely on the assumption that $F(t) = A(t)e^{i\varphi(t)}$ is an analytic function. We obtain

$$\int_\infty^\infty (\xi - \varphi'(t))^2 W_T(t,\xi)\,d\xi = -\pi A^2(t)\frac{d^2(\log A(t))}{dt^2}. \qquad (3.14)$$

If, in particular, $A(t) = 1$, the second member is zero, and $W_T(t,\xi)$ cannot be positive or zero unless it is concentrated on the curve $\xi = \varphi'(t)$, which represents the graph of the instantaneous frequency. Since $2\pi A^2(t) = \int_\infty^\infty W_T(t,\xi)\,d\xi$, the "variance" of it is equal to $-\frac{1}{2}\frac{d^2\log A(t)}{dt^2}$.

Here are two examples of the calculation of the instantaneous frequency.

Example 17 Suppose first that the original tone $T(t)$ is real, equal to 1 on the interval $[-t_1, t_1]$, and 0 outside this interval. The corresponding analytic tone is then

$$F(t) = T(t) + \frac{i}{\pi}\log\left|\frac{t+t_1}{t-t_1}\right|.$$

The phase $\varphi(t)$ of $F(t)$ is continuous on the whole real line, odd, equal to $\pi/2$ if $t \geq t_1$ (and thus $-\pi/2$ if $t \leq -t_1$), and strictly increasing on $[-t_1, t_1]$. The instantaneous frequency is 0 outside the interval $[-t_1, t-1]$, strictly positive on $(-t_1, t_1)$, equal to $2/\pi t_1$ at 0, and increases from $2/\pi t_1$ to $+\infty$ as t traverses the interval $[0, t_1)$.

As one could have guessed, the instantaneous Fourier analysis proposed by Ville is not even a local property. This means that knowing the tone in an arbitrary large interval centered at t_0 is not sufficient to calculate the instantaneous frequency at t_0. The operation responsible for this anomaly is the calculation of the analytic tone $F(t)$ associated with $T(t)$; as everyone knows, the kernel of the Hilbert transform decreases slowly at infinity. This discussion shows that the tones to which the Ville theory applies are necessarily academic tones (whose algorithmic structure does not change over time) or asymptotic tones whose behavior on a short time interval is equivalent to that of a normal tone over a much longer duration.

Example 18 The second example of a calculation of the instantaneous frequency is for the tone $T(t) = \cos t^2$, which is a chirp of infinite duration. The calculation of the corresponding analytic tone is interesting because it exhibits two different asymptotic behaviors depending on whether t tends to $+\infty$ or $-\infty$. Indeed, the Fourier transform of this analytic tone $F(t)$ is 0 for $\xi < 0$ and is equal to $\frac{1}{2}(\sqrt{i\pi}e^{-i\xi^2/4} + \sqrt{-i\pi}e^{i\xi^2/4})$ for $\xi > 0$. It follows that $F(t)$ is asymptotically equal to e^{it^2} when t tends to $+\infty$ and to e^{-it^2} when t tends to $-\infty$. The instantaneous frequency of $\cos t^2$ is thus equal to $2t + \varepsilon(t)$ when t tends to $+\infty$ and to $-2t + \varepsilon(t)$ when t tends to $-\infty$. In both cases, $\varepsilon(t)$ tends to 0 when $|t|$ tends to ∞.

3.3.3 Wigner–Ville transform of asymptotic tones

As we have already seen in Subsection 3.3.1, the Wigner–Ville transform can be generalized to the case where, instead of being a tone of finite energy, $T(t)$ is an arbitrary tempered distribution. We limit our discussion to three examples where $T(t) = e^{i\varphi(t)}$.

Example 19 We begin with the particular case where $\varphi(t) = \omega t$.

Note that $T(t)$ is an analytic tone only when $\omega \geq 0$, but the calculations that follow do not depend on this type of hypothesis.

The Wigner–Ville transform of $T(t) = e^{i\omega t}$ is $2\pi\delta_0(\xi - \omega)$, where $\delta_0(\xi)$ is the Dirac measure at the origin. Then the instantaneous frequency given by the Wigner–Ville transform is simply ω.

Example 20 Now, let $\varphi(t) = \omega t^2/2$. Then $T(t) = e^{i\omega t^2/2}$, for ω real, the Wigner–Ville transform of $T(t)$ is $2\pi\delta_0(\xi - \omega t)$. This is a distribution (in fact, a measure) supported by the line $\xi = \omega t$. The corresponding "instantaneous frequency" is ωt, and both members of 3.14 are 0 in this case. In fact, this statement is not correct because $e^{i\omega t^2/a}$ is not an analytic tone. However, as we have already observed, $e^{i\omega t^2/a}$ is asymptotic to an analytic tone when ω is strictly positive and when t tends to $+\infty$.

Example 21 Finally we come to the case where $T(t) = e^{i\omega t^3}$, i.e. $\varphi(t) = \omega t^3$ with $\omega > 0$. The Wigner–Ville transform of this function is easily calculated and is

$$\int_{-\infty}^{\infty} e^{i\frac{\omega \tau^3}{4+3\omega \tau t^2 - \omega \tau}}\, d\tau = 2\pi \left(\frac{4}{3\omega}\right)^{\frac{1}{3}} A\left(\left(\frac{4}{3\omega}\right)^{\frac{1}{3}}(3\omega t^2 - \omega)\right),$$

where

$$A(\omega) = \frac{1}{2\pi}\int_{-\infty}^{\infty} e^{i\frac{s^3}{3+\omega s}}\, ds$$

is the Airy function. Here again the Wigner–Ville transform of the function $e^{i\omega t^3}$ is "essentially" concentrated around the curve $\xi = 3\omega t^2$, which is the graph of the instantaneous frequencies. All of this must be put in quotation marks since $e^{i\omega t^3}$ is not an analytic tone. But the tone is asymptotically analytic because the Airy function decreases exponentially when ω tends to $+\infty$.

3.3.4 Return to the problem of optimal decomposition of tones into time-frequency atoms

As we indicated in the introduction to this chapter, the analysis of the energy distribution of a tone in the time-frequency plane was, for Ville, a precondition for his search for optimal decompositions in time-frequency atoms. Consider the example of the tone $T(t) = e^{i\omega t^2/2}$. In the time-frequency plane, its energy is concentrated on the line $\xi = \omega t$. If we try to decompose this function in time-frequency atoms of the kind advocated by Gabor, this comes down to covering the line $\xi = \omega t$ in an optimal way with "Heisenberg boxes" of area 1. We can think of these squares as leading to Gabor wavelets of the form $e^{i\omega k t} g(t - k)$. Finally, all of this leads to approximating $T(t) = e^{i\omega t^2/2}$ with the sum of the series

$$\sum_{-\infty}^{\infty} e^{\frac{-i\omega k^2}{2}} e^{\frac{i\omega k^2}{2}} g(t - k).$$

And this is a poor approximation.

The other major shortcoming of the Wigner–Ville transform is that it is not always positive. One might have thought, in light of the example of the tone

$$T(t) = e^{i\omega_1 t} g(t - t_1) + e^{i\omega_2 t} g(t - t_2),$$

that the places where the Wigner–Ville transform is positive correspond to the time frequency atoms and that the places where it oscillates were simply artifacts. But this is not the case, as we see from the example of the tone $T(t) = e^{-t}$ for $t \geq 0$; $T(t) = 0$ otherwise.

We are forced to conclude that the Wigner–Ville transform yields only *imperfect information* about the distribution of energy in the time-frequency plane. *We have no effective algorithm that allows us to find an atomic decomposition of a tone by using the Wigner–Ville transform.*

Chapter 4

Pitch granulation and ambiguity

4.1 Garbuzov zones: strife

4.1.1 An classical experiment

Since the time of Pythagoras it has been believed in Europe that the ideal values of musical intervals are equal to ratios of small integers. The possibility of reducing the problem of integer-ratio-intervals to phenomena of physical resonance can be viewed as an argument in favor of this belief. But only a relatively small number of European and Asian tunings contain close approximations to the small integer ratios. Numerous ethnographic musicological studies have demonstrated that e.g. African and Indonesian, as well as European folk music, also use intervals significantly different from the ratios of small integers. Moreover, it has been shown that intervals may vary even in different performances of the same musical piece.

The musicologist N. A. Garbuzov (1880-1955), c.f. [18], revolutionized the study of musical intervals suggesting a concept of musical "zones" in the 1940s. This theory can be characterized in the present scientific language as an information granulation in the sense of Zadeh, c.f. [190]. In [18], Table 4 and Table 5 (= our Table 4.1) are statistics of hundreds of measurements. To each note in a score, there is a set of possible tones with frequencies which form the Gar-

TABLE 4.1: Garbuzov zones [in cents]: Sequential and Simultaneous uncertainty

granule	melodic	harmonic
unison (octave)	(−12, 12)	(−30,30)
minor second	(48, 124)	(66,130)
major second	(160, 230)	(166,230)
minor third	(272, 330)	(266,330)
major third	(372, 430)	(372,430)
fourth	(472, 530)	(466,524)
tritone	(566, 630)	(566,630)
fifth	(672, 730)	(672,730)
minor sixth	(766, 830)	(766,830)
major sixth	(866, 930)	(866,924)
minor seventh	(966, 1 024)	(966,1 024)
major seventh	(1 066, 1 136)	(1 066,1 136)

buzov zone of this tone. The values may be chosen from a discrete set of numbers (the time constant functions, the theoretical situation) or are continuous functions (the typical real situation; every assemble play).

The set of frequencies may take values derived from one or more theoretical tunings. On the other hand, theoretical tone systems often use two or more crisp values for one Garbuzov zone. According to the more modern terminology used in the fuzzy sets and systems theory, these qualitative musical degrees we will call the *granule*. If there is only one value in each granule, we have no uncertainty. Clearly this situation is only a theoretical possibility which never occurs in real live music.

The choice of tone system is made on the basis of acoustical, psychological, and mathematical principles and relations in the frame of a given composition. Psychologically (i.e. after the sound was produced), the tone system in music is an example of the fact that the human perceptive mechanism uses various systems of uncertainty-based information coding. Particularly, N-tone systems with the octave periodicity, $N \geq 12$, for the 12 qualitative different music interval granules (e.g., Just Intonation and Pythagorean Tuning, [145], [128]) are well-known. The main requirement is that the subject's perceptive system would be able to encode the informational content unambiguously.

Uncertainty with regard to every tone system is relevant because there are ever present two or more different tone systems under consideration. For instance, in the case of the piano and the 12-tone Equal Temperament E_{12}, there is an interference among the played tones and their natural overtone systems. Since these tone systems are mutually incommensurable (theoretically—except of octaves), the interference valuable modifies the resulting sound. These modifications depend on many factors, e.g., the concrete composition, player, instrument, set of hall acoustics.

It is known that the sensitivity of human aural perception is 5–6 cents. The psychophysical boundary [to what listeners or players tend to expect] is 1–2 cents, professional musicians typically pick up differences between 10–20 cents. We will describe now the pioneering experiment of Garbuzov, classical today. Of course, there are more precise statistics today. Nevertheless, this experiment is enough to effectively demonstrate the idea.

In [18], we can find the numerical data of the following classical experiment. The first 12 measures of Air from J. S. Bach's Suite in *D* ("On the *G*-String") were interpreted by three famous Russian violinists of that time: Oistrach, Elman, and Cimbalist. The piece of Bach's Air was chosen because (1) the tempo is Slow (*Lento*, M 1/4 = 52), c.f. also Table 6.8; (2) it consists of a great number of pitches of various duration; (3) the piece has two parts, the second one is a repetition of the first one; (4) the piece is well-known, often recorded and interpreted; (5) it is played on violin (not fix-tuned instrument).

The recordings of Oistrach, Elman, and Cimbalist were numerically analysed with the accuracy of 5 cents. The results were as follows: Oistrach produced 65 intervals (25 *TET*, 19 *PYTH*, 15 *JI*, 30 *UI*); Elman produced 66 intervals (27 *TET*, 26 *PYTH*, 14 *JI*, 27 *UI*); Cimbalist produced 70 intervals (18 *TET*, 13 *PYTH*, 11 *JI*, 49 *UI*), where *TET* denotes intervals which can be identified as 12-*TET*, *PYT*—Pythagorean, *JI*—pure [Just Intoned], and *UI*—unidentified, not belonging to the considered tunings 12-*TET*, Pythagorean System, or Just Intonation. These three tone systems are, of course, not disjoint and the measured values were taken within the tolerance of 5 cents (due to temporary measuring apparatus). Further, the various number of intervals produced by the three violinists depended only on the various manners in which they played the trill in the second (fourteenth) measure and this fact was not important for the classifying of intervals to the three tunings. Moreover, the results were deformed with the fact that the accompaniment piano was tuned into E_{12}.

It is clear that we may consider also another set of theoretical tone systems to have no "unidentified" intervals or to have a lesser number of the mentioned theoretical tunings covering the set of all produced intervals. However, we may also say that there are three (Oistrach's, Elman's, and Cimbalist's) unique tone systems for Bach's Aria.

It is known that some string orchestra tend to play in Pythagorean Tuning, some in Just Intonation, and the large symphonic orchestras in Equal Temperament. It is also clear that there is no problem for two or more professional violinists to play the Bach's Air excellently together. In which tone system?

The effect of ambiguity of tuning (without any psychological interaction of players) was specially used when tuning some historical organs with two consoles which were each tuned to a different tuning.

In modern pipe organs, the effect is used, too. But in a different way. Some special two-row registers (called double voices) are constructed on the physical principle of beats, the interference of sound waves. For instance, such are the sound timbres (registers) named *Vox coelestis, Unda maris, Voce umana, Gemshornschwebung, Schwebend Harf*, c.f. [1].

The basic idea of Garbuzov zones (granules) is that the same musical quality is characterized not merely with a one single point frequency but either as a single point within a relatively large granule, c.f. Table 4.1.

Note that melodic and harmonic granules have uncertainty of different types. While melodic granules are unions of varieties of strict tunings (uncertainty of the type strife); harmonic granules represent uncertainty of the fuzzy type.

Further, Garbuzov asserted that the pitch granules appeared not as merely artificial convention and psychoacoustical phenomenon but, rather, as the natural development of musical culture. In Europe, this development led to aural selection of 12 pitch granules (called Garbuzov zones now) each of them having its own, particular character of individuality. In the Pacific Ocean region we can observe 10 granules within an octave. The observation of pitch granulation is interesting in music of nations which use the finite sets T with more than 12 basic values of pitches per octave (e.g., 22 in India, 53 in Turkey) and choose sets T (ragas, maquams) from these universum.

4.1.2 Psychological model

Classes T of tones (better and poetically said, a realms of tones) are various and different. The following metaphorical assertion seems rather true: the tone notion is a door which makes the theory of tone systems open and, in the same time, it is a connection to other objects and disciplines of science and art. Tones are usually coded into notes (e.g., one note codes one fuzzy set, one Fourier series, one wavelet, or one integer number) and notes are composed into musical scores. But also other codings are good (digital records, tabulatures, naumes, etc.).

Intervals belonging to one zone have unambiguous musical meaning. In European music, the basic zones have their own names, such as fifth, fourth, major third, etc., c.f. Table 4.1. The other interval zones are perceived as mixtures (fuzzy intersections) of basic musical zones. The question naturally arises of why basic zones have stable meaning while others do not. There are some psychological models which try to explain this, c.f. [145].

The purpose of projecting the subject's inner state onto a tone system is to transfer the emotional profile it represents from one subject to another. A diagram of this transfer is given in Figure 4.1.

The transfer of the state consists of sending a signal (the physical agent) encoding the given musician's inner state. In the perceptual system of the subject receiving the signal an automatic decoding process takes place: information is extracted from the physical agent. Then another automatic mechanism transfers the subject into the listener's inner state. A musical interval fulfills the function of this

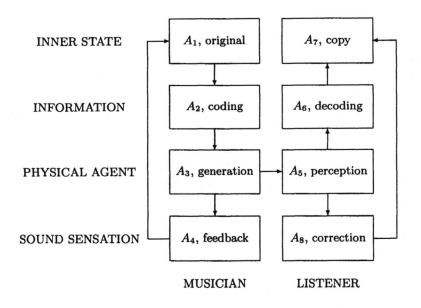

FIGURE 4.1: The projecting of the subject's inner state

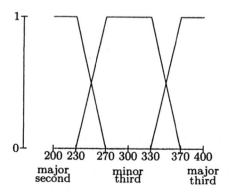

FIGURE 4.2: Garbuzov zones; the minor third

signal. Without diminishing the importance of such acoustic properties as loudness and timbre, the main factor distinguishing a musical interval from any other pair of sounds is its *informational content*. Formal considerations, however, demonstrate that the human perceptive mechanism does not need the very exact values (the ideal intervals). The important thing is that the deviations should not disturb the unambiguous restoration although this restoration can be different from the original. Statistical analysis has allowed us to determine the bounds for tolerable deviation and it turns out that those bounds coincide approximately with the Garbuzov zones for the European cultural zone.

In this way we have obtained a key to understanding the nature of intervals and their granules. As long as deviations from the subjective ideal value of an interval are small, the human perceptive system is able to decode information from the signal received. These deviations correspond to the plateau in Fig 4.2. But when deviations become bigger, exact decoding becomes difficult. Those deviations correspond to mixed intervals (overlapping slops of plateaus in Figure 4.2). It is important to note that the plateaus corresponding, e.g., to the fifth and to the fourth consist of tens of cents in the musical scale, while the psycho-acoustic theory "allows" deviations of only a few cents.

Therefore, we can see that in order to transfer emotional content from one subject to another one, there is no need for a musical system to contain the unambiguous values of physical intervals. The ideal integer-ratio values can be uncertainty-represented, shifted, varying within quite broad limits. This can be obscured by the fact that most music allows performers to deviate from "exact" values and use nonspecific (disagreement in choosing among several alternatives) intervals, as long as they remain inside Garbuzov's granules. Thus uncertainty substantially enriches the acoustical structure of music and makes each performance unique and individual. Also, uncertainty enables the orchestra to play.

Let us note that Garbuzov in the 1940s has also found experimental data demonstrating that rhythmic and dynamical structures in music also have a granulated structure. There is, similarly as concerning pitch granulation, no acoustical explanation for this phenomenon.

4.2   Geometric nets

4.2.1 $\mathbb{Z}$-chain condition

If we take into the account that tone is a dual object—both spiritual and physical—we need to use a more ingenious mathematical tool to reflect at least the information aspects. This apparatus provides the *concept of geometric net*.

Recall that a *net* with values in $\mathbb{L}$ is a function from I to $\mathbb{L}$, where I is a directed partially ordered set, c.f. [32]. Let $\mathbb{L} = ((0, \infty), \cdot, 1, \leq)$ be the usual multiplicative group with the usual order on $\mathbb{R}$. If $a \leq b$, $a, b \in (0, \infty)$, then b/a is an $\mathbb{L}$-length of the interval (a, b). Or, borrowing the usual musical terminology, we simply say that b/a is an interval.

The set $P \subset \mathbb{Z}$ in the following Definition 6 provides *information coding*, e.g., notes in a score (very often P is assumed to be $\mathbb{Z}$ in the Western music, but also P finite). To each $T \in \mathbb{T}$ there exists a unique (note) $p \in P$ but, due to psychological or spiritual aspects, possible more than one physical variants or theoretical decompositions of this tone. This ambiguity will be expressed with help the structure of the index set I of the geometric net in Definition 7. At the same time, the set P provides *the granulation* of the tone system. In the Western music there are used now 12 granules (qualitative degrees, Garbuzov interval zones) within one octave. Fuzziness of tone system will be expressed via the nature of quotients $X_1, \ldots, X_n$ of geometrical nets: they can be considered as crisp values, real positive numbers, or as fuzzy sets, i.e., functions in $[0, 1]^{\mathbb{R}}$.

Definition 6 Let $I = \{\phi(p) \in \mathbb{R}^n;\ p \in P\}$ be a lattice of n-tuples $\phi(p) = (\alpha_1(p), \ldots, \alpha_n(p))$, where $P \subset \mathbb{Z}$, $\alpha_i : P \to \mathbb{R}$ are real functions, $i = 1, 2, \ldots, n$. We say that the lattice I satisfies the $\mathbb{Z}$-chain condition if

(1) $\alpha_1, \ldots, \alpha_n$ are isotone functions, i.e. for every $i = 1, 2, \ldots, n$,

$$p_1 \leq p_2,\ p_1, p_2 \in P \text{ implies } \alpha_i(p_1) \leq \alpha_i(p_2),$$

(2) $\alpha_1(p) + \cdots + \alpha_n(p) = p,\ p \in P$.

Example 22 Let $P = \mathbb{Z}$, $n = 2$, define $I = \{(\alpha_1(z), \alpha_2(z)) \in \mathbb{Z}^2;\ z \in \mathbb{Z}\}$ as follows:
$(\alpha_1(z), \alpha_2(z)) = (z/2, z/2)$ if $z \in \mathbb{Z}$ is even (one couple of integers) and
$(\alpha_1(z), \alpha_2(z)) = (z - 1/2, z + 1/2), (z + 1/2, z - 1/2)$ if $z \in \mathbb{Z}$ is odd (two couples of integers).

Example 23 The following tone systems consist only of three elements each. All index sets I or sets P for these systems are "very poor".
CHIMES-3; Heavenly Chimes;
32/29, 1/2, 16/29
MBOKO(BOW)-3; African Mboko Mouth Bow (chordophone, single string, plucked);
0, 492, 625
HARM-3; Third octave of the harmonic overtone series;
5/4, 3/2, 7/4

Definition 7 Let $I = \{\phi(p) \in \mathbb{R}^n;\ p \in P\}$ be a lattice satisfying the $\mathbb{Z}$-chain condition, where $P \subset \mathbb{Z}$, $\phi(p) = (\alpha_1(p),\ \ldots,\ \alpha_n(p))$,

$\alpha_i : P \to \mathbb{R}$ are n real functions, $n \in \mathbb{N}$. Let $X_1 > 0, \ldots, X_n > 0$ be n positive real numbers (nonnegative functions, fuzzy sets, positive operators). The *geometric net* $\langle \Gamma, I \rangle$ of numbers (nonnegative functions, fuzzy sets, positive operators) *on the lattice I* is a net of numbers (nonnegative functions, fuzzy sets, positive operators) defined on the lattice I such that

$$\langle \Gamma, I \rangle = \{ X_1^{\alpha_1(p)} \ldots X_n^{\alpha_n(p)}; \ \phi(p) \in I, \ p \in P \}.$$

Re mark 2 It is easy to see that the notion of geometric net generalized the elementary notion of the *geometric progression* (equivalently, equal temperament in music) when $X_1 = \cdots = X_n \in \mathbb{R}^+$. According to this origin, we use the term *quotient* for objects $X_1, \ldots, X_n$ in Definition 7.

Re mark 3 $n = 1$ (geometric progression) is typical for equal temperaments, $n = 2$ for meantones, $n = 3$ is for the diatonic tone systems.

The lattice I structures the tone set $\mathbb{T}$ such that elements of the lattice I represent tones $T \in \mathbb{T}$ independently whether they are mentioned in the psychological or in the physical sense. The set $P \subset \mathbb{Z}$ provides coding of information. And if $X_1, \ldots, X_n$ are positive real numbers, the set $\{ X_1^{\alpha_1(p)} \ldots X_n^{\alpha_n(p)}; \ \phi(p) \in I, \ p \in P \} = \Omega(\mathbb{T}) = S$, tone system in the narrower sense, c.f. Preface, Definition 1.

Moreover, Definition 7 provides a straightforward extension of the tone system notion to fuzzy sets and positive operators. The additional concepts (S1)–(S3) to Definition 1, c.f. Preface, can be also satisfactorily "hidden" into the geometrical net. So, considering the notion of geometric nets as a possible *constructive definition of tone system*, the only question is how to deal with the doubtless necessary additional concept (S4), uncertainty measure, with respect to geometric nets.

Depending on the additional conditions on the set P, number n, functions $\alpha_1, \ldots, \alpha_n$, or values $X_1 > 0, \ldots, X_n > 0$, the consideration of a tone system $(\mathbb{T}, \Omega)$ with elements of $\Omega(T)$ in the form

$$X_1^{\alpha_1(p)} \ldots X_n^{\alpha_n(p)}; \ \phi(p) \in I, \ p \in P, \qquad (4.1)$$

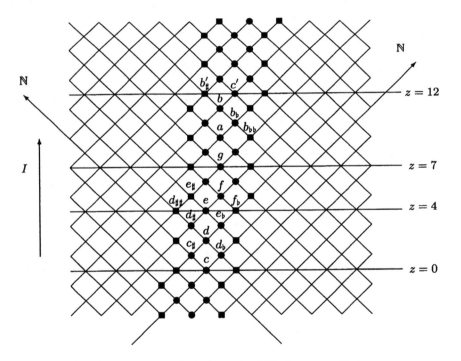

FIGURE 4.3: Lattice I for the 31-valued Pythagorean System.

gets substantial and non-trivial. In any case, geometric net is a universal tool for analytical description and consideration of tone systems.

With connection with the representation (4.1), we bring the following definition of a class of tones $\mathcal{T}$, the n-dimensional Euler music space $\mathbb{E}^n$.

Definition 8 For the sequence $X_1, X_2, \ldots, X_n = 2, 3, 5, 7, \ldots, p_n$, where p_n denotes the n-th prime number (not including 1), the set $\mathcal{T} = \mathbb{E}^n = \{2^a 3^b \ldots p_n^c;\ a, b, \ldots, c \in \mathbb{Q}\}$, is called the ($n$-dimensional) *Euler (musical) space* and $x = 2^a 3^b \ldots p_n^c \in \mathbb{E}^n$ the *Euler point*.

The following two assertions are obvious:

Lemma 8 *If $f : \mathbb{E}^n \to \mathbb{E}^n$ is a rational function, then f is a bijection, i.e. $f(x) = f(y)$ if and only is $x = y$, where $x, y \in \mathbb{E}^n$.*

Lemma 9

$$\{a \log 2 + b \log 3 + \cdots + c \log p_n;\ (a, b, \ldots, c) \in \mathbb{Q}^n\}$$

is a vector space over $\mathbb{Q}$.

Re mark 4 Note also that for a given tone system, in general, transforms of $X_1, \ldots, X_n$, change the partial order of the lattices I, c.f. Subsection 4.2.3.

4.2.2   Discrete geometric nets

For $I \subset \mathbb{Z}^n$ we will use also the following conventional notation:

$$X = (X_1, X_2, \cdots, X_n) \in \mathbb{L}^n, \quad \nu_{z,\cdot} = (\nu_{z,1}, \nu_{z,2}, \ldots, \nu_{z,n}) \in \mathbb{Z}^n,$$

$$|\nu_{z,\cdot}| = \nu_{z,1} + \cdots + \nu_{z,n}, \quad \nu_{z,\cdot} \le \nu_{z+1,\cdot} \Leftrightarrow \nu_{z,k} \le \nu_{z+1,k}$$

$$X^{\nu_{z,\cdot}} = X_1^{\nu_{z,1}} X_2^{\nu_{z,2}} \ldots X_n^{\nu_{z,n}},$$

where $k = 1, 2, \ldots, n$, $z, n \in \mathbb{Z}$. Rewrite Definition 7 for discrete geometric nets as follows:

Definition 9 Let $n \in \mathbb{N}$. Let $I \subset \mathbb{Z}^n$ be a lattice, $P \subset \mathbb{Z}$. Let $X \in \mathbb{L}^n$. We say that a net $\langle \Gamma, I \rangle$ is a *discrete geometric net* if there exist $\nu_{z,j} \in \mathbb{Z}$, $z \in \mathbb{Z}$, $j = 1, 2, \ldots, n$ such that

$$\langle \Gamma, I \rangle = \{X^{\nu_{z,\cdot}}; \nu_{z,\cdot} \in I, z \in P\}, \ P \subset \mathbb{Z},$$

$$\cdots \leq \nu_{0,\cdot} \leq \nu_{1,\cdot} \leq \cdots \leq \nu_{z,\cdot} \leq \cdots,$$

$$|\nu_{z,\cdot}| = z.$$

We will omit the adjective "discrete" when it will not lead to misunderstanding of the text.

Example 24 $I = \{(m_z, n_z); \ z \in \mathbb{Z}\} \subset \mathbb{Z}^2$, $(a, b) \leq (c, d)$ if and only if $a \leq c$ and $b \leq d$, where $(a, b), (c, d) \in \mathbb{Z}^2$. We have:

$$m_z + n_z = z, \ m_z \leq m_{z+1}, n_z \leq n_{z+1}, \ i \in \mathbb{Z}.$$

Let $X > 0, Y > 0$. Then

$$\{X^{m_z} Y^{n_z}; \ (m_z, n_z) \in I, z \in \mathbb{Z}\}$$

is a geometric net.

Thus, in Figure 4.3, there are depicted the lattices I for nets of the Pythagorean Systems *PYTH-17* (circles) and *PYTH-31* (both circles and squares). If we take $X = 253/243$ (limma) and $Y = 2\,187/2\,048$ (apotome), we obtain exactly values of *PYTH-17* and *PYTH-31*, e.g., $f = X^3 Y^2 = 4/3$, etc. See also Section 6.5.

Sometimes, when the concrete lattice I play no role in our explanation, we will use also the following simplified two notations of the n-quotient geometric net in $\mathbb{R}$:

$$\langle \Gamma_i \rangle = \langle X^{\nu_{i,\cdot}}; |\nu_{i,\cdot}| = i, \cdots \leq \nu_{0,\cdot} \leq \nu_{1,\cdot} \leq \cdots \leq \nu_{i,\cdot} \leq \cdots \rangle_{\nu_{i,\cdot} \in \mathbb{Z}^n},$$

or,

$$\langle \Gamma_i = X^{\nu_{i,\cdot}}; \ i \in \mathbb{Z} \rangle$$

keeping in the mind that there are possible more values $X^{\nu_{i,\cdot}}$ for one $i \in \mathbb{Z}$.

We say that the (n-quotient) geometric net is an M-granule (n-quotient) tone system if $X^{M,\cdot} = 2$ (octave). The M-granule tone system is scale (an M-granule n-quotient scale) if it is a chain, a sequence.

Now we introduce two useful notions derived from the notion of geometric net.

Definition 10 We say that two (discrete) tone systems S_1 and S_2 are *isomorphic* if they can be represented as geometric nets $\langle \Gamma, I \rangle$ and $\langle \Delta, I \rangle$, $\nu_{z,\cdot} \in I \subset \mathbb{Z}^n$, $X, Y \in \mathbb{L}^n$, and $P \subset \mathbb{Z}$, such that

$$\langle \Gamma, I \rangle = \{X^{\nu_{z,\cdot}}; \nu_{z,\cdot} \in I, z \in P\} \text{ and } \langle \Gamma, I \rangle = \{Y^{\nu_{z,\cdot}}; \nu_{z,\cdot} \in I, z \in P\}.$$

Definition 11 Let $\{X, \ldots, Y \in \mathbb{R};\ 1 < X \leq \cdots \leq Y\}$ be a set of N numbers. Let $m_0, m_1, \ldots, m_M;\ \ldots\ ;\ n_0, n_1, \ldots, n_M$ be $M \times N$ nonnegative integers, such that

$$0 = m_0 \leq m_1 \leq \cdots \leq m_M;\ \ldots;\ 0 = n_0 \leq n_1 \leq \cdots \leq n_M$$

and

$$m_z + \cdots + n_z = z,\ z = 0, 1, \ldots, M.$$

Then the set $S = \{X^{m_0} \ldots Y^{n_0}, X^{m_1} \ldots Y^{n_1}, \ldots, X^{m_M} \ldots Y^{n_M}\}$ is said to be an *M-granule N-interval scale* (i.e., it is a chain).

Particularly, a 12-granule N-quotient scale is a *12-granule 2-quotient (2,3)-scale* if

$$X^{m_{12}} Y^{n_{12}} = 2 \text{ and } X^{m_7} Y^{n_7} = 3/2$$

(equivalently,

$$X^{m_{12}} Y^{n_{12}} = 2 \text{ and } X^{m_{19}} Y^{n_{19}} = 3).$$

We say that S is a *M-granule N-quotient system* ((2,3)-*system*), if $S = \bigcup s$, where s are *M-granule N-quotient scales* ((2,3)-*scales*).

Example 25 E_{12} is a 12-granule 1-quotient scale which is not a 12-granule (2,3)-scale.

Example 26 Pythagorean System is a 12-granule 2-quotient (2,3)-system. The Pythagorean approximations $\mathcal{A}$ of Just Intonation, c.f. Table 9.9 is also a 12-granule 2-quotient (2,3)-system.

4.2.3 Various net representations of a tone system

For a given tone system, the index set I need not be unique. For instance, every mean-tone system can be represented at the same time as one chain as a "spiral of (generalized) fifths" (i.e. linearly ordered); as a net which is a many valued map, c.f. Figure 4.3; or linearly ordered according to the pitch of tones. Using the representation of a tone system via geometric nets, we implant information aspects into the consideration of tone systems. Explain this idea on Pythagorean System in more detail.

Denote by Π the class of tones $\mathcal{T}$ of Pythagorean System. We know that it can be defined as the set of all numbers of the form $2^{\alpha}3^{\beta}$, where $\alpha, \beta \in \mathbb{Z}$. So, it can be described as a geometric net defined on some lattices and generated with two quotients. In general, let for the suitable quotients $X > 0$ and $Y > 0$, there exists a lattice $I \subset \mathbb{Z}$ such that

$$\langle \Pi, I \rangle = \{ X^{\alpha_z} Y^{\beta_z}; \ (\alpha_z, \beta_z) \in I \subset \mathbb{Z}, z \in \mathbb{Z} \} \tag{4.2}$$

where for every $z \in \mathbb{Z}$,

$$\alpha_z + \beta_z = z, \tag{4.3}$$

$$\alpha_z \leq \alpha_{z+1}, \quad \beta_z \leq \beta_{z+1}, \tag{4.4}$$

and $\langle \Pi, I \rangle = \Pi$ as a set (not considering the lattice structure of I). Note, that we introduced a lattice I ambiguously. Choosing suitable different couples (X, Y), we may obtain different lattices I, i.e., different geometric nets $\langle \Pi, I \rangle$. For instance, the following $\langle \Pi, I_a \rangle$, $\langle \Pi, I_b \rangle$, $\langle \Pi, I_c \rangle$, $\langle \Pi, I_d \rangle$ are four geometric nets which are identical as the sets, but are equipped with various lattices I_a, I_b, I_c, I_d, respectively.

$$
\begin{aligned}
\langle \Pi, I_a \rangle &= \{ (256/243)^{\alpha_z} (2\,187/2\,048)^{\beta_z}; \ (\alpha_z, \beta_z) \in I_a, z \in \mathbb{Z} \}, \\
\langle \Pi, I_b \rangle &= \{ 2^{\gamma_z} 3^{\delta_z}; \ (\gamma_z, \delta_z) \in I_b, z \in \mathbb{Z} \}, \\
\langle \Pi, I_c \rangle &= \{ 2^{\varepsilon_z} (3/2)^{\eta_z}; \ (\varepsilon_z, \eta_z) \in I_c, z \in \mathbb{Z} \}, \\
\langle \Pi, I_d \rangle &= \{ (9/8)^{\theta_z} (256/243)^{\varkappa_z}; \ (\theta_z, \varkappa_z) \in I_d, z \in \mathbb{Z} \}.
\end{aligned}
$$
$$\tag{4.5}$$

The directions of the nets (i.e. lattices of integer numbers in our case) are given in all four cases with the equations (4.3), (4.4).

Musically, $\langle \Pi, I_a \rangle$ represents the set Π by semitones, $\langle \Pi, I_b \rangle$ by overtones, $\langle \Pi, I_c \rangle$ by the perfect fifths and octaves, $\langle \Pi, I_d \rangle$ by the whole tones and minor semitones. We can observe an very interesting psychological information moment. The geometric net $\langle \Pi, I_a \rangle$ gives the melodic structure of music (pitches of tones), $\langle \Pi, I_b \rangle$—overtone structure, $\langle \Pi, I_c \rangle$—harmonic structure ("the spiral of fifths"), and $\langle \Pi, I_d \rangle$—the diatonic structure. All these structures are present (in the same time, of course) in every musical composition. Clearly, these structures are not isomorphic in any sense (we cannot transform melody into harmony, etc.). In other words, transforms of quotients X, Y change the net supporting lattices and thus change the geometric nets.

Recall that the class of tones $\mathcal{T}$ (with the large entropy) is restricted to a set $\mathbb{T} \subset \mathcal{T}$ (with the small or no entropy). This same holds for geometric nets of these systems. In Figure 4.3, we can see the sublattices of the lattice I for nets of the 17-valued Pythagorean System Π_{17} (circles) and 31-valued Pythagorean System Π_{31} (both circles and squares), $\Pi_{17} \subset \Pi_{31} \subset \Pi$.

4.2.4   The harmony–melody uncertainty

Authors of [33] claim that the fuzziness, strife, and nonspecificity are rather all possible types of uncertainty. In Section 4.2.3 we discussed a type of uncertainty which is rather different one. It is given as a set of various strife-type structures of one object (in our case, the set of lattices $\{I_a, I_b, I_c, I_d\}$). The couple harmony–melody is a model example of this uncertainty in music. We cannot express harmony via melody or vice versa. On the other hand, harmony and melody cannot exist separately, they are ever present as a couple. Melody is a manifestation of harmony and vice versa. There is no hierarchy between harmony and melody. On the other hand, both structures are relatively independent, complementary, and nonconflicting. In every music we can ever find the harmony-melody couple. It is a join type of uncertainty.

Part II

Special Systems

Chapter 5

Equal temperaments

> *An individual number or tone is nothing. It begins to have a sense only in a concrete context. But, when a tone system is formalized, the initializing context is no more interesting.*
>
> Moisei Boroda

Equal Temperament E_{12} was already known to A. Werckmeister (this is obvious from his book "Erweiterte und verbesserte Orgel-Probe," 1698) and is commonly used in the present day. The sequence of real numbers $E_{12} = \{ \sqrt[12]{2^z};\ z \in \mathbb{Z}\}$ defines fully this tone system.

Mathematical problems about 12-tone Equal Temperament and music created in this system are relatively often stated, discussed and solved by mathematicians. Here is a small sample list of publications: [71], [76], [77], [78], [91], [90], [105], [106], [108], [109], [111], [20], [136], [28], [44], [152], [45], [164], [183], [184], [74], [2], [101], [142].

5.1 Algebraic language and harmony by Anna Romanowska

5.1.1 Formal language for the theory of harmony

Section 5.1 surveys the results of more than twenty years of discussion concerning the foundations of a formal language for the theory of harmony.[1] In 1976, R. Wille [185] proposed seven theses that should underline mathematical music theory and help to answer the question of the subsection title. Three of these theses will play an essential role in this section:

(1) Mathematics is the basis of a formal language for music theory;

(2) Mathematical language makes possible exact definitions of concepts in music theory;

(3) Mathematics furnishes complete and effective notation systems for concepts in music theory;

These three theses address the following problems of music theory in general, and the theory of harmony in particular:

(i) The lack of a precise language for concepts;

(ii) The ambiguity and vagueness of many concepts;

(iii) The lack of appropriate specifications, and even names, for many concepts that prove to be useful.

These deficiencies preoccupied many musicologists in the twentieth century. As the Polish musicologist M. Zalewski wrote in 1973 [68]: "Even a fleeting glance at the analyzes of the compositions of modern music, of the twentieth century classics or of the twelve-tone school, reveals what one might dare to call powerlessness, a lack of conceptual or terminological apparatus. Indeed, many theoreticians and musicologists have expressed this."

Zalewski's work [68] was one of several attempts to overcome these deficiencies. Mathematical formulations of some of his concepts were provided by Romanowska [165], [166], [168]. Another attempt of a similar type was undertaken by Wille [185], [186].

In this section, we will propose a language for the theory of harmony, using and unifying concepts of these three authors. This will serve to illustrate the three theses presented above. The "harmonic"

---

[1]The author of Section 5.1: Anna Romanowska, Dept. of Mathematics, Iowa State University, Ames, Iowa 50011, U.S.A.

content of a chord is described using certain sequences of natural numbers and their permutations, or else certain matrices of natural numbers and certain sets of similar matrices. They are obtained by introducing some equivalence relations on the set of all chords of a tone system. Only equal temperament tone systems are considered.

The paper only sets up a language describing an "harmonic content" of chords. It does not discuss the relations between chords, nor their meaning in different styles of music.

Though practicing musicians currently do not seem to pay much attention to theoretical considerations, the development and importance of computers and the Internet in the modern world may renew interest in a more precise foundation of the theory of harmony. The mathematics used in this Section are elementary, but not without subtleties. Readers are referred to e.g., [63].

5.1.2 Tone systems and chords

The central concept of music theory is the concept of a *tone*. And one of its most important characteristics is its *pitch*. In music one uses only a small choice of possible sounds, usually with determined pitches. The pitch is a psychological concept, depending on the frequency of the tones.

A higher frequency is perceived as a higher pitch. Most essential are the differences of pitches of two tones, described physically as ratios of their frequencies, and usually called intervals. Later, in Subsection 5.1.3, we will describe such differences in another way, as is usually done in music theory, and in such a way that intervals really have the properties of a distance. In particular, an interval with a frequency ratio equal to 2 is called an *octave*. Two tones a multiple of an octave apart are perceived as "the same." A set of tones chosen according to some rules (generally governing the intervals between consecutive tones) is called a *tone system*. Mathematically, we can consider it as a pair $(\mathbb{T}, \Omega)$ consisting of a set $\mathbb{T}$ of tones and an function, called the *pitch function*, $\Omega : \mathbb{T} \to \mathbb{R}^+$ assigning to each tone its frequency (e.g., in Hz). A system that contains along with each tone the tones an octave apart, and where the octave is "divided" into N equal parts, is called an *N-tone Equal Tempered*

System, and will be denoted by E_N. In this section we will consider only such systems. The 12-tone system E_{12} is the best known and most widely used. For theoretical reasons, it is convenient to consider the set $\mathbb{T}$ to be isomorphic to the set $\mathbb{Z}$ of integers. We frequently will identify the set $\mathbb{T}$ with the set $\mathbb{Z}$. In practice, obviously one restricts $\mathbb{T}$ to a finite subset. For two tones T_1 and T_2 of a system $(\mathbb{T}, \Omega)$, we say that T_2 is *higher* than T_1, and write $T_1 < T_2$, if $\Omega(T_1) < \Omega(T_2)$. In what follows, we assume that the tones of a system $(\mathbb{T}, \Omega)$ are ordered in this way.

Example 27 The system $(\mathbb{Z}, \Omega)$, $S = E_{12}$ has its pitch function $\Omega : \mathbb{Z} \to \mathbb{R}^+$ defined by $\Omega(z) = \omega_0 2^{\frac{z}{12}}$, where $\omega_0 = 440\,\mathrm{Hz}$, $z \in \mathbb{Z}$. Clearly, for each $z \in \mathbb{Z}$, $\Omega(z+1)/\Omega(z) = 2^{\frac{1}{12}}$. One can also consider this system as a pair $(\mathbb{T}, \Omega)$ with $\mathbb{T} = \{-48, \ldots, 36\}$.

Let us note that for any system $(\mathbb{Z}, \Omega)$, $S = E_N = \{ \sqrt[N]{2^z}; \ z \in \mathbb{Z} \}$, $N \in \mathbb{N}$, one has $\Omega(z) = \omega_0 \cdot 2^{\frac{z}{n}}$ for any z in $\mathbb{Z}$ and ω_0 is the reference frequency which depends on the choice of physical units. So, without loss of generality we can put $\omega_0 = 1\,\mathrm{NU}$ (i.e., new unit, where $1\,\mathrm{NU} = 440\,\mathrm{Hz}$).

By a "chord" in a system E_N, one usually understands a set of several tones played together or in a sequence, built according to certain rules. This justifies defining an *m-chord* C in a system $(\mathbb{T}, \Omega)$, $S = E_N$, as a finite subset of $\mathbb{T}$ with m elements. Denote by $C^m(E_N)$ the set of all m-chords of E_N and by $C(E_N) = \bigcup_{m \in \mathcal{N}} C^m(E_N)$ the set of all m-chords for all natural m (including the "empty" chord or a "rest"). The theory of harmony deals with chords of a given tone system and the relationships between them.

Harmonic analysis also requires the possibility of speaking about chords with some special selected notes, like a key note or a note in a melody. Such chords will be called *pointed*, and defined mathematically as pairs (C, T), where T is a selected element of a chord C. The "point" T can then be interpreted according to need. In this section we will interpret T always as a key note. The symbol $C_p^m(E_N)$ will denote the set of all pointed m-chords of the system E_N. Similarly the symbol $C_p(E_N)$ will denote the set of all pointed m-chords for all possible $m > 0$.

5.1.3 Structure of chords and pointed chords

The fact that tones of a system $(\mathbb{T}, \Omega)$, $S = E_N$, an octave apart are harmonically identical can be expressed mathematically by means of the equivalence relation θ defined on $\mathbb{T}$ as follows:

$$(T, T') \in \theta \Leftrightarrow \log_2 \Omega(T) - \log_2 \Omega(T') \in \mathbb{Z}.$$

One can extend this relation to an equivalence relation Υ on the set $C(E_N)$ by defining

$$(C, C') \in \Upsilon \Leftrightarrow \forall T \in C, \exists T' \in C'; T\theta T' \text{ and } \forall T' \in C', \exists T \in C; T\theta T'.$$

The Υ-classes are called *harmonies*. The set of all harmonies is denoted $C(E_N)^\Upsilon$. Examples are given by the harmonies C major and A minor of the system E_{12}. The representatives of the harmony C major consist of C major triads in all possible (closed or open) positions, and inversions. Let us note that each non-empty harmony has as a representative an ordered k-chord $(T_1, \ldots, T_k)$ such that for any T_i, T_j

$$\frac{1}{2} < \frac{\Omega(T_i)}{\Omega(T_j)} < 2,$$

i.e. the chord $(T_1, \ldots, T_k)$ is in a closed position. Let us call such a representative *ordered closed*.

An equivalence relation Υ_p similar to Υ can be defined on the set $C_p(E_N)$ of pointed chords:

$$((C, T), (C', T')) \in \Upsilon_p \Leftrightarrow (C, C') \in \Upsilon \text{ and } (T, T') \in \theta.$$

The Υ_p-classes are called *pointed harmonies*. The set of all pointed harmonies is denoted $C_p(E_N)^{\Upsilon_p}$. All C major triads in the first inversion, i.e. with a note E as the lowest tone, form representatives of the pointed harmony C major in the first inversion (i.e. with a third in the bass) in the system $\dot{E}_{12}$. As in the previous case, each pointed harmony has an ordered closed representative.

The harmonies and pointed harmonies of a system E_N are related as follows. Let $oc(h) \stackrel{\text{def}}{=} (T_1, \ldots, T_k)$ be an ordered closed representative of a pointed harmony $h = (\{T_1, \ldots, T_k\}, T_1)^{\Upsilon_p} \in C_p(E_N)^{\Upsilon_p}$.

Then the (ordered closed) chord $(T_i, \ldots, T_k, T_1, \ldots, T_{i-1})$ is called an $(i-1)$-*inversion* of $(T_1, \ldots, T_k)$ and the pointed harmony $h^{i-1} = (\{T_1, \ldots, T_k\}, T_i)^{\Upsilon_p}$ is called an $(i-1)$-*th inversion of the harmony* h. It is easy to see that the relation Ψ defined on the set $C_p(E_N)^{\Upsilon_p}$ by

$$(h, h') \in \Psi \Leftrightarrow \quad \forall oc(h'), \exists oc(h), \exists i; \ oc(h')\Upsilon(oc(h))^{i-1}$$
$$\text{and}$$
$$\forall oc(\omega), \exists oc(h'), \exists j; \ oc(\omega)\Upsilon(oc(h'))^{j-1},$$

is an equivalence relation and that

$$C(E_N)^{\Upsilon} = (C_p(E_N)^{\Upsilon_p})^{\Psi}.$$

Let r be a positive real number. An injective function $\tau_r : \mathbb{T} \to \mathbb{T}$, such that for each T in $\mathbb{T}$

$$\frac{\Omega(\tau_r(T))}{\Omega(T)} = r,$$

is called a *transposition* of the system E_N by r. The transposition τ_r moves each tone of $\mathbb{T}$ the same interval up or down.

Using the concept of a transposition, we will define one more equivalence relation Φ on the set $C(E_N)$:

$$(C, C') \in \Phi \Leftrightarrow \exists r \in \mathbb{R}^+; \ \tau_r(C) = C'.$$

Similarly for (C, T) and (C', T') in $C_p(E_N)$ we define

$$((C, T), (C', T')) \in \Phi_p \Leftrightarrow \exists r \in \mathbb{R}^+; \ \tau_r(t) = t' \text{ and } \tau_r(C) = C'.$$

Elements of the sets $C(E_N)^{\Phi}$ and $C(E_N)^{\Phi_p}$ of classes are called *chordal forms* and *pointed chordal forms* respectively. Intervals, like thirds, fifths, etc. are examples of chordal forms of the system E_{12}.

Now composition of the relations Υ and Φ will lead us to the central concept of this section. Since $\Upsilon \circ \Phi = \Phi \circ \Upsilon$ and $\Upsilon_p \circ \Phi_p = \Phi_p \circ \Upsilon_p$, the compositions $\Upsilon \circ \Phi$ and $\Upsilon_p \circ \Phi_p$ are equivalence relations on $C(E_N)$ and $C_p(E_N)$ respectively. Two chords of $C(E_N)$ are $\Upsilon \circ \Phi$-related if one can be obtained from the other by the octave identification and transposition. Elements of $C(E_N)^{\Upsilon \circ \Phi}$ are called *harmonic structures* and elements of $C_p(E_N)^{\Upsilon_p \circ \Phi_p}$ are called *pointed harmonic structures.* As examples of harmonic structures of E_{12}, one has major and minor triads. The first inversion of the major triad is an example of a pointed harmonic structure. As in the case of harmonies, (pointed) harmonic structures have ordered closed representatives. For example, the first inversion of the major triad has as representatives chords E, G, C' or B, D', G'.

Harmonic structures and pointed harmonic structures are related to each other similarly as harmonies and pointed harmonies. Let $\mathrm{oc}(s) \stackrel{\text{def}}{=} (T_1, \ldots, T_k)$ and $\mathrm{oc}(s') \stackrel{\text{def}}{=} (T'_1, \ldots, T'_l)$ be ordered closed representatives of the pointed harmonic structures

$$ s = (\{T_1, \ldots, T_k\}, T_1)^{\Upsilon_p \circ \Phi_p} $$

and

$$ s' = (\{T'_1, \ldots, T'_l\}, T'_1)^{\Upsilon_p \circ \Phi_p}. $$

Then the $(i\text{-}1)$-th *inversion* of s is the pointed harmonic structure $s^{i-1} = (\{T_1, \ldots, T_k\}, T_i)^{\Upsilon_p \circ \Phi_p}$. Define the following equivalence relation ψ on $C_p(E_N)^{\Upsilon_p \circ \Phi_p}$ similarly as Ψ:

$$ (s, s') \in \psi \Leftrightarrow \quad \forall \mathrm{oc}(s'), \exists \mathrm{oc}(s), \exists i; \ \mathrm{oc}(s') \Upsilon \circ \Phi(\mathrm{oc}(s))^{i-1} $$
$$ \text{and} $$
$$ \forall \mathrm{oc}(s), \exists \mathrm{oc}(s'), \exists j; \ \mathrm{oc}(s) \Upsilon \circ \Phi(\mathrm{oc}(s'))^{j-1}. $$

Then, as in the previous case,

$$ C(E_N)^{\Upsilon \circ \Phi} = (C_p(E_N)^{\Upsilon_p \circ \Phi_p})^{\psi}. $$

5.1.4 Representations of harmonic structures

The definitions of harmonic structures provided in the previous section capture what is essential in a chord from the harmonic point of view. Nevertheless they would be rather awkward to use. In this section we provide some simple ways of representing harmonic structures.

We first define an *interval* (or a *distance*) between two tones. Let (T_1, T_2) be an ordered closed representative of a pointed harmony of a system E_N. Then the interval between T_1 and T_2 is the number

$$ d(T_1, T_2) = \log_{2^{\frac{1}{n}}} \frac{\Omega(T_2)}{\Omega(T_1)} = n(\log_2 \Omega(T_2) - \log_2 \Omega(T_1)). $$

If $E_N = \Omega(\mathbb{T})$, then

$$d(T_1, T_2) = \log_{2^{\frac{1}{n}}} \frac{\Omega(T_2)}{\Omega(T_1)} = \log_{2^{\frac{1}{n}}} 2^{\frac{T_2 - T_2}{n}} = T_2 - T_1$$

is a positive integer. For two consecutive tones $z, z+1$ of the system E_{12} one has

$$d(z, z+1) = \log_{2^{\frac{1}{12}}} 2^{\frac{1}{12}} = 1.$$

Hence in E_{12}, the interval 1 corresponds to the equal temperament semitone. For any two T_i, T_j of an ordered closed chord with $T_i < T_j$, $d(T_i, T_j)$ is the number of semitones between T_i and T_j.

Now let $(T_1, \ldots, T_k)$ be an ordered closed representative of a pointed harmonic structure $(C, T_1)^{\Upsilon_p \circ \Phi_p}$ of a system E_N. Let T_0 be a tone of $\mathbb{T}$ an octave higher than T_1. And finally let Q be the set of finite sequences of natural numbers. Define the following mapping

$$f : C_p(E_N)^{\Upsilon_p \circ \Phi_p} \longrightarrow S;$$
$$(C, T_1)^{\Upsilon_p \circ \Phi_p} \mapsto \left(\log_{2^{\frac{1}{n}}} \frac{\Omega(T_2)}{\Omega(T_1)}, \ldots, \log_{2^{\frac{1}{n}}} \frac{\Omega(T_k)}{\Omega(T_{k-1})}, \log_{2^{\frac{1}{n}}} \frac{\Omega(T_0)}{\Omega(T_k)} \right)$$

Let $Q_N \overset{\text{def}}{=} f(C_p(E_N)^{\Upsilon_p \circ \Phi_p})$.

It is obvious that elements of Q_N are sequences $(a_1, \ldots, a_k)$ of positive integers less than N. Moreover, for each $(a_1, \ldots, a_k)$ in Q_N

$$\sum_{i=1}^{k} a_i = \log_{2^{\frac{1}{N}}} \left(\frac{\Omega(T_2)}{\Omega(T_1)} \cdot \frac{\Omega(T_3)}{\Omega(T_2)} \cdots \cdots \frac{2\Omega(T_1)}{\Omega(T_k)} \right) \log_{2^{\frac{1}{N}}} 2 = N.$$

The image $f((C, T_1)^{\Upsilon_p \circ \Phi_p})$ is called the *pointed interval structure* of the pointed harmonic structure $(C, T_1)^{\Upsilon_p \circ \Phi_p}$. The elements of Q_N are called *pointed interval structures of the system* E_N. Additionally, we may assume that the empty harmonic structure has an empty pointed interval structure denoted by $1 = (\ )$. For example the 1st inversion of the dominant seventh in E_{12} has the pointed interval structure $(3, 3, 2, 4)$.

The following propositions are easy consequences of the definitions above.

Lemma 10 *A sequence $(a_1, \ldots, a_k)$ of positive integers with $k \leq N$ is a pointed interval structure of a system E_N if and only if for all $i = 1, \ldots, k$*

$$a_i \in \{1, \ldots, N\} \quad and \quad \sum_{i=1}^{k} a_i = N.$$

Lemma 11
There is a bijective correspondence between the set $C_p(E_N)^{\Upsilon_p \circ \Phi_p}$ of pointed harmonic structures and the set Q_N of pointed interval structures.

The next step provides a passage from pointed to non-pointed interval structures.

Let $a = (a_1, \ldots, a_k)$ be in Q_N. Then for each $i = 1, \ldots, k$, the sequence $a^{i-1} \overset{\text{def}}{=} (a_i, \ldots, a_k, a_1, \ldots, a_{i-1})$ is also a pointed interval structure of E_N, and is called *the $(i-1)$-th inversion of a*. It is easy to see that the relation (ψ) defined on the set Q_N by

$$(a, b) \in (\psi) \Leftrightarrow \exists i; \ b = a^{i-1} \text{ for some } i$$

is an equivalence relation.

The elements $((a_1, \ldots, a_k)) \overset{\text{def}}{=} (a_1, \ldots, a_k)^{(\psi)}$ of the set $(Q_N) \overset{\text{def}}{=} Q_N^{(\psi)}$ of (ψ)-classes are called *interval structures* of the system E_N. The following theorem establishes the final relationship between harmonic structures and interval structures of the system E_N.

Theorem 12
There is a bijective correspondence between the set $C(E_N)^{\Upsilon \circ \Phi}$ of harmonic structures and the set (Q_N) of interval structures of the system E_N.

We omit the rather standard proof of this theorem. Note however the following:

$$C(E_N)^{\Upsilon \circ \Phi} = (C_p(E_N)^{\Upsilon_p \circ \Phi_p})^{\psi} \cong Q_N^{(\psi)}.$$

Harmonic and pointed harmonic structures can also be represented in a different way. First define the following mapping:

$$h : Q_N \to Q;$$

$$(a_1, \ldots, a_k) \mapsto \left(a_1, a_1 + a_2, \ldots, \sum_{i=1}^{k} a_i \right) \overset{\text{def}}{=} (A_1, A_2, \ldots, A_k).$$

For $a = (a_1, \ldots, a_k) \in Q_N$, the sequence $h(a)$ is called the *form of the pointed interval structure a* and the set $F_N = h(Q_N)$ is the set of forms of pointed interval structures of the system E_N. One has obvious counterparts of the following two lemmas.

Lemma 13 *A sequence $(A_1, \ldots, A_l)$ of positive integers with $l \leq N$ is the form of a pointed interval structure if and only if the sequence is strictly increasing and $A_k = N$.*

Lemma 14
There is a bijective correspondence between the set $C_p(E_N)^{\Upsilon_p \circ \Phi_p}$ of pointed harmonic structures and the set F_N of forms of pointed interval structures.

Proof.   The required bijection is given by the composition of the mappings f and ω.   □

The definition of an interval structure shows that we can consider it as the set of all cyclic permutations of any representative (a pointed interval structure). Sometimes it is more convenient to work with all representatives at the same time. One can do this by putting all of them in a matrix. However, one has much nicer properties, and an easier passage between interval structures and corresponding matrices, if instead of pointed interval structures one considers their forms arranged in a suitable way.

Let $\mathcal{M}$ be the set of square matrices of non-negative integers. Consider the following mapping $m : Q_N \to \mathcal{M}$;

$$a = (a_1, \ldots, a_k) \mapsto \begin{bmatrix} 0 & a_1 & a_1 + a_2 & \cdots & \sum_{i=1}^{k-1} a_i \\ \sum_{i=2}^{k-1} a_i & 0 & a_2 & \cdots & \sum_{i=2}^{k-1} a_i \\ & \cdots & \cdots & & \\ a_k & a_k + a_1 & a_k + a_1 + a_2 & \cdots & 0 \end{bmatrix}$$

The image $m(a)$ is called the *matrix of the pointed interval struc-ture* a. The set $\mathcal{M}_N \overset{\text{def}}{=} m(Q_N)$ is the set of matrices of the pointed interval structures of the system E_N.

Lemma 15 *A $k \times k$-matrix $\alpha = (a_{ij})$ with $k \leq N$ of non-negative integers belongs to the set $\mathcal{M}_N$ if and only if:*

(a) *it has zeros down the main diagonal;*

(b) *for each $i = 1, \ldots, k$, the sequence*

$$(a_{ii}, a_{ii+1}, \ldots, a_{ik}, a_{i1}, \ldots, a_{ii-1})$$

is strictly increasing;

(c) $a_{ij} + a_{ji} = N$ *for $j \neq i$;*

(d) $a_{ij} = a_{1j} - a_{1i}$ *for $j > i \neq 1$.*

We omit the easy proof based on elementary properties of ma-trices. Note only that for a given matrix α in the set $\mathcal{M}_N$, the corresponding pointed interval structure is recovered from its first row as

$$(a_{12}, a_{13} - a_{12}, \ldots, a_{1k} - a_{1k-1}, n - a_{1k}).$$

Inversions of pointed interval structures have an easy counterpart when dealing with their matrices. Indeed, given a pointed interval structure $a = (a_1, \ldots, a_k)$ in Q_N and its matrix $m(a)$ in $\mathcal{M}_N$, its i-th inversion

$$a^i = (a_{i+1}, \ldots, a_k, a_1, \ldots, a_i)$$

has as its matrix, the matrix α^i obtained from $m(a)$ by moving the top row down to the bottom and the left column across to the right i times. Mathematically, it can be expressed as follows

$$\alpha^i = E_{(i)}^{-1} \alpha E_{(i)},$$

whose $E_{(i)}$ is a suitable permutation $k \times k$-matrix. This leads to the last concept we need—the matrix counterpart of an interval struc-ture. In the set $\mathcal{M}_N$, define the relation $\widetilde{\psi}$ as follows

$$(\alpha, \beta) \in \widetilde{\psi} \Leftrightarrow \exists i \leq n; \ \beta = \alpha^i$$

The relation $\tilde{\psi}$ is an equivalence relation and one easily proves that for a, b in $Q_N, a(\psi)b$ if and only if $m(a)\tilde{\psi}m(b)$. This implies the following.

Theorem 16 *There is a one-to-one correspondence between the set (Q_N) of interval structures of the system E_N and the set of $\tilde{\psi}$-classes of $\mathcal{M}_N$.*

Remarks

As we have seen in the previous subsections, interval structures and their matrices provide a precise mathematical language to describe the harmonic aspects of chords in equal temperament tone systems E_N. Such concepts can also be defined for other tone systems, as for example systems $(\mathbb{T}, \Omega)$ such that for any T_2, T_1 in $\mathbb{T}$, one has $\Omega(T_2)/\Omega(T_1) \in \mathbb{Q}^+$, e.g., c.f. [167] and [110].

The set of (pointed) harmonic structures, as well as the corresponding isomorphic set of (pointed) interval structures and the set of their matrices, surprisingly all have quite rich (isomorphic) algebraic structures. It was fully described in [165], [166]. Not all these algebraic aspects have an interesting interpretation in music theory, so we will not repeat it here. We mention only some of them, and restrict ourselves mainly to the sets Q_N and (Q_N). (It is not difficult to describe isomorphic "algebras" of harmonic structures or matrices.)

Both the sets Q_N and (Q_N) have naturally defined partial orders. For $a = (a_1, \ldots, a_k)$ and $b = (b_1, \ldots, b_l)$ in Q_N define $a \leq b$ if there is a sequence $0 = T_0 < T_1 < \cdots < T_l = k$ such that for each $j = 1, \ldots, l$

$$b_j = \sum_{i=T_{j-1}+1}^{T_j} a_i.$$

It is easy to see that $\leq$ is a partial order on Q_N. Moreover the following holds.

Lemma 17 *The ordered sets $(Q_N, \leq)$ of (non-empty) pointed interval structures, and $(CgC_N, \leq)$ of the congruence relations of the N-element chain considered as a lattice, are order-isomorphic.*

The proof is rather standard, and we omit it here. As a consequence, we get the following.

Corollary 18 *The ordered set $(Q_N, \leq)$ is a Boolean lattice 2^{N-1}.*

The corresponding order relation on M_N is also easy to describe. For $a = (a_1, \ldots, a_k) \leq b = (b_1, \ldots, b_l)$, the matrix $m(a)$ is obtained form the matrix $m(b)$ by cancelling some rows and columns.

Though the Boolean operations are not preserved under the relation (ψ) on Q_N, the inversions and some further unary operations [as for example $-a = -(a_1, a_2, \ldots, a_k) = (a_k, a_{k-1}, \ldots, a_1)$] are preserved.

The order relation $\leq$ on the set (Q_N) of interval structures is defined by

(a) $\leq$ (b) $\Leftrightarrow$ for each c in (a) there is d in (b) with $c \leq d$.

5.2 Chord and tone rows enumerations

While H. Helmholtz, c.f. [26], represents the mathematical empiricism in its classical form, G. Pólya is one of "neo-empiricists." After him, mathematics is an inductive science which can be appropriately understood only as one of sciences which considers the nature, i.e. mathematics is a natural science, c.f. [47].

The calculation of the number of chords and tone rows using Pólya's Theorem and Burnside's Lemma can add a variety to the applications of these theorems usually given in combinatorics and algebra texts.

Let us deal with the tone system $E_N, n \in \mathbb{N}$. Analogously as in the previous Sections 5.1, we can consider the bijection $\omega = 2^{z/12}, \omega \mapsto z$ of E_N and the set $\mathbb{Z}$ of all integer numbers and suppose that the tone system consists of all integer numbers, i.e., $\mathbb{T} = \mathbb{Z}$. We equate octave tones, so our scale is mathematically $\mathbb{Z}_N$ with addition.

Ordinary Western music has $N = 12$. Debussy used a whole tone scale with $N = 6$. Hába used a quarter tone scale with $N = 24$.

A k-chord in E_N is an equivalence class (defined in the following subsubsection) of subsets of k elements each of $\mathbb{Z}_N$, and an N-tone *row* is an equivalence class (defined in the subsubsection about Tone rows) of permutations in S_N.

5.2.1   Chords

The equivalence relation is induced by one of two groups of permutations. We can use the transpositions (do not be confused with permutations which exchange two elements) $T^i : \mathbb{Z}_N \to \mathbb{Z}_N : a \mapsto i + a(\bmod N)$. A jazz guitarist generally uses chords which can be transposed easily so not so many fingerings must be memorized. Or we can use the larger group of transpositions (as above) and the inversion $I : a \mapsto -a \,(\bmod N)$. A typical group element looks like $T^i I : a \mapsto i - a \,(\bmod\ N)$. This use of the word "inversion" is standard in the serial music and must be distinguished from the usual meaning of octave transposition of one or more notes of a chord. For example, under the large group, a C major chord is equivalent to a C minor chord: $\{C, E, G\} = \{0, 4, 7\} \sim \{0, 3, 7\} \sim \{C, E_\flat, G\}$, the $\sim$ being via $T^i I$.

It is easy to see that these groups are cyclic groups C_N and the dihedral groups D_N, respectively, and we are thus dealing with what is known as the two-color necklace problem, either one-sided or two-sided, [58, p. 162]. That is, we have a circular necklace of notes to choose from, and we chose the ones in the chord by coloring them one color and the rest another color.

Polya's Theorem is the appropriate tool for this situation: a group acting on a domain set D, inducing a group action on the set of functions into some range set R. It suffices to let G be a group of permutation of D. G then acts on the set of all functions R^D in a natural way, inducing an equivalence relation.

To keep track of what's important in such problem, we use enumerators, polynomials or formal power series whose coefficients are the numbers of objects of specified types. The enumerator (*the uncertainty measure!*) of the range here can be whatever we like, say $\sum_{r \in R} w(r)$, where w, the "weight," is any function from R to a commutative ring containing the rationals. The weight of a function from D to R is then defined to be the product of the weights of its image values, and the enumerator or inventory of a set of functions is the sum of the weights of the individual functions. Usually this set of functions is a set of equivalence class representatives when we have an equivalence relation on R^D.

In our situation of chords, $R = \{0, 1\}$, $w(0) = 1$, $w(1) = x$, and the weight of a function from D to this R is x^k, where k is just the size of the subset of D mapped to 1. This is how we enumerate subsets of chords.

The importance of Polya's theorem is that it relates the inventory of equivalence classes of functions to the "store enumerators," the enumerator of the range, via the cycle index P_G of the group G acting on D, which is defined by

$$P_G(t_1, t_2, \dots) = \frac{1}{|G|} \sum_{g \in G} t_1^{a_1} t_2^{a_2} \cdots ,$$

where $a_i =$ the number of i - cycles in g.

Polya's Enumeration Theorem, [42, p. 148], states that the enumerator or invertory equivalence classes under G of functions in R^D is

$$P_G \left(\sum_{r \in R} w(r), \sum_{r \in R} w(r)^2, \dots, \sum_{r \in R} w(r)^k, \dots \right).$$

The cycle indices of C_N and D_N are well known, [58, pp.149–150]:

$$P_G(t_1, t_2, \dots, t_N) = \frac{1}{N} \sum_{j|n} \phi(j) t_j^{N|j} = \frac{1}{N} \Phi,$$

and

$$P_D(t_1, t_2, \dots, t_N) = \begin{cases} \frac{1}{2n}[\Phi + nt_1 t_2^{(N-1)/2}], & \text{if } n \text{ is odd,} \\ \frac{1}{2n}[\Phi + \frac{N}{2} t_1^{(N/2)-1}(t_1^2 + t_2)], & \text{if } n \text{ is even.} \end{cases}$$

Here we use $\phi(j)$, which is the Euler ϕ-function, the number of positive integers less than j which are relatively prime to j, and Φ, which is a short for the sum in the first equation.

Our enumerator is just $1 + x$, and $t_j = 1 + x^j$. Hence the number of k-chords is the coefficient of x^k in $P(1 + x, 1 + x^2, \dots, 1 + x^N)$. For C_N, a little rearrangement produces

$$\sharp \text{ of } k\text{--chords} = \frac{1}{N} \sum_{j|(N,k)} \phi(j) \binom{N/j}{k/j} = \frac{1}{N} \Phi_N(k),$$

TABLE 5.1: Number of hexagons in 12 tone music

k	0	1	2	3	4	5	6	7	8	9	10	11	12
	1	1	6	12	29	38	50	38	29	12	6	1	1

where $\Phi_N(k)$ is short for the summation. For D_N, the additional term gives

$$\sharp \text{ of } k\text{--chords } = \begin{cases} \frac{1}{2N}\left(\Phi_N(k) + N\binom{(N-1)/2}{k/2}\right), N \text{ odd}, \\ \frac{1}{2N}\left(\Phi_N(k) + N\binom{N/2}{k/2}\right), \text{ both } N, k \text{ even}, \\ \frac{1}{2N}\left(\Phi_N(k) + N\binom{(N/2)-1}{k/2}\right), \text{ otherwise}. \end{cases}$$

The situation in twelve-tone music is $N = 12$ with dihedral symmetry, and we obtain Table 5.1.

Thus, there are 50 hexagons in twelve-tone music and only 6 intervals, a fact well known in music theory, [16].

5.2.2   Tone rows

Here our equivalence classes are induced by the group generated by
(1) transposition T;
(2) inversion I;
(3) retrogradation R,
where

$$T : S_N \to S_N : (a_1, \ldots, a_N) \mapsto (a_1 + 1, \ldots, a_N + 1)(\bmod N),$$

$$I : (a_1, \ldots, a_N) \mapsto (a_1, 2a_1 - a_2, \ldots, 2a_1 - a_N + 1)(\bmod N),$$

$$R : (a_1, \ldots, a_N) \mapsto (a_N, \ldots, a_1)(\bmod N).$$

This group is not well known; however the entire structure of the group is not really needed. We can take a cue from music theorists, who regard transposition as such a basic transformation that they do not work with all permutations, but work with equivalence classes of permutations under transposition, those beginning with 0 being regarded as class representatives. Hence our set is now the set of $(N-1)!$ permutations of $\{1, \ldots, N-1\}$ with a prefix of 0, and our

TABLE 5.2: Number of N-tone rows

	N odd	N even
e	$(N-1)!$	$(N-1)!$
I	0	0
R	0	$(N-2)(N-4)\cdots(4)$
IR	$(N-1)(N-3)\dots(2)$	$(N/2)(N-2)(N-4)\cdots(2)$

group is generated by R and I. On this new set, $RI = IR$, so we have the Klein four group V acting.

Burnside's Lemma, [42, p. 136], is the appropriate tool in this setting. It states that for a group G acting on D, the number of equivalence classes is

$$\frac{1}{|G|} \sum_{g \in G} (\sharp \text{ of elements of } D \text{ fixed by } g).$$

The computation proceed as follows, c.f. Table 5.2. In order to avoid triviality, we assume that $N > 3$.

Thus the number of N - tone rows is

$$[(N-1)! + (N-1)(N-3)\cdots(2)]/4$$

if N is odd; or

$$[(N-1)! + (N-2)(N-4)\cdots(2)(1+N/2)]/4$$

if N is even, respectively. For example, there are $9\,985\,920$ twelve-tone rows.

5.2.3 Computation of numbers of fixed elements

1. $I(0, a_2, \dots, a_N) = (0, -a_2, \dots, -a_N) \sim (0, a_2, \dots, a_N)$ implies $a_i \equiv -a_i (\text{mod } N)$ for $i = 2, \dots, n$ since no transposition is allowed because the first element 0 is fixed. There is at most one nonzero solution to $x \equiv -x (\text{mod } N)$, but that is not enough to fill out the permutation.

2. $R(a_1, \ldots, a_N) = (a_N, \ldots, a_1) \sim (a_1, \ldots, a_N)$ implies that a t exists such that $a_1 \equiv a_N + t, a_2 \equiv a_{N-1} + t, \ldots, (\text{mod } N)$. If N is odd, the middle element is fixed and no transposition is allowed: $t = 0$. But then $a_N = a_1$, a contradiction. If N is even, the first and last congruences imply that $2t \equiv 0$; hence $t = 0$ or $t = N/2$. The first is impossible just as when N is odd, but the other gives fixed permutation. Since $a_1 = 0, a_N = N/2$. For a_2, we can choose any of $N - 2$ elements, and this determines a_{N-1}. For a_3 we have $N - 4$ choices, etc.

3. $IR(a_1, \ldots, a_N) = (-a_N, \ldots, -a_1) \sim (a_1, \ldots, a_N)$ implies that a t exists such that $a_1 + a_N \sim t, a_2 + a_{N-1} = t, \ldots$. The last congruence for N odd is $2a_{(N+1)/2} \equiv t(\text{mod } N)$. Clearly it is not important that we fix the first element as 0; we could fix $2a_{(N+1)/2}$ as 0 and obtain the same count. Thus we may assume that $t = 0$, $2a_{(N+1)/2} = 0$. This allows $N - 1$ choices for a_1, with a_N thus terminated, $N - 3$ choices for a_2, etc. For N even, we fix $a_1 = 0, a_N \neq t \neq 0$. A little thought shows that t must be odd in order for us to complete the permutation. If $t = 2k$, then there is no mate for k in the permutation. There are thus $N/2$ choices for $t = a_N$. For a_2, there are $N - 2$ choices, with a_{N-1} determined thereby, etc.

5.3 Measuring equal temperaments

5.3.1 Continued fraction method

E_{12}, the 12-tone Equal Temperament, is a commonly well-known theoretical tone system. It will serve us as a starting point when considering other tone systems. It is also a common element of the majority of modern tone systems or tone systems families. An easy generalization of E_{12} is the following

Definition 12 The *N-equal tempered* tone system

$$E_N = \{\omega_z \in \mathbb{R}; \ \omega_z = \sqrt[N]{2^z}, z \in \mathbb{Z}\}, \ N \in \mathbb{N}.$$

The following theorem belongs rather to the mathematical folklore. Concerning the notions of temperature and mistuning ($\tau_{3/2}$ and $\mu_{3/2}$, respectively), c.f. Chapter 7, Subsection 7.1.4.

Theorem 19 *Denote by* $\{E_{N_k}\}_{N_k=1}^{\infty}$ *the sequence of equal tempered tone systems such that* $|\mu_{3/2}(E_{N_k})|_{N_k=1}^{\infty}$ *strictly decreases to 0 as* $N_k \to \infty$, *where* $\{\mu_{3/2}(E_{N_k})\}_{N_k=1}^{\infty}$ *is a subsequence of the sequence of all fifth mistunings* $\{\mu_{3/2}(N)\}_{N=1}^{\infty}$ *of the equal tempered tone systems* $\{E_N\}_{N=1}^{\infty}$. *Then*

$$\{E_{N_k}\}_{N_k=1}^{\infty} = \{E_1, E_2, E_5, E_{12}, E_{41}, E_{53}, E_{306}, E_{665}, \ldots\}.$$

Proof. Consider the approximations

$$\frac{3}{2} \approx 2^{x/N_k},$$

where N_k denotes the number of equal steps per octave and x the order number of the tempered fifth, respectively. We have:

$$\frac{x}{N_k} \approx \log_2 \frac{3}{2} = \cfrac{1}{1+\cfrac{1}{1+\cfrac{1}{2+\cfrac{1}{2+\cfrac{1}{3+\cfrac{1}{1+\cfrac{1}{5+\cfrac{1}{2+\cfrac{1}{10}\cdots}}}}}}}}$$

This continuous fraction yields the following sequence:

$$\{a(N_k)\}_{N_k=1}^{\infty} = \left\{1, \frac{1}{2}, \frac{3}{5}, \frac{7}{12}, \frac{24}{41}, \frac{31}{53}, \frac{179}{306}, \frac{389}{665}, \cdots\right\},$$

$$\lim_{N_k \to \infty} a(N_k) = \log_2 \frac{3}{2}.$$

The corresponding fifth temperature sequence $\{\tau_{3/2}(E_{N_k})\}_{N_k=1}^{\infty}$ is as follows:

$$\{\tau_{3/2}(E_{N_k})\}_{N_k=1}^{\infty} = \{ \quad 1 + 0.010\,48, 1 - 0.001\,135,$$
$$1 + 0.000\,278\,9, 1 - 0.000\,039\,4,$$
$$1 + 0.000\,003\,18, 1 - 0.000\,000\,07, \ldots\}.$$

We see that $|\tau_{3/2}(E_{N_k})| \downarrow 0$ as $N_k \to \infty$.   □

We can verify that the third temperatures are greater than $81/80$ in the case $\tau_{3/2}(E_N) > 1$. For $\tau_{3/2}(E_{N_k}) < 1$, the N_k-tone equal tempered tone systems are then E_{12}, E_{53}, E_{665}, ... The E_{12} is well known, the E_{665} (the tone system of J. Sumec) is not very interesting since the number 665 is too large from the practical viewpoint. The E_{53} is used in Turkish national music and can be considered also as a temperament of a very interesting Petzval's cyclic Tone system of the 2nd type, c.f. Example 63.

Analogously to the proof of Theorem 19, we can compute continuous fractions for other musical intervals, e.g., $\log_2 6/5$, $\log_2 5/4$, $\log_2 7/4$, and obtain separately decreasing sequences of mistunings and corresponding E_{N_k}. However, we would like to have measures that let us compare musical temperaments of E_N, $n \in \mathbb{N}$ in a more complex view.

5.3.2   Increasing number of keys per octave

In Table 5.3, there are twelve powers of $\sqrt[12]{2^z}$, $z = 0, 1, 2, \ldots, 11$, in the first column. In the second one, there are just intonation values, and in the third column, we can find the temperature of this tone system expressed in cents. Notice how remarkably close most of them are to the ratios of small whole numbers. If the distribution were random we would expect an average temperature of 25 cents. Instead the average temperature is only approximately 10 cents (for the chosen set of rationals). There arises a question: *is there something really special about E_{12}?*

In Subsection 5.3.1 we compared E_N for various $N \in \mathbb{N}$ taking into the account only one interval, $3/2$, the perfect fifth. A following trivial measure can be used

$$\sigma(S) = \frac{1}{\mu_{3/2}(S)}, \tag{5.1}$$

where $\mu_{3/2}(S) = \min_{\omega \in S} \left| 1 - \frac{\omega}{3/2} \right|$, $S \in \mathfrak{S}_E$, where $\mathfrak{S}_E = \{E_N; N \in \mathbb{N}\}$ is the class of all equal temperaments.

TABLE 5.3: Temperature 12-*TET/JI* (in cents)

E_{12} ratio ϕ	Just ratio f	ϕ/f
1.000	$1/1 = 1.000$	0
1.059	$16/15 \approx 1.066$	-12
1.122	$9/8 = 1.125$	-4
1.189	$6/5 = 1.200$	-16
1.256	$5/4 = 1.250$	$+14$
1.335	$4/3 \approx 1.333$	$+2$
1.414	$7/5 = 1.400$	$+17$
1.498	$3/2 = 1.500$	-2
1.587	$8/5 = 1.600$	-14
1.682	$5/3 \approx 1.667$	$+16$
1.782	$16/9 \approx 1.778$	$+4$
(1.782	$7/4 = 1.750$	$+31$)
1.888	$15/8 = 1.875$	$+12$

Example 28 Try

$$\sigma(S) = \frac{\frac{1}{9} + \frac{1}{25} + \frac{1}{49}}{\frac{\mu_{3/2}^2(S)}{9} + \frac{\mu_{5/4}^2(S)}{25} + \frac{\mu_{6/5}^2(S)}{25} + \frac{\mu_{7/4}^2(S)}{49}}.$$

For illumination, execute the practical search for $S = E_N$, $N = 5, 6, \ldots, 60$ (we stop the evaluations on the number 60 since each frequency would fall in a ± 10 cent range from some frequency $f \in E_{60}$). If there was no disadvantage in having more notes per octave, then E_{12} is only slightly special with regard to harmonies measured by σ. Several scales with less than 60 notes do better, c.f. Figure 5.1, [194].

Example 29 Of course we expect the temperatures to get smaller as the number of divisions increases and hence their size decreases. We modify the harmonic mean in Example 28 as follows:

$$\sigma_N^{(q)}(S) = \frac{1}{N^q} \cdot \sigma(S), \qquad q \in \mathbb{N}.$$

For $q = 1$, the only ones equal to or better than E_{12} on this criterion are E_{31}, E_{41} and E_{53}, c.f. Figure 5.2, [194].

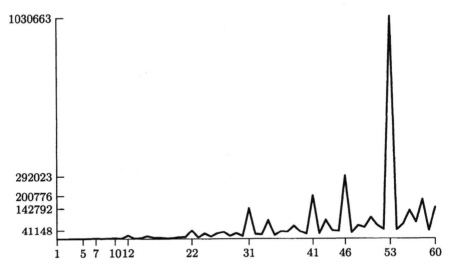

FIGURE 5.1: Measuring I of E_N, $N = 5, 6, \ldots, 60$

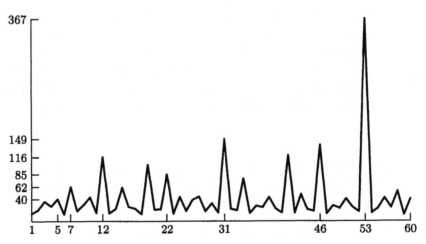

FIGURE 5.2: Measuring II of E_N, $N = 5, 6, \ldots, 60$

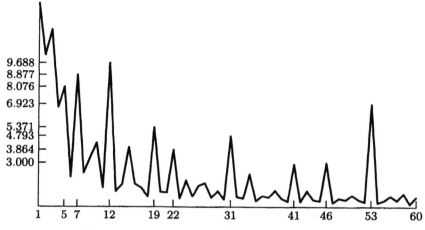

FIGURE 5.3: Measuring III of E_N, $N = 5, 6, \ldots, 60$

TABLE 5.4: Tempered fifths of E_N, $N = 1, 2, \ldots, 72$

E_7	E_{47}	E_{40}	E_{33}	E_{26}	E_{45}	E_{64}
675.8	689.4	690.0	691.0	692.4	693.4	693.8
E_{19}	E_{69}	E_{50}	E_{31}	E_{43}		
694.8	695.6	696.0	696.8	697.6	698.2	698.6
E_{12}	E_{65}	E_{53}	E_{41}	E_{70}	E_{29}	E_{46}
700.0	701.6	701.8	702.4	702.8	703.4	704.4
E_{63}	E_{17}	E_{56}	E_{39}	E_{61}	E_{22}	E_{71}
704.8	705.8	707.2	707.6	708.2	709.0	709.8
E_{49}	E_{27}	E_{59}	E_{32}	E_{37}	E_{42}	E_5
710.2	712.2	711.8	712.6	713.6	714.2	720.0

Example 30 For $q = 2$, the situation is graphed in Figure 5.3.

5.3.3 Fifths of equal temperaments

Concerning the equal tempered tone systems E_N, $N = 1, 2, \ldots, 72$, we give the overview of their tempered fifths in Table 5.4.

If an equal temperament does not appear here, it is because its tempered fifths are either:

- much too narrow (2, 4, 9, 11, 16, 23) or

- much too wide (1, 3, 6, 8, 13, 18) or

- it has more than one chain of fifths:

  - 2 chains (4, 6, 10, 14, 24, 34, 38, 44, 52, 54, 58, 62),
  - 3 chains (15, 21, 36, 51, 57, 66),
  - 4 chains (20, 28, 48, 68),
  - 5 chains (25, 35, 60),
  - 6 chains (30, 72).

Incidentally, the fifth size for a multi-chain equal temperament is the same as that for the equal temperament obtained by dividing by the number of chains; e.g. E_{72} has the same fifth size as $E_{72/6} = E_{12}$. Those with multiple chains may have more going for them than their fifth size, but we do not address this here.

5.4 Equal temperaments in practice

5.4.1 Music of nations

There are numerous very old drawings showing assemble play which requires the use of some tone system. For instance, in some reliefs in Ninive, there are depicted playing, singing, and dancing music quartets: drum, lute, harp, and cymbal. We have only very poor knowledge about tone systems of Summer, Babylon, Assyria, and other old nations.

Africa and Vietnam There are equal temperaments used in Africa and Vietnam:

Example 31

TET-5; 5-tone Equal Temperament: Southern Uganda;
240, 480, 720, 960, 1 200

TET-7; 7-tone Equal Temperament: Khmer scale of South Vietnam (c.f. [10]), Lower Zambezi Valley, Eastern Angola;
171, 242, 514, 685, 857, 1028, 1 200

Besides E_5, E_7, and E_{12}, E_{10} (which we deal with in other chapters because they are important for the information granulation), other equal temperaments are mostly served as classes T of tones. The systems used in this way are E_{22} (India); E_{17}, E_{19}, E_{24}, E_{31}, E_{53} (Middle Asia); E_7 (Siam).

Siam In Siamese 7-TET 5-valued modes, we see the relation with gamelan music, c.f. Section 8.2.3.

Example 32

BARANG-5; Siamese 7-TET mode Barang: $\sqrt[7]{2^i}$, $i = 2, 3, 4, 5, 7$;
342.857, 514.286, 685.714, 857.143, 1 200.000

LIMA-5; Siamese 7-TET mode Lima:: $\sqrt[7]{2^i}$, $i = 1, 2, 4, 5, 7$;
171.429, 342.857, 685.714, 857.143, 1 200.000

NEM-5; Siamese 7-TET mode Nem: $\sqrt[7]{2^i}$, $i = 1, 3, 4, 5, 7$;
171.429, 514.286, 685.714, 857.143, 1 200.000

The present Middle Asia tone systems (the Arabic, Persian, Turkish, Hebrew, etc. music) consists of choosing 7–13 valued sets $\mathbb{T}$ (called *modes* or *maquams*; singular: maquam, original plural: *maqamat*) from a finite (17, 19, 24, 31, or 53 values within an octave) "acoustically dense" equally tempered or Pythagorean set T .
We now explain maqams of the Turkish and Arabic systems.

Turkish systems According to music theory, the structure of the *Turkish* maqams is such that the octave is divided into 53 parts (about the acoustic reasons to take the number 53 we have spoken earlier. Moreover, there is an interesting Petzval's tone system, c.f. Chapter 7). The frequency multiplier for successive notes in equally tempered systems is given by $2^{n/N}$, where $N = 53$. The values in Table 5.5 and Table 5.6 are the n values, given as number of increments between successive notes. All have been reduced to within one octave and ascending and descending variants are combined, etc.
Notice the high frequency of occurrence of intervals such as 9 (i.e., $2^{9/53} = 1.124\,9 \approx 9/8$), 9+8 ($2^{17/53} = 1.249\,0 \approx 5/4$), 9+5 ($2^{14/53} = 1.200\,9 \approx 6/5$), and 9+8+5 ($2^{22/53} = 1.333\,4 \approx 4/3$). These simple

TABLE 5.5: Turkish maqams

Rast	9	8	5	9	9	4	4	5		
Nahawand	9	4	9	9	4	9	9			
Hicaz, Uzzal, Humayun	5	12	5	9	4	4	5	9		
Hicazkar	5	12	5	9	5	3	5	4	5	
Yegah	9	8	5	9	8	1	4	4	5	
Sultaniyegah, Ruhnevaz	9	4	9	9	4	9	4	5		
Ferahnuma, Askefza	4	9	9	9	4	9	9			
Sedaraban	5	12	5	9	5	8	4	5		
Huseyniasiran	8	5	9	8	5	9	9			
Suzidil	5	12	5	9	4	4	5	4	5	
Acemasiran	9	9	4	4	5	9	9	4		
Sevkefza	9	5	4	4	4	5	5	13	4	
Iraq	5	9	8	5	9	9	4	4		
Evic, Segah	5	9	8	5	4	5	9	4	4	
Ferahnak	5	9	8	1	4	4	5	9	4	4
Evicara	5	13	4	9	5	13	4			
Mahur	9	9	4	9	9	4	5	4		
Suzidilara	9	5	4	4	4	5	9	4	5	4
Buzurk	9	9	4	4	5	9	4	4	5	
Suzinak	5	4	4	4	5	9	4	9	9	
Zirguleli Suzinak	5	12	5	5	4	4	4	5	9	
Kurdilihicazkar	4	1	4	4	4	5	9	4	4	5	9
Nihavend	9	4	9	5	4	4	5	8	5	
Neveser	9	5	12	5	5	12	5			
Nikriz	9	5	12	5	9	4	4	5		

TABLE 5.6: Turkish maqams (continued)

Maqam													
Huseyni, Muhayyer	8	5	9	9	4	4	5	9					
Gulizar, Beyati	8	5	9	5	4	4	4	5	9				
Ussak, Acem	8	5	9	9	4	9	9						
Kurdi	4	4	1	4	9	4	5	4	9	9			
Buselik	9	4	9	5	4	4	4	5	4	5			
Arazbar	8	5	9	5	3	1	4	4	5	9			
Zirgule	5	12	5	9	4	1	8	4	5				
Sehnaz	5	12	5	9	4	1	3	5	4	5			
Saba, Sunbule	8	5	5	13	4	9	9						
Kucek	8	5	5	13	4	4	5	5	4				
Eski Sipihr	8	5	5	1	3	9	4	4	5	4	5		
Dugah	4	4	5	4	1	13	4	9	9				
Hisar	8	5	9	4	5	4	1	8	4	5			
Yeni Sipihr	5	3	5	4	5	4	5	4	1	3	5	4	5
Nisaburek	9	8	5	9	5	8	9						
Huzzam	5	9	5	12	5	9	4	4					
Mustear	9	5	8	9	5	9	4	4					
Maye (Yeni Maye)	5	9	8	1	4	9	9	8					
Vechi Arazbar	1	4	9	8	5	9	9	3	5				
Nisabur	8	5	9	4	9	9	4	5					
Cargah I	5	13	4	9	5	12	5						
Cargah II	9	9	4	9	9	9	4						
Araban	5	8	13	5	5	8	4	5					
Urmawi	9	8	5	9	9	5	8						

ratios are approximated much better than in the Western 12-tone system.

Apparently, the Turkish theorist further divide each of the 53 steps into 20 parts, making for a 1 060-part octave, comparable to the Western system of 1 200 cents.

Arabic systems   There are at least two sets of *Arabic maqams*. One is the "commercial" 24-tone equally tempered and another is a 17-tone Pythagorean set, c.f. *PYTH-17*. Modern popular music generally is played in 24-tone equal temperature while strict traditionalists and some musicians employ the more sublime 17 system. Maquams are listed in Table 5.7 and Table 5.8 (given as number of increments).

Music in the E_{24} Arabic maqams can be written in western musical staff notation with the adoption of the quarter pitch notation. Western music has a 12-tone octave, so adding the quarter tones makes 24. With the quarter step (half-flat) marker, the 5-bar Western staff can thus be used.

In Example 33 we bring values of some maqams, both Turkish and Arabic, Pythagorean or equally tempered:

Example 33

AL-FARABI-19; Arabic scale by Al Farabi;
256/243, 12/11, 9/8, 32/27, 8 192/6 561, 81/64, 2816/2 187, 4/3, 1 024/729, 16/11, 3/2, 128/81, 32 768/19 683, 27/16, 891/512, 16/9, 4 096/2 187, 64/33, 2

AL-KINDI-14; Arabic mode by Al-Kindi;
256/243, 9/8, 32/27, 8 192/6 561, 81/64, 4/3, 1 024/729, 3/2, 128/81, 32 768/19 683, 27/16, 16/9, 4 096/2 187, 2

AL-MAUSILI-11; Arabic mode by Ishaq al-Mausili, ? - 850 AD;
256/243, 9/8, 32/27, 81/64, 4/3, 1 024/729, 3/2, 128/81, 27/16, 16/9, 2

ARABIC(NO.2)-12; From [17]. Try C or F minor;
100, 150, 300, 400, 500, 600, 700, 18/11, 900, 1 000, 1 100, 1 200

AVICENNA(CHRO/NO.2)-7; Dorian Mode, a 1:2 Chromatic, 4 + 18 + 8 parts;
200/3, 1 100/3, 500, 700, 2 300/3, 3 200/3, 1 200

TABLE 5.7: Arabic maqams in the 24-tone Equal Temperament

Maquam											
Rast	4	3	3	4	4	3	3				
Sikah	3	4	4	3	3	4	3				
Miha'il Musaqa's mode	4	3	3	4	3	3	4				
Mohajira	3	4	3	4	3	4	3				
Ushshaq Turki	3	3	4	4	2	4	4				
Bayati	3	3	4	4	2	1	3	4			
Saba, Mansuri	3	3	2	6	2	4	4				
Suznak (Soznak)	4	3	3	4	2	6	2				
Mahur, Dilkashidah, Rast	4	3	3	4	4	2	1	3			
Yakah	4	3	1	2	1	3	4	2	1	3	
Qarjighar	3	3	4	2	6	2	4				
Huzam	3	4	2	2	2	2	2	4	3		
Hizam	3	4	2	6	2	4	3				
Hijaz	2	6	2	4	2	1	3	4			
Mustaár	3	4	4	2	1	3	4	3			
Iraq	3	2	6	2	4	4	3				
Bastanikar	5	2	4	2	4	4	3				
Farah Faza, Nakriz	4	2	6	2	4	2	1	3			
Hayyan	4	2	1	3	2	2	2	2	3	1	2
Basandida	4	2	1	3	2	2	4	2	1	3	
Shawq Afza	4	4	2	1	3	2	2	4	2		
Shawq Tarab	2	1	3	2	2	4	2	4	2	2	
Jabburi	3	1	2	1	4	4	2	4	4		
Nawa	2	4	4	4	3	3	4				
Higaz-kar	2	5	3	4	2	5	3				
Suár	3	4	4	2	4	4	3				
Aug-ara	3	6	1	5	2	6	1				
Buselik	4	1	5	4	2	6	2				
Neuter	4	2	6	2	2	5	3				

TABLE 5.8: Arabic maqams in the 17-tone Pythagorean System

Safi al-Din's maquam								
Ussaq	3	3	1	3	3	1	3	
Nawa	3	1	3	3	1	3	3	
Abu Salik	1	3	3	1	3	3	3	
Rast	3	2	2	3	2	2	3	
Iraq	2	3	2	2	3	2	2	1
Isfahan	3	2	2	3	2	2	2	1
Zirafkend	2	2	3	2	2	1	3	2
Buzurk	2	3	2	2	1	3	2	2
Zankulah	3	2	2	2	3	2	3	
Rahawi	2	3	2	2	2	3	3	
Husaini	2	2	3	2	2	3	3	
Higazi	2	2	3	2	3	2	3	

Maquam														
al-Kindi	1	2	1	1	1	1	1	2	1	1	1	1	1	2
Ishaq al-Mausili	1	2	1	2	1	1	2	1	2	1	3			

AVICENNA-19; Arabic scale by Ibn Sina;
256/243, 1 024/945, 9/8, 32/27, 8 192/6 561, 81/64, 4/3, 48/35, 729/512,
4 096/2 835, 3/2, 128/81, 512/315, 27/16, 16/9, 64/35, 243/128,
129 140 163/67 108 864, 2

AWARD-24; [11], *vol. 5, p. 37, after Mans.ur 'Awad;*
40/39, 20/19, 40/37, 10/9, 8/7, 20/17, 40/33, 5/4, 40/31, 4/3, 48/35, 24/17,
16/11, 3/2, 20/13, 30/19, 60/37, 5/3, 12/7, 30/17, 20/11, 15/8, 60/31, 2

BEY-24; Idris Rag'ib Bey, [11], *vol. 5, p. 40;*
1 000/969, 1 000/931, 167.000, 9/8, 257.000, 25/21, 369.000, 432.000, 200/153,
4/3, 500/363, 605.000, 500/343, 3/2, 767.000, 808.000, 2 000/1 209, 940.000,
970.000, 16/9, 1 038.000, 1 000/537, 25/13, 2

BOZOURK-8; Bouzourk;
65 536/59 049, 8 192/6 561, 4/3, 262 144/177 147, 3/2, 27/16, 4 096/2 187, 2

CAIRO-26; [11], *vol. 5., p. 42. Congress of Arabic Music, Cairo,*
1932;
625/607, 5 000/4 739, 400/367, 1 000/891, 1 250/1 087, 2 000/1 689, 500/419,
400/327, 5 000/3 989, 2 500/1 937, 4/3, 250/183, 10 000/7 111, 10 000/6 881, 3/2,
2 500/1 631, 1 000/631, 1 000/627, 2 500/1 529, 500/297, 10 000/5 789, 500/279,
200/109, 250/133, 125/64, 2

COLLENGETTES-24; R.P. Collengettes, from p. 23 of [11], *vol 5. 24 tone Arabic system;*
36/35, 256/243, 12/11, 9/8, 81/70, 32/27, 27/22, 81/64, 729/560, 4/3, 48/35, 1 024/729, 16/11, 3/2, 54/35, 128/81, 18/11, 27/16, 243/140, 16/9, 64/35, 243/128, 64/33, 2

IRAN(DIAT)-7; Iranian Diatonic from Dariush Anooshfar, Safi-a-ddin Armavi's scale from 125-TET;
220.800, 441.600, 489.600, 710.400, 931.200, 979.200, 1 200.000

KIMBALL-18; Buzz Kimball 18-tone just scale;
25/24, 135/128, 10/9, 9/8, 75/64, 5/4, 81/64, 4/3, 25/18, 45/32, 3/2, 25/16, 5/3, 27/16, 225/128, 16/9, 15/8, 2

LEBANON-7; Lebanese scale;
150, 300, 500, 700, 800, 1 000, 1 200

PERSIAN-17; Persian Tar Scale, from Dariush Anooshfar, Internet Tuning List 2/10/94;
256/243, 27/25, 9/8, 32/27 243/200, 81/64, 4/3, 25/18 36/25, 3/2, 128/81, 81/50, 27/16, 16/9 729/400, 243/128, 2

TURKISH-24; Ra'uf Yaqta Bey, 24 of 53 tones, Theoretical Turkish gamut;
256/243, 2 187/2 048, 65 536/59 049, 9/8, 32/27, 19 683/16 384, 8 192/6 561, 81/64, 2 097 152/1 594 323, 4/3, 1 024/729, 729/512, 262 144/177 147, 3/2, 128/81, 6 561/4 096, 32 768/19 683, 27/16, 8 388 608/4 782 969, 16/9, 4 096/2 187, 243/128, 1 048 576/531 441, 2

ZIRAFKEND-8; Arabic Zirafkend mode;
65 536/59 049, 32/27, 4/3, 262 144/177 147, 128/81, 32 768/19 683, 4 096/2 187, 2

Example 34

AL-DIN/ROUANET-6; Zirafkend Bouzourk from both Rouanet and Safi al-Din;
14/13, 7/6, 6/5, 27/20, 3/2, 2

AD-DIK-24; Amin Ad-Dik, [11], *vol 5, p.42;*
1 053/1 024, 256/243, 12/11, 9/8, 147/128, 32/27, 27/22, 5/4, 9/7, 4/3, 48/35, 1 024/729, 81/56, 3/2, 49/32, 128/81, 18/11, 27/16, 26/15, 9/5, 11/6, 15/8, 35/18, 2

AVICENNA-7; Soft diatonic of Avicenna (Ibn Sina);
10/9, 8/7, 4/3, 3/2, 5/3, 12/7, 2

AVICENNA(CHRO)-7; Dorian mode of a chromatic genus of Avicenna;
36/35, 8/7, 4/3, 3/2, 54/35, 12/7, 2

AVICENNA(CHRO/NO.3)-7; Avicenna's Chromatic permuted;
10/9, 35/27, 4/3, 3/2, 5/3, 35/18, 2

AVICENNA(DIAT)-7; Dorian mode of a soft diatonic genus;
14/13, 7/6, 4/3, 3/2, 21/13, 7/4, 2

AVICENNA(ENH)-7; Dorian mode of an Enharmonic genus;
40/39, 16/15, 4/3, 3/2, 20/13, 8/5, 2

IRAQ-8; Iraq 8-tone scale, Ellis;
394/355, 8 192/6 561, 4/3, 623/421, 591/355, 16/9, 513/260, 2

ISFAHAN-8; Isfahan;
394/355, 8 192/6 561, 4/3, 3/2, 591/355, 16/9, 513/260, 2

KIMBALL-53; Buzz Kimball 53-tone just scale;
18/17, 17/16, 16/15, 14/13, 13/12, 12/11, 11/10, 17/15, 8/7, 7/6, 20/17, 13/11,
6/5, 17/14, 11/9, 16/13, 5/4, 14/11, 22/17, 13/10, 17/13, 4/3, 11/8, 18/13,
7/5, 24/17, 17/12, 10/7, 13/9, 16/11, 3/2, 26/17, 20/13, 17/11, 11/7, 8/5, 13/8,
18/11, 28/17, 5/3, 22/13, 17/10, 12/7, 7/4, 30/17, 20/11, 11/6, 24/13, 13/7,
15/8, 32/17, 17/9, 2

SAFIYU(MAJ)-6; Singular Major from Safi al-Din;
14/13, 16/13, 4/3, 56/39, 3/2, 2

Here are arabic tetrachordal scales (two octaves), the interaction with Greek culture.

Example 35

AL-FARABI-7; Al-Farabi Syn Chrom;
16/15, 8/7, 4/3, 3/2, 8/5, 12/7, 2

AL-FARABI(CHRO)-7; Al-Farabi's Chromatic permuted;
16/15, 56/45, 4/3, 3/2, 8/5, 28/15, 2

AL-FARABI(DIAT/NO.1)-7; Al-Farabi's Diatonic;
8/7, 64/49, 4/3, 3/2, 12/7, 96/49, 2

AL-FARABI(DIAT/NO.2)-7; Permuted form of Al-Farabi's reduplicated 10/9 diatonic genus;
10/9, 6/5, 4/3, 3/2, 5/3, 9/5, 2

AL-FARABI(DOR/NO.1)-7; Dorian mode of Al-Farabi's 10/9 Diatonic;
27/25, 6/5, 4/3, 3/2, 81/50, 9/5, 2

AL-FARABI(DOR/NO.2)-7; Dorian mode of Al-Farabi's Diatonic;
49/48, 7/6, 4/3, 3/2, 49/32, 7/4, 2

AL-FARABI(TETRA)-12; Al-Farabi's tetrachord division,
including extra 2 187/2 048 and 19 683/16 384;
256/243, 18/17, 2 187/2048, 162/149, 54/49, 9/8, 32/27, 81/68, 19 683/16 384,
27/22, 81/64, 4/3

AL-HWARIZMI-6; Al-Hwarizmi's tetrachord division;
9/8, 81/70, 81/68, 27/22, 81/64, 4/3

AL-KINDI-6; Al-Kindi's tetrachord division;
256/243, 2 187/2048, 9/8, 32/27, 81/64, 4/3

MOHAJIRA/BAYATI-7; Mohajira + Bayati (Dudon) 3 + 4 + 3
Mohajira and 3 + 3 + 4 Bayati tetrachords;
150, 350, 500, 700, 850, 1 000, 1 200

MOHAJIRA-7; Mohajira (Dudon) Two 3 + 4 + 3 Mohajira tetrachords;
150, 350, 500, 700, 850, 1050, 1 200

MUSAQA-7; Egyptian scale by Miha'il Musaqa;
200, 350, 500, 700, 850, 1 000, 1 200

RAST-MOHAJIRA-7; Rast + Mohajira (Dudon) 4 + 3 + 3 Rast
and 3 + 4 + 3 Mohajira tetrachords;
200, 350, 500, 700, 850, 1 050, 1 200

SAFIYU(DIAT)-7; Safi al-Din's Diatonic, also the strong form of
Avicenna's 8/7 diatonic;
19/18, 7/6, 4/3, 3/2, 19/12, 7/4, 2

SAFIYU(DIAT/NO.2)-7; Safi al-Din no.2 Diatonic, a 3/4 tone diatonic like Ptolemy's Equable Diatonic;
64/59, 32/27, 4/3, 3/2, 96/59, 16/9, 2

SEGAH-7; Arabic Segah (Dudon) Two 4 + 3 + 3 tetrachords;
200, 350, 500, 700, 900, 1050, 1 200

SEGAH(PERSIAN)-7; Persian Segah (Dudon) 4 + 3 + 3 and 3 +
4 + 3 degrees of 24-TET;
200, 350, 500, 700, 850, 1 050, 1 200

5.4.2   Present equal temperament inspired systems

Unlike the ordinary 12-TET, the present exponents of equal temperaments typically use refretted guitars, tubulongs, and/or digital synthesizers to realize their compositions.

In general, they seem more technologically-oriented, more prone to use "alternative controllers" like the Power Glove and the *MIDI* Theremin and the Buchla Lightning, and less interested in Partch's doctrine of "corporeality" than members of the American just intonation movement. Many proponents of equal temperaments prefer to work exclusively in one particular tone system: typically E_{19}, E_{22}, E_{24}, E_{31} or E_{53} which can be performed on traditional European instruments, as well as a few equal temperaments proposed by various European theorists (A. Hába, I. Wyschnegradsky, A. D. Fokker, J. Yasser).

In central New York, J. Reinhard performs divisions of the whole-tone (i.e., E_{24}, E_{36}, E_{48}, E_{96}, etc.), E_{19} and E_{31}, and others.

J. Mandelbaum has since 1961 established himself as another notable figure in microtonality; his 1961 thesis *Multiple division of the octave and the tonal resources of* E_{19} is the most thorough discussion of E_{19} yet penned. Mandelbaum composes for traditional acoustic instruments (example: the composition Three Xenophonies) as well as the 31-tone Netherlands electronic organ called the archiphone, and recently one of his operas played to sell-out crowds (it included several microtonal sections). His 1961 "Preludes for 19-tone Piano" use 2 pianos with black keys shared in common as 5 out of E_{19} along with 2 different sets of 7 out of E_{19}: this set of microtonal piano compositions remains an acknowledged classic.

Example 36

TET19-12; 12 out of E_{19}, Mandelbaum's dissertation;
63.158, 189.474, 252.632, 378.947, 505.263, 568.421, 694.737, 757.895, 884.211, 947.368, 1 073.684, 1 200.000

MERCATOR-19; 19 out of E_{53}, Mandelbaum's dissertation, p. 331;
67.925, 113.208, 181.132, 271.698, 339.623, 384.906, 430.189, 498.113, 566.038, 611.321, 679.245, 769.811, 815.094, 883.019, 950.943, 996.226, 1 086.793, 1 154.717, 1 200.000

In Boston, J. Maneri, and E. Sims compose E_{72} music for traditional instruments.

A more technologically elaborate current of microtonal music can be found at M.I.T and Berklee College of Music, where R. Boulanger works in exotic equal temperaments and non-octave scales (E_{60} and the 13th root of 3, i.e. the Bohlen–Pierce scale) using the CSOUND acoustic compiler, the Mathews radio drum and various *MIDI* synthesizers; nearby, E. Mullen performs cyberpunk music in E_{19} and the 13th root of 3. Randy Winchester has composed many remarkable synthesizer studies in almost all of the equal temperaments from E_6 through E_{23}.

Meanwhile, St. Louis, Missouri, and environs are dominated by E_{31}. This is largely due to the powerful influence of L. Gerdine, who served as dean of Webster College for many years and who translated A. D. Fokker's "New Music With 31 Tones" back in 1973.

In Texas, a number of composers regularly work outside of 12. G. Morrison has worked for years in E_{10}, and more recently in a non-octave scale with 13.6 tones per octave (or 88 cents per step).

Example 37

MORRISON(CET88)-14; step 88 cents, by G. Morrison;
88, 176, 264, 352, 440, 528, 616, 704, 792, 880, 968, 1 056, 1 144, 1 232.000

Since the 1950s, E. Wilson has categorized and catalogued many different equal temperaments and developed generalized Bosanquet-type keyboards and notations for them. Wilson was one of the first Americans to refret guitars to a variety of different equal temperaments: inspired by his reading of A. Novaro's refretted E_{15} and E_{12} tempered guitars, Wilson's guitar-refretting efforts influenced I. Darreg and from there spread far and wide across the United States. Wilson has also discovered and characterized many different new classes of just intonation scales, and his 1967 discovery of the Combination Product Set method of generating just scales has been called "a giant step forward" for just intonation theory. Since the 1960s, Wilson has explored aliquot scales and *JI* tone systems based on many different numerical series.

Example 38

DARREG(ENNEA)-9; I. Darreg's Mixed Enneatonic, a mixture of chromatic and enharmonic;

50, 100, 200, 500, 700, 750, 800, 900, 1 200

DARREG(MIX)-9; I. Darreg's Mixed Enharmonic and Chromatic Scale;

50, 100, 200, 500, 700, 750, 800, 900, 1 200

Since inventing his first electronic musical instrument in 1939, I. Darreg wrote and lectured and made hundreds of comparison-and-contrast cassettes showing the characteristics of a vast array of different microtonal tone systems, from just intonation to equal temperament and beyond. In 1975, Darreg made the crucial discovery that *each microtonal tone system boasts its own unique "sound" or "mood"*, and that a composer could tailor the "mood" of the composition by choosing the equal temperament or just intonation system: for a calm mood, E_{31}, for a brilliant steely mood, E_{17}; for a strange outer-space sound, the 11-limit just intonation with plenty of 11-ratios; and for a vibrant propulsive "mood," Pythagorean just intonation used polyphonically; etc.

Example 39 *Used restrictions from E_N, $N \in \mathbb{N}$:*

TET19-5; 5 out of 19-TET;

252.632, 505.263, 757.895, 1 010.526, 1 200.000

TET19-7; 7 out of 19-TET, major;

189.474, 378.947, 505.263, 694.737, 884.211, 1 073.684, 1 200.000

TET19-8; 8 out of 19-TET;

126.316, 315.789, 442.105, 568.421, 757.895, 884.211, 1 010.526, 1 200.000

TET19-10; 10 out of 19-TET. For 9 out of 19 discard degree 3;

126.316, 252.632, 315.789, 442.105, 568.421, 694.737, 821.053, 947.368, 1 073.684, 1 200.000

TET19-11; 11 out of 19-TET;

126.316, 189.474, 315.789, 442.105, 568.421, 631.579, 757.895, 884.211, 1 010.526, 1 073.684, 1 200.000

TET19-13; 13 out of 19-TET;

126.316, 189.474, 315.789, 378.947, 505.263, 568.421, 694.737, 757.895, 884.211, 947.368, 1 073.684, 1 136.842, 1 200.000

TET19-14; 14 out of 19-TET;
63.158, 189.474, 252.632, 315.789, 442.105, 505.263, 568.421, 694.737, 757.895, 821.053, 947.368, 1 010.526, 1 136.842, 1 200.000

TET31-19; 19 out of 31-TET;
77.419, 116.129, 193.548, 270.968, 309.677, 387.097, 464.516, 503.226, 580.645, 619.355, 696.774, 774.194, 812.903, 890.323, 967.742, 1 006.452, 1 083.871, 1 161.290, 1 200.000

TET31-20; 20-tone mode of 31-TET;
77.419, 116.129, 193.548, 270.968, 309.677, 387.097, 425.806, 503.226, 580.645, 619.355, 696.774, 735.484, 774.194, 851.613, 890.323, 967.742, 1 006.452, 1 083.871, 1 161.290, 1 200.000

TET36-24; 12 and 18-TET mixed;
200/3, 100, 400/3, 200, 800/3, 300, 1 000/3, 400, 1 400/3, 500, 1 600/3, 600, 2 000/3, 700, 2 200/3, 800, 2 600/3, 900, 2 800/3, 1 000, 3 200/3, 1 100, 3 400/3, 1 200

TET50-19; 19 out of 50-TET;
72, 120, 192, 264, 312, 384, 456, 504, 576, 624, 696, 768, 816, 888, 960, 1 008, 1 080, 1 152, 1 200

TET60-24; 12 and 15-TET mixed;
80, 100, 160, 200, 240, 300, 320, 400, 480, 500, 560, 600, 640, 700, 720, 800, 880, 900, 960, 1 000, 1 040, 1 100, 1 120, 1 200

TET/ENNEA-9; Quasi-Equal Enneatonic, each tetrachord has 125 + 125 + 125 + 125 cents;
125, 250, 375, 500, 700, 825, 950, 1 075, 1 200

COUL-20; Tuning for a 3-row symmetrical keyboard, M. Op de Coul, 1989;
100.000, 131.283, 200.000, 268.717, 300, 400.000, 431.283, 500.000, 25/18, 600.000, 700.000, 731.283, 800.000, 868.717, 900.000, 1 000.000, 1 031.283, 1 100.000, 1 168.717, 1 200.000

ARISTO(DIAT/HEM/CHRO)-7; Diat. + Hem. Chrom. Diesis, Another genus of Aristoxenos, Dorian Mode;
75, 300, 500, 700, 775, 1 000, 1 200

ARISTO(DIAT/SOFT/CHRO)-7; Diatonic + Soft Chromatic Diesis, Another genus of Aristoxenos, Dorian Mode;
200/3, 300, 500, 700, 2 300/3, 1 000, 1 200

DIAT/SOFT-7; Soft Diatonic genus 5 + 10 + 15 parts;
250/3, 250, 500, 700, 2 350/3, 950, 1 200

DIAT/SOFT(2)-7; Soft Diatonic genus with equally divided Pyknon, Dorian Mode;
125, 250, 500, 700, 825, 950, 1 200

HANSON-19; 19 out of 53-TET by L. H. Hanson, 1978;
67.925, 135.849, 203.774, 249.057, 316.981, 384.906, 452.830, 498.113, 566.038,
633.962, 701.887, 769.811, 815.094, 883.019, 950.943,
1 018.868, 1 086.793, 1 132.076, 1 200.000

HEM/CHRO(2)-7; 1:2 Hemiolic Chromatic genus 3 + 6 + 21 parts;
50, 150, 500, 700, 750, 850, 1 200

MARION-19; Scale with two different TET step sizes;
53.996, 107.993, 161.990, 215.986, 269.983, 323.979, 377.976, 431.972, 485.969,
539.965, 593.962, 647.959, 3/2, 784.963, 867.970, 950.978, 1 033.985, 1 116.993,
1/,200.000

MIX/ENNEA3-9; A mixture of the hemiolic chromatic and diatonic genera, 75 + 75 + 150 + 200 c;
75, 150, 300, 500, 700, 775, 850, 1 000, 1 200

MIX/ENNEA(4)-9; Each "tetrachord" contains 67 + 67 + 133 + 233;
200/3, 400/3, 800/3, 500, 700, 2 300/3, 2 500/3, 2 900/3, 1 200

MIX/ENNEA(5)-9; A mixture of the intense chromatic genus and the permuted intense diatonic;
100, 200, 400, 500, 700, 800, 900, 1 100, 1 200

MIXED(9-3)-9; A mixture of the hemiolic chromatic and diatonic genera, 75 + 75 + 150 + 200 c;
75, 150, 300, 500, 700, 775, 850, 1 000, 1 200

MIXED(9-4)-9; Mixed enneatonic 4, each "tetrachord" contains 67 + 67 + 133 + 233;
200/3, 400/3, 800/3, 500, 700, 2 300/3, 2 500/3, 2 900/3, 1 200

MIXED(9-5)-9; A mixture of the intense chromatic genus and the permuted intense diatonic;
100, 200, 400, 500, 700, 800, 900, 1 100, 1 200

MIXED(9-6)1-9; Mixed 9-tonic 6, Mixture of Chromatic and Diatonic;

100, 200, 300, 500, 700, 800, 900, 1 000, 1 200

MIXED(9-7)-9; Mixed 9-tonic 7, Mixture of Chromatic and Dia-tonic;

100, 300, 400, 500, 700, 800, 1 000, 1 100, 1 200

MIXED(9-8)-9; Mixed 9-tonic 8, Mixture of Chromatic and Dia-tonic;

200, 300, 400, 500, 700, 900, 1 000, 1 100, 1 200

PROG/ENNEA-9; Progressive Enneatonic, 50+100+150+200; in each half (500;);

50, 150, 300, 500, 700, 750, 850, 1 000, 1 200

QUASI-5; Quasi-Equal 5-tone in 24-TET, steps: 5, 5, 4, 5, 5;

250, 500, 700, 950, 1 200

QUASI-9; Quasi-Equal Enneatonic, Each "tetrachord" has 125 + 125 + 125 + 125;

125, 250, 375, 500, 700, 825, 950, 1 075, 1 200

SMITH-19; 19 out of 612-TET by Roger K. Smith, 1978;

84.314, 154.902, 203.922, 266.667, 315.686, 386.275, 470.588, 498.039, 582.353, 652.941, 701.961, 764.706, 813.726, 884.314, 968.627,

1 017.647, 1 080.392, 1 150.980, 1 200.000

VRIES-18; Leo de Vries 19/72 Through-Transposing-Tonality 18-tone scale;

200/3, 400/3, 200, 800/3, 950/3, 2 500/3, 450, 516.667, 1 750/3, 1 900/3, 700, 2 300/3, 2 500/3, 950, 3 050/3, 3 250/3, 1 150, 1 200

Example 40 *Non-octave periodic equal temperaments:*

CARLOS(ALPHA)-18; W. Carlos' Alpha scale with the step $\sqrt[9]{3/2}$, 2 periods;

78, 156, 234, 312, 390, 468, 546, 624, 702, 780, 858, 936, 1 014, 1 092, 1 170, 1 248, 1 326, 1 404

CARLOS(ALPHA/P)-36; W. Carlos' Alpha prime scale with the step $\sqrt[18]{3/2}$, 2 periods;

39, 78, 117, 156, 195, 234, 273, 312, 351, 390, 429, 468, 507, 546, 585, 624, 663, 702, 741, 780, 819, 858, 897, 936, 975, 1 014, 1 053, 1 092, 1 131, 1 170, 1 209, 1 248, 1 287, 1 326, 1 365, 1 404

CARLOS(BETA)-22; W. Carlos' Beta scale with $\sqrt[11]{3/2}$, 2 periods;

63.800, 127.600, 191.400, 255.200, 319.000, 382.800, 446.600, 510.400, 574.200,

638.000, 701.800, 765.600, 829.400, 893.200, 957.000,
1 020.800, 1 084.600, 1 148.400, 1 212.200, 1 276.000, 1 339.800,
1 403.600

CARLOS(BETA/P)-44;
W. Carlos' Beta prime scale with $\sqrt[22]{3/2}$, *2 periods;;*
31.900, 63.800, 95.700, 127.600, 159.500, 191.400, 223.300, 255.200, 287.100,
319.000, 350.900, 382.800, 414.700, 446.600, 478.500, 510.400, 542.300, 574.200,
606.100, 638.000, 669.900, 701.800, 733.700, 765.600, 797.500, 829.400, 861.300,
893.200, 925.100, 957.000, 988.900, 1 020.800, 1 052.700, 1 084.600, 1 116.500,
1 148.400, 1 180.300, 1 212.200, 1 244.100, 1 276.000, 1 307.900, 1 339.800, 1 371.700,
1 403.600

BOHLEN/PIERCE-13; *13-tone equal division of 3/1;*
146.304, 292.608, 438.913, 585.217, 731.521, 877.825, 1 024.130, 1 170.434,
1 316.738, 1 463.042, 1 609.347, 1 755.651, 3

CET105-13; *Equal temperament with very good 6/5 and 13/8;*
105.000, 210.001, 315.001, 420.001, 525.002, 630.002, 735.002, 840.003, 945.003,
1 050.003, 1 155.004, 1 260.004, 11/5

CET133-13; *Equal temperament, step* $\sqrt[13]{e}$;
133.172, 266.344, 399.516, 532.687, 665.859, 799.031, 932.203, 1 065.375, 1 198.547,
1 331.719, 1 464.890, 1 598.062, 1 731.234

CET140-24; *Equal temperament, step* $\sqrt[24]{7}$;
140.368, 280.736, 421.103, 561.471, 701.839, 842.207, 982.574, 1 122.942, 1 263.310,
1 403.678, 1 544.045, 1 684.413, 1 824.781, 1 965.150, 2 105.517, 2 245.885, 2 386.253,
2 526.621, 2 666.988, 2 807.356, 2 947.724, 3 088.092, 3 228.459, 7

KIDJEL(CET181)-16; *6.625 TET. The 16/3 is the so-called Kidjel Ratio promoted by Kidjel in 60's;*
181.128, 362.256, 543.383, 724.511, 905.639, 1 086.767, 1 267.895, 1 449.023,
1 630.150, 1 811.278, 1 992.406, 2 173.534, 2 354.663, 2 535.790, 2 716.918, 16/3

CET195-7; *Equal temperament, step* $\sqrt[7]{11/5}$;
195.001, 390.001, 585.002, 780.002, 975.003, 1 170.004, 11/5

CET45-11; *Equal temperament, step* $\sqrt[11]{4/3}$;
45.277, 90.554, 135.830, 181.107, 226.384, 271.661, 316.938, 362.215, 407.491,
452.768, 4/3

CET63-30; *Equal temperament, step* $\sqrt[30]{3}$, *stretched 19-TET;*
63.399, 126.797, 190.196, 253.594, 316.993, 380.391, 443.790, 507.188, 570.587,

633.985, 697.384, 760.782, 824.181, 887.579, 950.978, 1 014.376, 1 077.775, 1 141.173, 1 204.572, 1 267.970, 1 331.369, 1 394.767, 1 458.166, 1 521.564, 1 584.963, 1 648.361, 1 711.760, 1 775.158, 1 838.557, 3

CET90-17; Scale with limma steps;
256/243, 65 536/59 049, 16 777 216/14348907, 360.900, 451.125, 541.350, 631.575, 721.800, 812.025, 902.250, 992.475, 1 082.700, 1 172.925, 1 263.150, 1 353.375, 1 443.600, 1 533.825

CARLOS(GAMMA)-35; Step $\sqrt[11]{5/4}$ or $\sqrt[20]{3/2}$;
35.100, 70.200, 105.300, 140.400, 175.500, 210.600, 245.700, 280.800, 315.900, 351.000, 386.100, 421.200, 456.300, 491.400, 526.500, 561.600, 596.700, 631.800, 666.900, 702.000, 737.100, 772.200, 807.300, 842.400, 877.500, 912.600, 947.700, 982.800, 1 017.900, 1 053.000, 1 088.100, 1 123.200, 1 158.300, 1 193.400, 1 228.500

PI-13; $\sqrt[13]{\pi}$;
152.446, 304.892, 457.337, 609.783, 762.229, 914.675, 1 067.121, 1 219.566, 1 372.012, 1 524.458, 1 676.904, 1 829.350, 1 981.796 $\approx \pi$

PHI-17; Equal temperament, step $\sqrt[17]{\pi + 1}$;
98.011, 196.021, 294.032, 392.042, 490.053, 588.064, 686.074, 784.085, 882.096, 980.106, 1 078.117, 1 176.128, 1 274.138, 1 372.149, 1 470.159, 1 568.170, 1 666.181

STOCKHAUSEN-25; Stockhausen's 25-tone TET scale;
111.453, 222.905, 334.358, 445.810, 557.263, 668.715, 780.168, 891.620, 1 003.073, 1 114.526, 1 225.978, 1 337.431, 1 448.883, 1 560.336, 1 671.788, 1 783.241, 1 894.694, 2 006.146, 2 117.600, 2 229.052, 2 340.505, 2 451.957, 2 563.410, 2 674.862, 5

Remarks and ideas for exploring

1 Compute continuous fractions for 5/4, 6/5 and 7/4, and corresponding sequences of numbers of steps in equal temperaments with octave periode.

2 Let A_3, A_5, A_6, A_7 be classes of equal temperaments from the previous exercise corresponding 3/2, 5/4, 6/5, and 7/4, respectively. Find the intersections of these sets.

3 Explore graphs of $\sigma(S)$ in Example 28 and Example 29 with weighting factors other than 1/9, 1/25, and 1/49.

Chapter 6

Mean tone systems

6.1 Pythagorean System

Pythagorean System was established about five hundred years B.C. and used in Western music up to the 14th century. For this system, $\mathcal{T} = \{2^p 3^q; \ p, q \in \mathbb{Z}\} \overset{\text{def}}{=} \Pi$. Depending on the sampling algorithms, c.f. Figure 6.11, there are known several tone systems called ambiguously also Pythagorean Systems, Pythagorean Tunings, or Pythagorean Scales such that $\mathbb{T} \subset \mathcal{T}$ and having various cardinality per octave (Π_5, Π_7, Π_{31}, etc.).

In Sections 6.1.1–6.1.3, we will stress the interplay between two tone systems—Pythagorean and Equal Temperament E_{12}. We show that the Pythagorean System Π_{12} (or more valued) can be canonically represented as discrete sets of the plane and there is a natural parallel projection of Pythagorean System to the points of Equal Temperament. This fact implies particularly that when performing a composition on two instruments simultaneously in both Pythagorean System and Equal Temperament, then the Garbuzov zones can be considered as segments on parallel lines in the plane.

6.1.1 Pythagorean minor and major semitones

In this section we describe the structure of Pythagorean System $\Pi_{17} \subset \Pi$ (the system *PYTH-17* in our colection of tone systems in this book)

PYTH-17; 17-tone Pythagorean System Π_{17};
256/243, 2 187/2 048, 9/8, 32/27, 19 683/16 384, 81/64, 4/3, 1 024/729, 729/512,
3/2, 128/81, 6 561/4 096, 27/16, 16/9, 59 049/32 768, 243/128, 2
in the sense of Definition 9. So, we are looking for rational numbers
X, Y such that $X^m Y^n = 2^p 3^q$, where $m, n, p, q \in \mathbb{Z}$. We have:

$$m_0 + n_0 = 0$$
$$m_1 + n_1 = 1$$
$$\vdots$$
$$m_7 + n_7 = 7$$
$$\vdots$$
$$m_{12} + n_{12} = 12,$$
$$0 = m_0 \le m_1 \le \cdots \le m_{12},$$
$$0 = n_0 \le n_1 \le \cdots \le n_{12}.$$

Theorem 20 *The unique (up to the order of X, Y) pair of rational
intervals for 12-granule 2-quotient $(2, 3)$-scales is*

$$(X, Y) = (256/243, 2\,187/2\,048).$$

This concrete values of X, Y can be obtained from Definition 9,
c.f. [180], and they are well-known as the *minor* and *major Pythagorean semitones*, respectively.
    Possible numbers m_i, n_i, $i = 0, 1, \ldots, 12$, are:

$$
\begin{aligned}
0 + 0 &= 0, \\
1 + 0 &= 1, \\
1 + 1 &= 2, \\
2 + 1 &= 3, \\
2 + 2 &= 4, \\
3 + 2 &= 5, \\
4 + 2 &= 6, \\
4 + 3 &= 7, \\
5 + 3 &= 8, \\
5 + 4 &= 9, \\
6 + 4 &= 10, \\
6 + 5 &= 11, \\
7 + 5 &= 12.
\end{aligned}
$$

Now we express the numbers of Π_{17} in the sense of Definition 9 and denote them by $C = 1$, $D_\flat = X$, $C_\sharp = Y$, $D = XY$, $E_\flat = X^2Y$, $D_\sharp = XY^2$, $E = X^2Y^2$, $F = X^3Y^2$, $G_\flat = X^4Y^2$, $F_\sharp = X^3Y^3$, $G = X^4Y^3$, $A_\flat = X^5Y^3$, $G_\sharp = X^4Y^4$, $A = X^5Y^4$, $B_\flat = X^6Y^4$, $A_\sharp = X^5Y^5$, $B = X^6Y^5$, $C' = X^7Y^5$.

It is evident that the sum of exponents $1 + 0 = 1$ for $D_\flat$ is the same as $0 + 1 = 1$ for $C_\sharp$, likewise $2 + 1 = 3$ for $E_\flat$ as $1 + 2 = 3$ for $D_\sharp$, $4 + 2 = 6$ for $G_\flat$ as $3 + 3 = 6$ for $F_\sharp$, $5 + 3 = 8$ for as $A_\flat$ $4 + 4 = 8$ for $G_\sharp$, and $6 + 4 = 10$ for $B_\flat$ as $5 + 5 = 10$ for $A_\sharp$. Thus we have obtained the following theorem:

Theorem 21 *Let*

$$\Pi_{r,s,t,u,v} = \{C, r, D, s, E, F, t, G, u, A, v, B, C'\},$$

where $r = C_\sharp, D_\flat$; $s = D_\sharp, E_\flat$; $t = F_\sharp, G_\flat$; $u = G_\sharp, A_\flat$; $v = A_\sharp, B_\flat$.
Then

(1) $\Pi_{r,s,t,u,v}$ *are 12-granulated 2-quotient $(2,3)$-scales,*

(2)

$$\Pi_{17} = \bigcup_{r,s,t,u,v} \Pi_{r,s,t,u,v}.$$

Theorem 22

$$\Pi_{17} = \{(X^4Y^3)^k(X^4Y^2)(\text{mod } X^7Y^5) \in \Pi; \ k = 0, 1, 2, \ldots, 16\}.,$$

Proof. Evidently, $X^7Y^5 = 2/1 = 2, X^4Y^3 = 3/2$. Further, it is $X^4Y^2 = G_\flat$, $(X^4Y^2) \cdot (X^4Y^3) \cdot (X^7Y^5)^{-1} = X = D_\flat$, $X \cdot (X^4Y^3) = X^5Y^3 = A_\flat$, $(X^5Y^3) \cdot (X^4Y^3) \cdot (X^7Y^5)^{-1} = X^2Y = E_\flat$, $(X^2Y) \cdot (X^4Y^3) = X^6Y^4 = B_\flat$, $(X^6Y^4) \cdot (X^4Y^3) \cdot (X^7Y^5)^{-1} = X^3Y^2 = F$, $(X^3Y^2) \cdot (X^4Y^3) \cdot (X^7Y^5)^{-1} = 1 = C$, $1 \cdot (X^4Y^3) = G$, $(X^4Y^3) \cdot (X^4Y^3) \cdot (X^7Y^5)^{-1} = XY = D$, $(XY) \cdot (X^4Y^3) = X^5Y^4 = A$, $(X^5Y^4) \cdot (X^4Y^3) \cdot (X^7Y^5)^{-1} = X^2Y^2 = E$, $(X^2Y^2) \cdot (X^4Y^3) = X^6Y^5 = B$, $(X^6Y^5) \cdot (X^4Y^3) \cdot (X^7Y^5)^{-1} = X^3Y^3 = F_\sharp$, $(X^3Y^3) \cdot (X^4Y^3) \cdot (X^7Y^5)^{-1} = Y = C_\sharp$, $Y \cdot (X^4Y^3) = X^4Y^4 = G_\sharp$, $(X^4Y^4) \cdot (X^4Y^3) \cdot (X^7Y^5)^{-1} = XY^2 = D_\sharp$, $(XY^2) \cdot (X^4Y^3) = X^5Y^5 = A_\sharp$. $\square$

Corollary 23 Pythagorean System Π_{17} is a geometric net of 17 rationals with the quotient $3/2$ considered modulo 2 (in the multiplicative group $(0, +\infty)$). Combine the results of this section and consequently collect Table 6.3, where $X = Y_1$, $Y = Y_2$; in the fifth column, there are values in cents; in the sixth column, there is a musical denotation.

6.1.2   Images in the plane

For $X, Y \in \mathbb{Q}$, denote by $\mathbb{Q}_{X,Y} = \{X^\alpha Y^\beta;\ \alpha, \beta \in \mathbb{Z}\}$. Consider the map $\vartheta : \mathbb{Q}_{X,Y} \to \mathbb{Z}^2, \vartheta(X^\alpha Y^\beta) = (\alpha, \beta)$. Then

$$\vartheta((X^\alpha Y^\beta)(X^\gamma Y^\delta)) = (\alpha + \gamma, \beta + \delta),$$
$$\vartheta((X^\alpha Y^\beta)^\gamma) = (\gamma\alpha, \gamma\beta), \alpha, \beta, \gamma, \delta \in \mathbb{Z}.$$

Lemma 24 *The map $\vartheta : \mathbb{Q}_{X,Y} \to \mathbb{Z}^2$ is an isomorphism.*

Proof.   It is sufficient to show that $\vartheta(a) = \vartheta(b)$ implies $a = b$. Indeed, then

$$\vartheta(a) = \vartheta(X^{\alpha_1} Y^{\beta_1}) = \vartheta(X^{\alpha_2} Y^{\beta_2}) = \vartheta(b),$$

$$(\alpha_1, \beta_1) = (\alpha_2, \beta_2)$$

which implies

$$\alpha_1 = \alpha_2, \beta_1 = \beta_2, X^{\alpha_1} Y^{\beta_1} = X^{\alpha_2} Y^{\beta_2}. \qquad \square$$

Embed $\mathbb{Z}^2$ identically into the complex plane $\mathbb{C}$. Denote this injection by η. Denote by ϑ_*^{-1} the extended map $\vartheta_*^{-1} : \mathbb{C} \to \mathbb{R}_{X,Y}$, where $\mathbb{R}_{X,Y} = \{X^\alpha Y^\beta;\ \alpha, \beta \in \mathbb{R}\}$. So we have the following commutative diagram, c.f. Figure 6.1, which defines the embedding $\zeta : \mathbb{Q}_{X,Y} \to \mathbb{R}_{X,Y}$.

In Figure 6.2 we see the image of Pythagorean System Π_{17} in the isomorphism ϑ.

Theorem 25 *Let $\pi : \mathbb{C} \to p$ be the projection of the complex plane $\mathbb{C} = \{(\alpha, \beta) \in \mathbb{R}^2\}$ into the line*

$$p : 5\alpha - 7\beta = 0$$

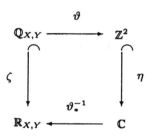

FIGURE 6.1: The embedding ζ

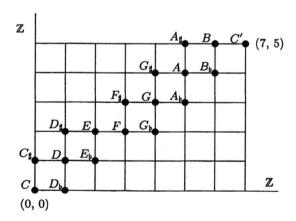

FIGURE 6.2: ϑ-image of Pythagorean System Π_{17}

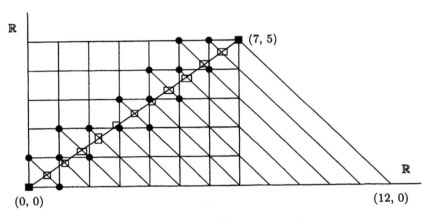

FIGURE 6.3: ζ-image of E_{12} and Pythagorean System Π_{17}

along the line

$$q : \alpha + \beta = 1.$$

Then

$$E_{12} = \vartheta_*^{-1}(\pi(\eta(\vartheta(\Pi)))).$$

Proof.   Cf. Figure 6.3. Denote by

$$
\begin{aligned}
w_0 &= & \tfrac{0}{12} \cdot (7,5), \\
w_1 &= & \left(\tfrac{7}{12}, \tfrac{5}{12}\right) = \tfrac{1}{12} \cdot (7,5), \\
w_2 &= & 2 \cdot \left(\tfrac{7}{12}, \tfrac{5}{12}\right) = \tfrac{2}{12} \cdot (7,5), \\
&\vdots& \\
w_{12} &= & 12 \cdot \left(\tfrac{7}{12}, \tfrac{5}{12}\right) = \tfrac{12}{12} \cdot (7,5).
\end{aligned}
$$

the individual points of the line q.
    We have:

$$
\begin{aligned}
w_0 &= & \pi(\eta(\vartheta(C))), \\
w_1 &= & \pi(\eta(\vartheta(D_\sharp))) = \pi(\eta(\vartheta(D_\flat))), \\
w_2 &= & \pi(\eta(\vartheta(D))), \\
&\vdots& \\
w_{12} &= & \pi(\eta(\vartheta(C'))),
\end{aligned}
$$

and

$$
\begin{aligned}
\vartheta_*^{-1}(w_0) &= (X^7 Y^5)^0 &= 1, \\
\vartheta_*^{-1}(w_1) &= (X^7 Y^5)^{\frac{1}{12}} &= \sqrt[12]{2}, \\
\vartheta_*^{-1}(w_2) &= (X^7 Y^5)^{\frac{2}{12}} &= \sqrt[12]{2^2}, \\
&\vdots \\
\vartheta_*^{-1}(w_{12}) &= (X^7 Y^5)^{\frac{12}{12}} &= 2.
\end{aligned}
$$

□

The image of E_{12} in Theorem 25 can be obtained also with the projection of the images of the whole-tone (9/8) scales: $(C, D, E, F_\sharp, G_\sharp, A_\sharp)$ and $(D_\flat, E_\flat, F, G, A, B)$ into the line

$$p : 7\alpha - 5\beta = 0$$

along the line

$$q : \alpha + \beta = 1.$$

The proof of the following theorem is easy. Here values of the boundaries of the Garbuzov zones are understood to be expressed in relative frequencies.

Theorem 26 *(Fuzzy) boundaries of the Garbuzov granules in the map ϑ_* (in the complex plane with coordinates (α, β)) are the following lines:*

$$
1 = \frac{\alpha}{\log_x R} + \frac{\beta}{\log_y R}, \quad 1 = \frac{\alpha}{\log_x Q} + \frac{\beta}{\log_y Q},
$$

where (R, Q) are the Garbuzov granule boundaries, c.f. Table 4.1.

In Figure 6.4, we c.f. the Garbuzov zone for the minor third (c.f. Table 6.3) in the map ϑ_*.

6.1.3 Fuzziness and beats

Suppose that a musical piece (e.g., the J. S. Bach's Air) is played simultaneously by two violinists both in Π_{17} and E_{12} but with no psychological interaction, and we are interested in the result of such "an operation." Moreover, we simplify the situations as follows: we

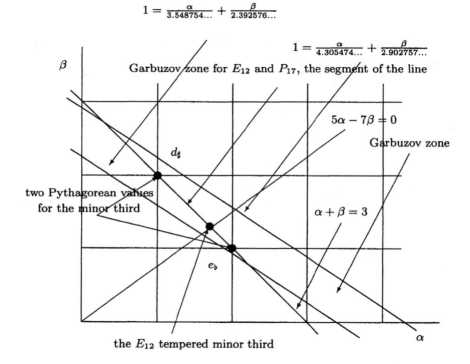

FIGURE 6.4: The Garbuzov zone (granule) of the minor third

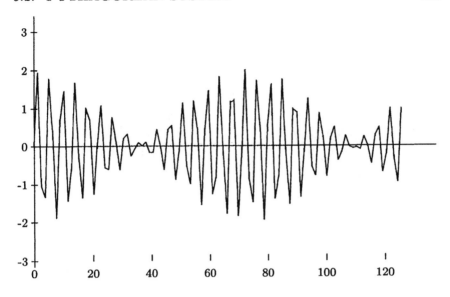

FIGURE 6.5: Interference of two sine waves, equal amplitudes

consider only the theoretical values of basic tones and do not mention overtones which arise automatically.

In Figure 6.5, we can see such the interference of the E_{12} and Π_{17} fifths both with the equal intensity of the sound, $\sin(\alpha \cdot 3/2) + \sin(\alpha \cdot \sqrt[12]{2^7})$. In Figure 6.6, the intensity rate of the sounds is 1:2, $\sin(\alpha \cdot 3/2) + (1/2) \cdot \sin(\alpha \cdot \sqrt[12]{2^7})$. The resulting sound has an altering intensity and also pitch. For some frequency couples the kind of altering will get stable after a short time (rational ratios, harmonic tone intervals, e.g., $5:4$), for some not (irrational ratios, inharmonic tone intervals, e.g., $(3/2): \sqrt[12]{2^7}$).

For the proof of the following theorem, c.f. [35] [34].

Theorem 27 *Denote by $\mathfrak{K}$ the set of all solutions of the equation*

$$\sin\left(\alpha \cdot 3/2\right) + \sin\left(\alpha \cdot \sqrt[12]{2^7}\right) = 0.$$

Denote by Δ the set of all distances between the neighbor points in $\mathfrak{K} \subset (0, \infty)$.

Then Δ is a uniform distributed dense subset of the segment $[\min \Delta, \max \Delta]$.

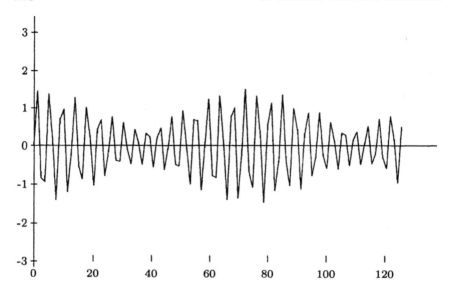

FIGURE 6.6: Interference of two sine waves, various amplitudes

Combining Theorem 25, Theorem 26, Theorem 27, and taking into account that the interference sound is a continuous nonperiodical function of time, we obtain the following:

Theorem 28 *The interference intensity of the sound of the Π_{17} and E_{12} fifths is a continuous nonperiodical (multi) function defined on a segment on the line $\alpha + \beta = 7$ in the map ϑ_*.*

The qualitative similar results we obtain when considering the remaining 11 Garbuzov musical granules. We have, c.f. Figure 6.4.

Theorem 29 *Garbuzov granules of the interference of Π_{17} and E_{12} considered in the map ϑ_* are segments on the lines $\alpha + \beta = i$, $i \in \mathbb{N}$.*

6.2 Pythagorean System worldwide

6.2.1 Indian systems

The Hindu 22-value set of tones called *Sruti*, c.f. e.g. Table 6.1, is based both on ascending fifths and fourths (= descending fifths)

TABLE 6.1: Hindu Srutis, *INDIA(NO.2)-22*

Cents	Cycle	Ratio	Sruti Name	Abbr.
0	$(3/2)^0$	1/1	Shadjam	Sa
24	$(3/2)^{12}$	256/243	Komal Suddha Rishabham	Ri
112	$(\mathcal{D} \cdot 256)/243$	16/15	Suddha Rishabham	Ri
114	$9/(8 \cdot \mathcal{D})$	10/9	TriSruti Rishabham	Ri
204	$(3/2)^2$	9/8	ChatuSruti Rishabham	Ri
294	$(3/2)^{-3}$	32/27	Komal Sadarana Gandharam	Ga
316	$\mathcal{D} \cdot 32)/27$	6/5	Sadarana Gandharam	Ga
318	$81/(64 \cdot \mathcal{D})$	5/4	Antara Gandharam	Ga
408	$(3/2)^4$	81/64	Tivra Antara Gandharam	Ga
498	$(3/2)^{-1}$	4/3	Suddha Madyamam	Ma
520	$(\mathcal{D} \cdot 4)/3$	27/20	Tivra Suddha Madyamam	Ma
590	$(3/2)^6/\mathcal{D}$	45/32	Prati Madyamam	Ma
600	$\mathcal{D} \cdot (3/2)^{-6}$	64/45	Tivra Prati Madyamam	Pa
702	$(3/2)^1$	3/2	Panchamam	Fa
792	$(3/2)^{-4}$	128/81	Komal Suddha Dhaivatam	Dha
814	$(\mathcal{D} \cdot 128)/81$	8/5	Suddha Dhaivatam	Dha
816	$27/(16 \cdot \mathcal{D})$	5/3	TriSruti Dhaivatam	Dha
906	$(3/2)^3$	27/16	ChatuSruti Dhaivatam	Dha
996	$(3/2)^{-2}$	16/9	Komal Kaisiki Nishadam	Ni
1018	$(\mathcal{D} \cdot 16)/9$	9/5	Kaisiki Nishadam	Ni
1020	$243/(128 \cdot \mathcal{D})$	15/8	Kakali Nishadam	Ni
1110	$(3/2)^5$	243/128	Tivra Kakali Nishadam	Ni
1 200	–	2	Shadjam	Sa

modified by Didymus comma $\mathcal{D} = 81/80 \approx 22$cents, or *Pramana* in the Indian version. The acoustic importance of number 22 is obvious from Chapter 1.

The peculiarity of Indian music is that there is a variety of Sruti sets. In our terminology, each of them plays the role of the universe $\mathcal{T}$ of tones. Some of Sruti sets are based also on other prime factors than (1), 2, 3, and 5.

Example 41

DEVARAJAN-22; Derived by Devarajan of Madurai;
256/243, 16/15, 10/9, 9/8, 32/27, 6/5, 5/4, 81/64, 4/3, 27/20, 45/32, 64/45, 3/2, 128/81, 8/5, 5/3, 27/16, 16/9, 9/5, 15/8, 243/128, 2

INDIA(NO.1)-22; Indian shruti;
256/243, 16/15, 10/9, 9/8, 32/27, 6/5, 5/4, 81/64, 4/3, 27/20, 45/32, 729/512,
3/2, 128/81, 8/5, 5/3, 27/16, 16/9, 9/5, 15/8, 243/128, 2

INDIA(NO.2)-22; Indian Sruti, with tritone schisma lower: 64/45;
256/243, 16/15, 10/9, 9/8, 32/27, 6/5, 5/4, 81/64, 4/3, 27/20, 45/32, 64/45, 3/2,
128/81, 8/5, 5/3, 27/16, 16/9, 9/5, 15/8, 243/128, 2

INDIA(NO.3)-22; From [17], after [26], the approximating ratios;
34/33, 35/33, 12/11, 9/8, 22/19, 35/29, 5/4, 40/31, 4/3, 11/8, 17/12, 16/11, 3/2,
17/11, 35/22, 59/36, 27/16, 7/4, 38/21, 15/8, 60/31, 2

HAHN-22; Indian Shruti, Paul Hahn proposal;
25/24, 16/15, 10/9, 9/8, 75/64, 6/5, 5/4, 32/25, 4/3, 27/20, 45/32, 36/25, 3/2,
25/16, 8/5, 5/3, 27/16, 16/9, 9/5, 15/8, 48/25, 2

PERKIS-22; Indian Sruti, Perkis;
36/35, 18/17, 12/11, 9/8, 36/31, 6/5, 5/4, 9/7, 4/3, 26/19, $\sqrt{2}$, 13/9, 52/35,
26/17, 167/106, 13/8, 99/59, 26/15, 52/29, 115/62, 52/27, 2

Indian music is 12 granule and the Indian tone systems, called
ragas, are diatonic restrictions of Sruti sets. Thus there are usually 2 semitones and 5 whole tones between tones per octave (e.g.:
$W_1, W_2, X_3, W_4, W_5, X_6, W_7$) and the order is important. But it
does not matter from which tone we start. Often the increasing
and decreasing forms of a raga contain other values of frequencies,
however, the structure of whole tones and semitones is saved (i.e.,
$W_8, X_9, W_{10}, W_{11}, X_{12}, W_{13}, W_{14}$).

Example 42

INDIA(INV/ROT)-12; Inverted and rotated North Indian gamut;
128/125, 16/15, 6/5, 5/4, 32/25, 4/3, 3/2, 8/5, 128/75, 15/8, 48/25, 2

INDIA-12; North Indian Gamut, modern Hindustani gamut;
16/15, 9/8, 6/5, 5/4, 4/3, 45/32, 3/2, 8/5, 27/16, 9/5, 15/8, 2

INDIA(ROT)-12; Rotated North Indian gamut;
25/24, 16/15, 75/64, 5/4, 4/3, 3/2, 25/16, 8/5, 5/3, 15/8, 125/64, 2

*INDIA(MA-GRAMA)-7; Indian mode Ma-grama (Sa-Ri-Ga-Ma-Pa-
Dha-Ni-Sa);*
9/8, 5/4, 45/32, 3/2, 27/16, 15/8, 2

INDIA (A)-7; Observed Indian mode;
183, 342, 533, 685, 871, 1 074, 1 200

INDIA (B)-7; Observed Indian mode;
183, 271, 534, 686, 872, 983, 1 200

INDIA (C)-7; Observed Indian mode;
111, 314, 534, 686, 828, 1 017, 1 200

INDIA (D)-7; Observed Indian mode (Ellis);
174, 350, 477, 697, 908, 1 070, 1 200

INDIA (E)-7; Observed Indian mode;
90, 366, 493, 707, 781, 1 080, 1 200

INDIA (SA-GRAMA)-7; Indian mode Sa-grama (Sa-Ri-Ga-Ma-Pa--Dha-Ni-Sa), inverse of Didymus' diatonic, tetrachordal scale (2 periods);
9/8, 5/4, 4/3, 3/2, 27/16, 15/8, 2

RAJASTAN-6; A folk scale from Rajasthan, the sixth is missing;
9/8, 5/4, 4/3, 3/2, 15/8, 2

Many tone systems from Middle Asia can be viewed as an intersection of the influence of the Indian systems with the Arabic cultural zone. Arabic systems also use many "universal spaces" T with finite sets of tones per octave. But the number of tones is different than 22 (and they do not use 22).

6.2.2 Europe

Boethius The Middle Ages period is very poor concerning the very development of tone systems. The Pythagorean System was developed in antiquity and prevailed through the whole Middle Ages. In Figure 6.7, we see the Boethius (?480–?524) scheme of the Pythagorean whole tone, c.f. [128].

If we denote by $R = 4/3$ the *perfect fourth* and by $O = 2$ the *octave*, then $Q = O/R = 3/2$ is the *perfect fifth*, $W = Q/R = 9/8$ the *Pythagorean whole tone, tonos*; $X = R/W^2 = 256/243$ the *minor Pythagorean semitone, limma*; $Y = W/X = 2 187/2 048$ the *major Pythagorean semitone, apotome*; $Z = \sqrt{X} = \sqrt[3]{2}/4$ *diaschisma*, $K = Y/X = 531 441/524 288$ *comat*, the *Pythagorean comma*; $H =$

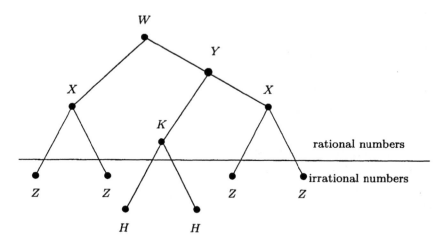

FIGURE 6.7: Boethius scheme of the Pythagorean whole tone

$\sqrt{K} = \sqrt{531\,441/524\,288}$ schisma. We see that the Boethius music theory does not avoid irrational numbers.

Marchetto   Another notable person of the Middle Ages period was Marchetto from Padua. Marchetto divided the whole tone (9/8) into five parts, dieses. But there is no consensus among present scholars as to how it was done. J. Herlinger consider a division of the 9/8 whole tone into 5 *equal* parts, i.e. $(9/8)^{1/5}$, c.f. [138], [27].

Another valuable musicologist, J. L. Monzo (mentioned in the Preface) argued that the corresponding passage of Marchetto's book, *Lucidarium*, is confusing and asserts that the Marchetto's system is much more mathematically elaborated and ingenious as follows. If the string were divided up into 3 parts, then the first of those parts divided into 3 parts, and so on two more times. The procedure would be diagrammed, c.f. Figure 6.8. The most surprising thing to notice is that the *Syntonic Comma*, 81/80, is explicitly measured off as both the first part and the first Diesis. If this is correct, it would mark the *Lucidarium* as one of the earliest explicit notings of the role of the Syntonic Comma in the harmonic practice of medieval European music.

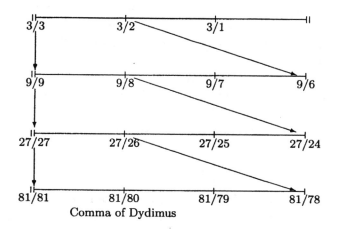

FIGURE 6.8: Marchetto's dividing of the string

At present, Pythagorean tone systems are mostly used in Europe when interpreting Gregorian chants.

Let us pick out one moment about Pythagorean tone systems and uncertainty which we can observe only in Europe and not in India, China, or elsewhere. In Figure 6.11 we see that the Pythagorean chromatic decreasing scale, *-reduced scale, and chromatic increasing scale have 12 tones each. All they are chains with different "shifts" on the "spiral of fifths".

6.2.3   Tone systems of China

Here are some quotes from D. Lentz, c.f. [40], pp. 27–28.

"Two theoretical systems evolved in China, one derived from the cyclic pentatonic and the other from the division of string lengths. They are found combined in the highest form of Ch'in music. The Cyclic Pentatonic, arrived at mathematically, is important in Chinese musical thought. The conception of the twelve liis (tones) dates back to the Han dynasty. They are formed by building empirical fifths in a manner similar to that used in Pythagorean tone system. Methods of arriving at these fifths included the use of twelve tubes. Levis indicates that a stopped bamboo tube 230 millimeters long,

8.12 millimeters in diameter, and vibrating at 366 vibrations per second was the Yellow Bell (Huang Chong), the standard established by the Bureau of Weights and Measures in 239 B.C. Tube number two was made two-thirds the length of the reference tube; number three was made equal to two-thirds of number two and then doubled to bring the tone within the ambit of an octave. This process was repeated for each of the twelve tubes. The fifths produced by these tubes were small compared to Western fifths. Various musicologists place them between 670 and 680 cents ... Fifths of varying sizes are produced on different pipes when the end-correction factor is not considered, thus not fitting a theoretical system. These convert to a standard when duplicated. This procedure of duplication is found in China along with the theoretical fifth, and although one cannot find positive documentation of it for the gamelans of Java and Bali, it is highly possible that it became a factor in the varying sizes of the fifth there too."

Note that the spiral of 23 fifths produces $23 \cdot 678 = 15\,594$ cents and the comma after 13 octaves is $1\,200 \cdot 13 - 15\,594 = 6$ cents, c.f. Meantone systems, Chapter 6.

"In music for the ch'in, a zither-type instrument, the seven strings are tuned to the cyclic pentatonic. Each string has frets or nodes dividing it into the following lengths: $1/2, 1/3, 2/3, 1/4, 3/4, 1/5,$ $4/5, 1/6, 5/6, 1/8,$ and $7/8$. Each string employs the principle of Just Intonation, but many microtonic intervals result when this theory of string length division is combined with the above cyclic pentatonic procedure, which is used to tune the seven open strings. There is no counterpart in Western tempered music. Many Oriental stringed instruments use frets which are movable. The accuracy of placing the fret, which is done by ear, can greatly affect the pitch and thus produce noticeable deviances in a system using natural intervals of microtonic size. The tones of the present-day scales of Japanese koto and gekkin music evolved with variance from the ch'in principle."

Western musicians think of the fifth as being an interval of 700 to 702 cents. The deviation of only two cents between the Just (Pythagorean) and tempered fifth $\sqrt[12]{2^7}$ is so small that these sizes are accepted as being the true fifth. This is not the case in Oriental music. The conception of the fifths is in many cases very different.

For the most part they are smaller than the Western fifth. In Chinese music, another common theoretical fifth of 693 cents (as contrasted with the cyclic fifth of 678 cents above) results from combining three of the characteristic large seconds derived from string-length division. Each of the large seconds has the value of $8/7 \approx 231$ cents.

Example 43

BAMBOO-23; Meantone scale with fifth average 678 cents from Chinese bamboo tubes;

48, 102, 156, 204, 258, 312, 366, 414, 468, 522, 570, 624, <u>678</u>, 726, 780, 834, 882, 936, 990, 1 044, 1 092, 1 146, 1 200

CHIN-LU-12; Chinese Lu (by Huai-nan-dsi), Han era. K. Reinhard: Chinesische Musik;

18/17, 9/8, 6/5, 54/43, 4/3, 27/19, 3/2, 27/17, 27/16, 9/5, 36/19, 2

LING-LUN-12; Scale of Ling Lun;

2 187/2 048, 9/8, 19 683/16 384, 81/64, 177 147/131 072, 729/512, 3/2, 6 561/4 096, 27/16, 59 049/32 768, 243/128, 2

TI-TSU-7; Observed Chinese flute scale;

178, 339, 448, 662, 888, 1 103, 1 196 (1 200?)

CHIN-SHENG-7; Observed scale, Chinese sheng or mouth organ;

210, 338, 4/3, 715, 908, 1 040, 1 200

HIRADOSHI(NO.1)-5; Observed Japanese pentatonic koto scale;

185, 337, 683, 790, 1 200

HIRADOSHI(NO.2)-5; Observed Japanese pentatonic koto scale;

9/8, 6/5, 3/2, 8/5, 2

6.3 On two algorithms in music acoustics

When tuning key instruments in music, two basic algorithms are known. The first one is based on the fifths, the second one on the major thirds. In their pure form, these algorithms lead to two different classical tone systems—to the Pythagorean System and Praetorius Tuning (1/4-comma meantone). In this section we describe these two basic tuning algorithms.

Side c

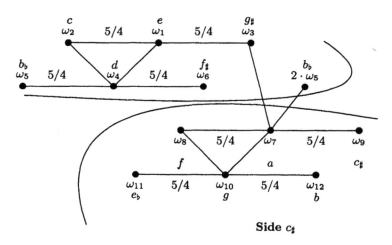

FIGURE 6.9: Algorithm of Prætorius System

6.3.1 Description of the algorithms

Algorithm based on the major thirds Describe the algorithm based on the major thirds. Starting with a frequency ω_1 and both divide and multiply ω_1 by $5/4$, we obtain the relative frequencies ω_2, ω_3, respectively. The next step, the frequency ω_4, we obtain by the geometric mean $\sqrt{\omega_1\omega_2}$, etc., c.f. Figure 6.9. If we replace the sequence $\omega_1, \omega_2, \cdots, \omega_{12}$ by the obvious musical notation (we will use small letters for Prætorius System), we obtain the sequence:

$$e, c, g_\sharp, d, b_\flat, f_\sharp, a, f, c_\sharp, g, e_\flat, b.$$

All other frequencies of Prætorius System we obtain via the so called *octave equivalency*, i.e. multiplying the described 12 values by 2^i, i is an integer. The relative values of Prætorius System within the interval $[1,2], c = 1$, are in Table 6.2. The set of pipes of an organ is divided usually into two parts (often, two boxes)—"Side c" and "Side $c_\sharp$." The reason for this arrangement is the algorithm of Prætorius tone system which prevailed in the 17th century, the golden age of

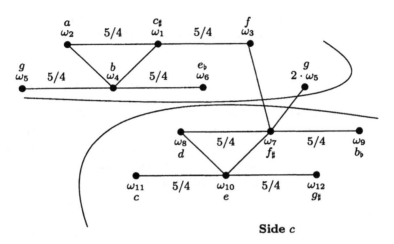

Side $c_\sharp$

FIGURE 6.10: Algorithm of Praetorius System (continued)

organ. The algorithm based on the thirds is/was less popular than the algorithms based on the fifths and now is commonly unknown among professional tuners.

Practically, tuners tuned exactly the third in the third octave over the pitch standard (c and e'' in this case, i.e. $e''/c = 2^2 \cdot 5/4$), then divided this interval into 4 equal fifths. So, they obtained c, g, d, a, e. Then they complete the rest 8 tones with thirds $5/4$. What interesting is that we can start from an arbitrary frequency of the resulting set to obtain the same tone system. In Figure 6.10 there is the same scheme starting from a. We may also do the geometric means of thirds as we wish and again obtain the same frequencies.

Algorithm based on the fifths Describe the algorithm based on the fifths. Start with a frequency $\omega_1 \in [1, 2)$. The n-th step, the frequency ω_n is evaluated as follows:

$$\omega_n = \omega_{n-1}\frac{3}{2} \text{ if } \omega_{n-1} \cdot \frac{3}{2} < 2$$

or
$$\omega_n = \omega_{n-1}\frac{3}{2}\cdot\frac{1}{2} \text{ if } \omega_{n-1}\cdot\frac{3}{2} \geq 2.$$

If $\omega_7 = 1$, then replacing $\omega_1, \omega_2, \cdots, \omega_{17}$ by the obvious music notation (we will use the capital letters for Pythagorean System), we obtain the sequence

$$G_\flat, D_\flat, A_\flat, E_\flat, B_\flat, F, C, G, D, A, E, B, F_\sharp, C_\sharp, G_\sharp, D_\sharp, A_\sharp.$$

Comparing with Praetorius tone system algorithm, the Pythagorean tone system algorithm is sensitive on the initial value (if $\omega_7 \neq 1$, we obtain another result). Again, we enlarge these frequencies to the whole halfline via the octave isomorphism (multiplying each value by 2^z, $z \in \mathbb{Z}$). The values within the interval $[1, 2)$, $C = 1$, are in Table 6.3.

Pythagorean System is mentioned here as 17-valued (also: 5-, 7-, 12- valued, c.f. Figure 6.11). We reduce the 17-valued Pythagorean System to 12-values, (and $Y_1 = 256/243$, $Y_2 = 2\,187/2\,048$ in Definition 15), the values are marked by $*$ in Table 6.3. Another approach to Pythagorean System is presented, e.g., in [115].

6.3.2   Isomorphism of the fifth and third tunings

Pythagorean System has two semitones, expressed via the unique rationals:

$$Y_1 = 256/243, \quad Y_2 = 2\,187/2\,048$$

(the minor Pythagorean semitone, *diesis* and the major Pythagorean semitone, *apotome*), and contains only harmonic music intervals (intervals expressed by rationals), c.f. Table 6.3 and [180].

Praetorius System (also known as the 1/4-comma meantone system), [1], contains both harmonic (octaves, major thirds, and minor sixths) and inharmonic music intervals, c.f. Table 6.2.

Excluding octaves, Equal Temperament contains only inharmonic intervals and is represented by the geometric net (= geometric progression) $\langle \sqrt[12]{2^i} \rangle$ with the quotient $\sqrt[12]{2}$, the *equal tempered semitone*.

For information about semitones (and other constructing intervals) of the the diatonic scales in general, c.f. [124].

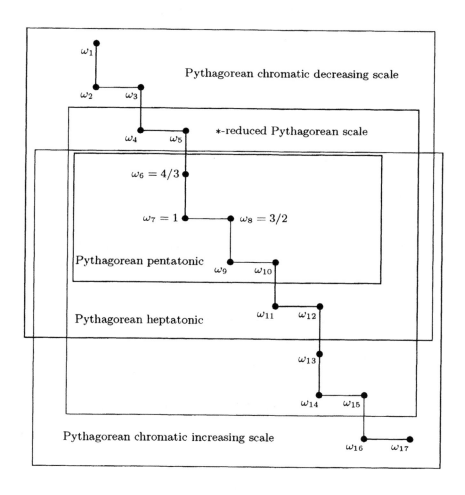

FIGURE 6.11: Algorithm of Pythagorean System

In the present section there is shown the existence and unique-
ness of semitones of Praetorius System and an isomorphism between
Pythagorean and Praetorius Systems.

Theorem 30 *The semitones of Praetorius System are as follows:*

$$\sqrt[4]{78\,125}/16, 8/\sqrt[4]{3\,125}.$$

More precisely, we have the following theory.

Definition 13 *Let $k = 0, 1, 2, ..., 11, 12$.*
We say that a matrix $(\nu_{i,j})_{12,k}^{1,2} \in \mathbb{N}^2 \times \mathbb{N}^2$ is a $(12, k)$-matrix, if

$$0 \le \nu_{k,\cdot} \le \nu_{12,\cdot}, \quad |\nu_{i,\cdot}| = i, i = 12, k.$$

The following definition is based on the algorithm of Praetorius
System.

Definition 14 We say that a 2-quotient geometric net $\langle \Gamma, I \rangle$ is $\mathcal{M}$-
generated by a $(12, 4)$-matrix if for some index lattice I and some
$X_1 > 0$, $X_2 > 0$,

$$\nu_{0,\cdot} = 0, \nu_{1,\cdot} = \frac{7}{4}\nu_{4,\cdot} - \frac{\nu_{12,\cdot}}{2}, \nu_{2,\cdot} = \frac{\nu_{4,\cdot}}{2}, \nu_{3,\cdot} = \frac{\nu_{12,\cdot}}{2} - \frac{3}{4}\nu_{4,\cdot\cdot},$$

$$\nu_{5,\cdot} = \frac{\nu_{12,\cdot}}{2} - \frac{\nu_{4,\cdot}}{4}, \nu_{6,\cdot} = \frac{3}{2}\nu_{4,\cdot\cdot}, \nu_{7,\cdot} = \frac{\nu_{4,\cdot}}{4} + \frac{\nu_{12,\cdot}}{2}, \nu_{8,\cdot} = 2 \cdot \nu_{4,\cdot\cdot},$$

$$\nu_{9,\cdot} = \frac{3}{4}\nu_{4,\cdot} + \frac{\nu_{12,\cdot}}{2}, \nu_{10,\cdot} = \nu_{12,\cdot} - \frac{\nu_{4,\cdot}}{2}, \nu_{11,\cdot} = \frac{5}{4}\nu_{4,\cdot} + \frac{\nu_{12,\cdot}}{2}$$

and for $i \in \mathbb{Z}\backslash\{0, 1, 2, \ldots, 12\}$, there exists $p \in \mathbb{N}, 0 \le p < 12$, and
$q \in \mathbb{Z}$, such that $\nu_{i,\cdot} = q\nu_{12,\cdot} + \nu_{p,\cdot}$.

The following definition is based on the algorithm of Pythagorean System (it defines the Pythagorean System Π_{12} when Y_1, Y_2 are
diesis and apotome, respectively).

Definition 15 *We say that a 2-quotient geometric net $\langle \Gamma, I \rangle$ is $*$-*
generated by a $(12, 7)$-matrix if for some index lattice I and some

$Y_1 > 0$, $Y_2 > 0$ the first 12 of Γ_i is evaluated as follows (reordering increasingly)

$$\{(Y_1^{\nu_{7,1}} Y_2^{\nu_{7,2}})^k (\mathrm{mod}\ Y_1^{\nu_{12,1}} Y_2^{\nu_{12,2}})\}_{k\,=3,\dots,14},$$

and for $i \in \mathbb{Z}\backslash\{0,1,2,\dots,12\}$, there exists $p \in \mathbb{N}, 0 \le p < 12$, and $q \in \mathbb{Z}$, such that $\nu_{i,\cdot} = q\nu_{12,\cdot} + \nu_{p,\cdot}$.

Theorem 31 According to the symmetry, there exists a unique pair of algebraic numbers (X_1, X_2) for $\mathcal{M}$-geometric nets which yields Praetorius System, where

$$X_1 = \frac{5^{7/4}}{2^4} = \sqrt[4]{78\,125}/16 \approx 1.044\,906,$$

$$X_2 = \frac{2^3}{5^{5/4}} = 8/\sqrt[4]{3\,125} \approx 1.069\,984.$$

Proof.   Suppose that there exist $X_1 > 0$, $X_2 > 0$ in Definition 14. Then it is easy to see that Definition 14 implies Praetorius System if both X_1, X_2 are algebraic.

Prove that there exist unique algebraic $X_1, X_2 > 0$ satisfying Definition 1. To show this, consider the following equation system

$$\begin{aligned}
X_1^{\nu_{12,1}} X_2^{\nu_{12,2}} \delta &= 2 \\
X_1^{\nu_{7,1}} X_2^{\nu_{7,2}} \delta &= \sqrt[4]{5} \\
X_1^{\nu_{4,1}} X_2^{\nu_{4,2}} \delta &= \tfrac{5}{4}
\end{aligned}$$

where δ is an arbitrary real number (a parameter, a shift when tuning) and values 2, $\sqrt[4]{5}$, $\tfrac{5}{4}$ are given by Praetorius System, [1], (the octave, the minor third and the Pretorius fifth). We have:

$$\begin{aligned}
\nu_{12,1} \log X_1 + \nu_{12,2} \log X_2 + \log \delta &= \log 2 \\
\nu_{7,1} \log X_1 + \nu_{7,2} \log X_2 + \log \delta &= \log \sqrt[4]{5} \\
\nu_{4,1} \log X_1 + \nu_{4,2} \log X_2 + \log \delta &= \log \tfrac{5}{4}.
\end{aligned}$$

Consider $\nu_{12,1}, \nu_{4,1} \in \mathbb{N}$, such that

$$0 \le \nu_{1,1} \le \frac{7}{4}\nu_{4,1} - \frac{\nu_{12,1}}{2}$$

$$\leq \frac{\nu_{4,1}}{2} \leq \frac{\nu_{12,1}}{2} - \frac{3}{4}\nu_{4,1}$$

$$\leq \nu_{4,1} \leq \frac{\nu_{12,1}}{2} - \frac{\nu_{4,1}}{2} \leq \frac{3}{2}\nu_{4,1} \leq \frac{\nu_{4,1}}{4} + \frac{\nu_{12,1}}{2}$$

$$\leq 2 \cdot \nu_{4,1} \leq \frac{3}{4}\nu_{4,1} + \frac{\nu_{12,1}}{2} \leq \frac{-\nu_{4,1}}{2} + \nu_{12,1},$$

$$\leq \frac{5}{4}\nu_{4,1} + \frac{\nu_{12,1}}{2} \leq \nu_{12,1} \leq 12.$$

The condition $\nu_{2,1} = \nu_{4,1}/2 \leq 4$ implies possibilities $\nu_{4,1} = 2, 4$ ($\nu_{4,1} \neq 0$ since otherwise $\nu_{1,1} < 0$). To $\nu_{4,1} = 2$ we have possibilities $\nu_{12,1} = 3, 5, 7$ (the last two are symmetric). To $\nu_{4,1} = 4$ we have possibilities $\nu_{12,1} = 2, 6, 8, 10$. From these cases only the unique pair $(\nu_{12,1}, \nu_{4,1}) = (5, 2)$ (symmetrically $(7, 2)$) is such that it may $\mathcal{M}$-generate a 2- geometric net.

Solve the equation system above for $\nu_{12,1} = 5, \nu_{4,1} = 2$. By Definition 13 and Definition 14, the determinant of this equation system is

$$\begin{vmatrix} \nu_{12,1} & 12 - \nu_{12,1} & 1 \\ \nu_{7,1} & 7 - \nu_{7,1} & 1 \\ \nu_{4,1} & 4 - \nu_{4,1} & 1 \end{vmatrix} = \begin{vmatrix} \nu_{12,1} & 12 & 1 \\ \nu_{7,1} & 7 & 1 \\ \nu_{4,1} & 4 & 1 \end{vmatrix} =$$

$$= 3\nu_{12,1} - 8\nu_{7,1} + 5\nu_{4,1}$$
$$= \nu_{12,1} + 5\nu_{4,1} - 8(\nu_{4,1}/4 + \nu_{12,1}/2)$$
$$= 3\nu_{4,1} - \nu_{12,1} = 1.$$

Consequently,

$$(X_1, X_2, \delta) = \left(\frac{5^{7/4}}{2^4}, \frac{2^3}{5^{5/4}}, 1 \right) =$$

$$\left(\sqrt[4]{78\,125}/16, 8/\sqrt[4]{3\,125}, 1 \right) \approx (1.044\,906, 1.069\,984, 1.0).$$

We obtained also, that $\delta = 1$ (there is no shift of the fundamental tone when tuning).   □

Combine these results with the algorithms above and collect Table 6.2, *MEAN(1/4-DIDY)-12* (in the fifth column, there are values in cents; in the sixth column, there is a musical denotation).

TABLE 6.2: Praetorius System

$X_2^0 X_1^0$	$2^0 5^0$	1	1.000000	0.000	c
$X_2^0 X_1^1$	$2^{-4} 5^{7/4}$	$\sqrt[4]{78\,125}/16$	1.044907	76.049	$c_\sharp$
$X_2^1 X_1^1$	$2^{-1} 5^{1/2}$	$\sqrt[2]{5}/2$	1.118034	193.157	d
$X_2^2 X_1^1$	$2^2 5^{-3/4}$	$4/\sqrt[4]{125}$	1.196279	310.265	$e_\flat$
$X_2^2 X_1^2$	$2^{-2} 5^1$	$5/4$	1.250000	386.314	e
$X_2^3 X_1^2$	$2^1 5^{-1/4}$	$2/\sqrt[4]{5}$	1.337481	503.422	f
$X_2^3 X_1^3$	$2^{-3} 5^{1/2}$	$\sqrt[2]{125}/8$	1.397542	579.471	$f_\sharp$
$X_2^4 X_1^3$	$5^{1/4}$	$\sqrt[4]{5}$	1.495349	696.578	g
$X_2^4 X_1^4$	$2^{-4} 5^2$	$25/16$	1.562500	772.627	$g_\sharp$
$X_2^5 X_1^4$	$2^{-1} 5^{3/4}$	$\sqrt[4]{125}/4$	1.671851	889.735	a
$X_2^6 X_1^4$	$2^2 5^{-1/2}$	$4/\sqrt[2]{5}$	1.788854	1006.843	$b_\flat$
$X_2^6 X_1^5$	$2^{-2} 5^{5/4}$	$\sqrt[4]{3\,125}/4$	1.869186	1082.892	b
$X_2^7 X_1^5$	2^1	2	2.000000	1200.000	c'

The music interval $e_\flat/g_\sharp$ is the famous so-called *wolf fifth*.

The following theorem is very interesting since: (1) Pythagorean and Praetorius Systems are obtained by two very different algorithms; (2) it leads to many consequences in music theory. The assertion follows when comparing Table 6.2 and Table 6.3.

Theorem 32 *Praetorius and Pythagorean System* Π_{12} *(c.f. Definition 14 and Definition 15) are isomorphic (in the sense of Definition 13).*

6.3.3   Pipe organs with subsemitones, 1468–1721

One[1] of the most interesting chapters in the history of keyboard instruments began with the wish to exceed the limits of certain restricted temperaments, that only allowed a certain number of usable keys. Pythagorean tone systems based on pure fifths, and 1/4-comma meantone temperament, based on pure thirds, posed restrictions in two different ways (for the relation between these tone systems, see the previous sections). Devices were developed to exceed

---

[1]The author of Subsection 6.3.3 and Appendix B is Ibo Ortgies, Göteborg University, Dept. of Musicology, GOArt, P. O. Box 200, SE-405 30 Göteborg, Sweden

TABLE 6.3: Pythagorean System Π_{17}

$Y_1^0 Y_2^0$	$2^0 3^0$	1	1.000 000	0.000	C	*
$Y_1^1 Y_2^0$	$2^8 3^{-5}$	256/243	1.053 498	90.225	$D_\flat$	
$Y_1^0 Y_2^1$	$2^{-11} 3^7$	2 187/2 048	1.067 871	113.685	$C_\sharp$	*
$Y_1^1 Y_2^1$	$2^{-3} 3^2$	9/8	1.125 000	203.910	D	*
$Y_1^2 Y_2^1$	$2^5 3^{-3}$	32/27	1.185 186	294.135	$E_\flat$	*
$Y_1^1 Y_2^2$	$2^{-14} 3^9$	19 683/16 384	1.201 355	317.595	$D_\sharp$	
$Y_1^2 Y_2^2$	$2^{-6} 3^4$	81/64	1.265 625	407.820	E	*
$Y_1^3 Y_2^2$	$2^2 3^{-1}$	4/3	1.333 333	498.045	F	*
$Y_1^4 Y_2^2$	$2^{10} 3^{-6}$	1 024/729	1.404 664	588.270	$G_\flat$	
$Y_1^3 Y_2^3$	$2^{-9} 3^6$	729/512	1.423 828	611.730	$F_\sharp$	*
$Y_1^4 Y_2^3$	$2^{-1} 3^1$	3/2	1.500 000	701.955	G	*
$Y_1^5 Y_2^3$	$2^7 3^{-4}$	128/81	1.580 247	792.180	$A_\flat$	
$Y_1^4 Y_2^4$	$2^{-12} 3^8$	6 561/4 096	1.601 807	815.904	$G_\sharp$	*
$Y_1^5 Y_2^4$	$2^{-4} 3^3$	27/16	1.687 500	905.865	A	*
$Y_1^6 Y_2^4$	$2^4 3^{-2}$	16/9	1.777 778	996.090	$B_\flat$	*
$Y_1^5 Y_2^5$	$2^{-15} 3^{10}$	59 049/32 768	1.802 032	1019.550	$A_\sharp$	
$Y_1^6 Y_2^5$	$2^{-7} 3^5$	243/128	1.898 438	1109.775	B	*
$Y_1^7 Y_2^5$	2^1	2	2.000 000	1 200.000	C'	*

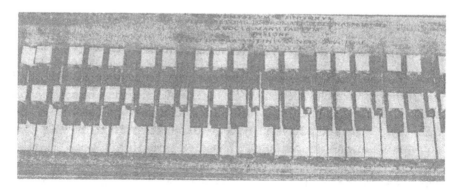

FIGURE 6.12: Keyboard MDCVI A.D., 31 tones per octave

these limits, either with split upper keys, "subsemitonia," c.f. Figure 6.12 or with mechanical systems to switch between the different pitches, c.f. Example 51. The desire to exceed these limits was motivated by various reasons. Throughout the time line of this article transposition was necessary for organists to provide the proper tones in the liturgy, to play transcriptions in different applied pitches and to accompany different ensembles. Most common were transposition by fourths, fifths, or whole tones, but also by minor thirds. Another important role was played by the renaissance idea to reconstruct antique classical Greek thought. This was applied to all sciences and arts and in this spirit the Italian theorists and musicians tried to reinvent the Greek diatonic, enharmonic and chromatic musical modes. These reconstruction attempts resulted in many instruments which had more than 12 keys per octave, even experimenting with just intonation and equal temperaments (i.e., equal temperaments with more than 12 keys per octave, e.g., 19 or 31 keys as the most usable equal tempered systems). Research on "enharmonic" keyboard instruments in general is carried out for his habilitation by M. Kirnbauer, Basel, Switzerland. Equal temperaments were based on 12, 19 or 31 keys per octave, based on different close-to-pure intervals: the fifth, the minor third and the major third.). The application of a limited number, c.f. Appendix B, presents only organs, with a maximum of 5 split keys per octave) of split keys per octave was certainly encouraged by this experimentation.

Though today commonly assumed to have been used in meantone temperaments, evidence has been presented by L. F. Tagliavini and Ch. Stembridge, that split keys were also used to exceed the limits of Pythagorean temperament as well. Stembridge refers in [176], p. 319, that a depiction of "an organ keyboard in an intarsia in Assisi, Basilica of San Francesco, shows an extra natural key between *E* and *F*. Since other details in this intarsia, such as the proportion of the pipes, suggest accuracy on the part of the artist (Apollonio da Ripatransone, 1471), this extra key, which is shown in both octaves of the two-octave keyboard, cannot easily be dismissed as an error. It would seem to the author [Stembridge] that the purpose of such a key might have been to provide a solution for the wolf-fifth in Pythagorean tone system—albeit in an unusual place (i.e., a second

e or f)." Thus the practice of split keys may be older than the commonly accepted date of the mid 15th century.

Pythagorean temperament was used in various forms until the 16th century. 1/4-comma meantone temperamentwas described first by B. Ramis de Pareja in 1482, but was probably used in practice even earlier, and would become the most common temperament system for keyboard instruments in western Europe between the end of the 15th and the early 18th century. The list in Appendix B is collected from the literature and presents more than 70 organs which had split keys. For a deeper analysis, the collection presented here might be too small. But a few working hypotheses can be drawn. It is evident, that Italy was the center of the subsemitonia development from the mid 15th century to around 1600. The only Austrian organ mentioned shows Italian influence. The picture in Spain remains somewhat vague, since it seems unclear whether split keys were actually realized or whether they were only proposed by builders. Considering the close political ties between Spain and Naples in that time and the trade relations between Granada and Northern Italy it would be surprising if there had been no split keys in Spanish instruments. For France there are good reasons to assume that split keys were the exception, if they ever existed at all. The 17th century shows adaptation in Germany and the surrounding countries. While in Italy the development cannot be connected to a certain organ building school, it clearly does in Germany. Two traditions stand out: G. Fritzsche and the Manderscheidt-family. Fritzsche, court organ builder at the Saxon Electoral Court in Dresden, seems to have built the first successful organ with split keys in Germany, in the Schloßkirche in Dresden, 1612. Fritzsche is said to have traveled to Prague at the end of 1610, where he might have had access to instruments like the "Clavicymbalum Vniversale, seu perfectum," [163], pp. 63–66. This harpsichord with split keys in all octaves for all semitones and 2 additional keys for $e\sharp$ (between e and f) and for $B\sharp$ (between b and c) might have been used for an expanded meantone temperament or a 19-tone-equal temperament. Modern reconstructions have been made by the harpsichord makers K. Hill, G. Rapids, in 1984, by W. Jurgenson and D. Wraight, 1987. Praetorius also described on p. 66 an organ positive at the court of the Archduke Ferdinand (the later Hapsburg Emperor, 1619–1637, Ferdinand II.) in Graz, Austria, made by an unknown organ builder of Italian origin, which has the same keyboard design, therefore this organ would have had 77 keys between C and c'''.

Since Fritzsche worked closely connected with Italian-influenced musicians at the Dresden and Braunschweig courts, such as H. L. Hassler, H. Schütz and Michael Praetorius, and after his move to Hamburg, with the Hanseatic organists Jacob Praetorius and H. Scheidemann. It seems obvious that this circle was strongly supportive of the idea of split keys. Michael Praetorius especially was in the forefront of the development, culminating in his design for the organ of the great church "Beatae Mariae Virginis" in Wolfenbüttel, built by Fritzsche 1620–1624, which would have 4 split keys in the c'-octave.

The mathematician O. Gibelius published his "Propositiones mathematico-musicae" (in German) in 1666 in the North German town Minden. In his manual of tuning he is speaking about the two additional tones, $d_\sharp$ and $a_\flat$, as Gibelius was a pupil of H. Grimm, who himself wrote in 1634 a treatise on tuning. Grimm held since 1631 the position of an organist in Braunschweig, where several organs with split keys of G. Fritzsche must have been well known to him. H. Schütz made Gibelius compliments for his work in a private letter, which Gibelius quotes. Gibelius complains that $d_\sharp$ and $a_\flat$ up to "today" are not in all organs and instruments (stringed keyboard instruments—many German organ builders called themselves "Orgel- und Instrumentenmacher," referring to their building stringed keyboard instruments, especially clavichords) and that not all organists would know how to use the split keys properly, not to speak about even more split keys per octave.

The Manderscheidt family, working in the middle of the century in the area around Nürnberg (Germany) and later in Switzerland used split keys rather regularly in organ positives, but also in a few bigger organs.

Though it cannot be assumed that German organ builders were the first to bring split keys to the Netherlands, Great Britain and Sweden, there is one, late example (Uppsala, Sweden, 1710), where a non-German organ builder working within the German tradition applied split keys. It was the Hagerbeer-family, coming from Ost-

friesland, North West Germany, in the 1640s, whose designs for Den
Haag (1641) and Alkmaar (1643-1645/6) with $e_b/d_\sharp$ and $b_b/a_\sharp$, leaving out the $g_\sharp/a_b$, enable the transposition a whole tone upwards,
because of the relatively low Dutch organ pitch.

The design, to be found in a number of Italian organs, goes in the
opposite direction: $c_\sharp/d_b$ and $g_\sharp/a_b$, leaving out $e_b/d_\sharp$. Both designs
enable also transposition by minor thirds upwards (Hagerbeer) or
downwards (Italian organs).

Bernard Smith (Schmidt), probably from Germany, is the only
English organ builder known to have used split keys. After the middle of the 18th century England shows an interesting development of
organs with a mechanical switch to shift between different adjacent
enharmonic keys. "The Russell Collection of Early Keyboard Instruments," Edinburgh (Great Britain) possesses such an organ built in
1765 by Thomas Parker, which was restored in 1998.

Parker's organ had levers at the console, two for each of the three
manuals. One gave $c_\sharp$ and e_b at rest, $c_\sharp$ and $d_\sharp$ when moved to the
left, and db and e_b when moved to the right. The other gave $g_\sharp$ and
b_b at rest, a_b and b_b when moved to the left, and $g_\sharp$ and $a_\sharp$ when
moved to the right." A similar organ was built by Parker in London
(Great Britain) for the Foundling Hospital as late as 1768.

The instruments listed in Appendix B might not be more than
the tip of the iceberg, but having such an amount of evidence, we
can be sure that further research will reveal that there must have
been existed many more instruments with these features.

For illumination, c.f. the keyboard, MDCVI A.D., 31 tones per
octave, Figure 6.12.

6.3.4   The Petzval's keyboard

J. M. Petzval's contributions to the tone systems theory was a theory
of temperatures based on the tempered fifths. The invention (and
also a practical realization) bounded with his tone systems is a "rational" keyboard (before 1870). On these ideas his keyboard which
is rather more known in the present day constructed Paul Jankó,
the Petzval pupil. Note that the most famous one in the present
time is an analogous Bosanquet's keyboard which is, maybe, of a bit

younger data.

The structure of the Petzval's (theoretically–infinite) two-dimensional keyboard is identical with the lattice structure of meantones, c.f. Figure 6.2, Figure 4.3, Figure 7.4, and Figure 7.5. Keyboards of this construction have a the property that we may transpose compositions (in the Petzval's Tone Systems and specially–in meantones and Fokker systems) into every key and, moreover, to use the identical fingering.

Note that the scheme of the Fokker's keyboard (enlarged to 31-keys per octave, 1940–1945, is mirror-symmetrical to the Petzval's keyboard.

Example 44 Let us have the sequence of tempered fifths ..., $C_\flat$, $G_\flat$, $D_\flat$, $A_\flat$, $E_\flat$, $B_\flat$, F, C, G, D, A, E, B, $F_\sharp$, $C_\sharp$, $G_\sharp$, $D_\sharp$, $A_\sharp$, $E_\sharp$, ... Design this set in the plane as follows (two octaves):

$$
\begin{array}{llllllllllll}
C_\sharp & D_\sharp & E_\sharp \\
C & D & E & F_\sharp & G_\sharp & A_\sharp & B_\sharp \\
C_\flat & D_\flat & E_\flat & F & G & A & B & c_\sharp & d_\sharp & e_\sharp \\
& F_\flat & G_\flat & A_\flat & B_\flat & c & d & e & f_\sharp & g_\sharp & a_\sharp & b_\sharp \\
& & & & & c_\flat & d_\flat & e_\flat & f & g & a & b \\
& & & & & & & & f_\flat & g_\flat & a_\flat & b_\flat
\end{array}
$$

There are fingering schemes (the numbers 1, 2, 3, 4, 5 denote the right hand fingers, 1 denotes the thumb)

• for every minor accord:

$$
\begin{array}{cccc}
1 & . & . & . \\
. & 3 & . & 5
\end{array}
$$

• for every major scale:

$$
\begin{array}{cccccccc}
1 & 2 & 3 & . & . & . & . & . \\
. & . & . & 1 & 2 & 3 & 4 & . \\
. & . & . & . & . & . & . & 1
\end{array}
$$

• for every minor scale (aiolish):

$$
\begin{array}{cccccccc}
1 & 2 & . & . & . & . & . & . \\
. & . & 3 & 1 & 2 & . & . & . \\
. & . & . & . & . & 3 & 4 & 1
\end{array}
$$

6.4   Overview of meantones

6.4.1   Era of meantones

In Europe after the ancient Greeks, the further development of tone systems went on after a relatively long time period. One major contribution to the theory of tone systems brought G. Zarlino (1517–1590), the composer and music theoretician in Venice, who in 1560 proposed inverting the Major and Minor tones of the upper group in Aristoxenus's scale to relieve the monotony of having two identically tuned halves of the scale. This gives us the following intervals, c.f. Table 9.7. This is, effectively, the pure tone scaling, resp. the major (and minor) diatonic scale! His idea took root and meant the end of tone systems with the tetrachord period in Europe. Since that time the development continued with the octave period. In the present time, the pure tone system is commonly used in Europe in the folklore music. But it was not Zarlino's last word! He suggested to save the algorithm of Pythagorean System and to reduce every fifth by two sevenths of the Didymus comma in order to lose the Pythagorean comma.

This method of redistributing commas was further refined by Francis Salinas, a blind mathematician and musician in Naples. If $\Theta_{3/2} = \tau_{3/2} \cdot 3/2$ is a tempered fifth with the temperature $\tau_{3/2} \in \mathbb{R}^+, 80/81 < \tau_{3/2} < 81/80$, then the basic formula describing meantones is

$$\left(\tau_{3/2} \cdot \frac{3}{2}\right)^{11} \cdot \mathcal{W} = 2^7, \qquad (6.1)$$

where $\mathcal{W}$ is called the *wolf* fifth, c.f. Figure 6.13.

Example 45 If $\tau_{3/2} = \sqrt[4]{80/81}, \sqrt[5]{80/81}, \sqrt[6]{80/81}$, then we obtain the 1/4-, 1/5-, 1/6- comma meantones, respectively.

Here we have the start of Meantone temperaments, the purpose of which is to redistribute the intervals such that the principle ones (in various meantones: fourths, fifths, major thirds, minor thirds, etc.) remain fairly true and others retuned so as to still fulfil their function within the scales.

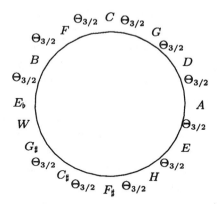

FIGURE 6.13: Meantones; $\mathcal{W}$—wolf, $\Theta_{3/2}$—tempered fifth

The most popular mean tone system during about 400 years was the 1/4-*comma mean-tone system*, which is used also today when playing the historical instruments and music, c.f. Section 6.3.3 and Appendix B. Its mathematical description is assigned to the mathematician, organist and composer M. Praetorius (Schultze) (1571–1621), c.f. Section 6.3 and Subsection 6.3.3. The 1/4-comma meantone is specific with the property that major thirds and minor sixths are exact, i.e. 5/4 and 8/5, respectively.

Solving one problem, another arises. One of the most problematical intervals is the fifth $E_\flat/G_\sharp$. It is too sharp and its inversion too flat. This interval is known historically as the *wolf*. There are also other wolves but this one is the most disturbing. The wolf intervals are probably the main reason that composers of the period avoided using keys with a large number of accidentals. For example, Mozart rarely, if ever, composed any works in $D_\flat$, $F_\sharp$, $A_\flat$, and B Major or $C_\sharp$, $E_\flat$, F, $F_\sharp$ and $G_\sharp$ Minor as these keys make wolf tones stick out like a sore thumb. Curiously he also avoided B Minor which is all the more odd when you consider that his favorite key was D Major, closely followed by C and $B_\flat$ Major.

One of the meantones is E_{12}, the 12 *Tone Equal Temperament*.

Here $\tau_{3/2} = 1/\sqrt[12]{\mathcal{K}}, \mathcal{W} = \sqrt[12]{2^7}$ in the formula (6.1), where $\mathcal{K} = 3^{12}/2^{19} = (3/2)^{12}/2^7$ is Pythagorean comma. Both Zarlino and Salinas knew about E_{12} but disliked the severe mistuning inherent in the thirds and sixths. But, progress moves ever on. A French mathematician, physician, and musician Marin Mersenne (1588–1648) was probably the first person who explicitly (about 1620) calculated the basis of the equal temperature—the equal tempered semitone, $\sqrt[12]{2} \approx 1.059\,463$. Although, some people accredit Simon Stevin (1548–1620), the organ tuner at the workshop of Andreas Werckmeister, physician, mathematician and technician, with the discovery somewhat earlier in 1608. There is also evidence to suggest that a Chinese gentleman by the name of Chu Tsai-yu worked it out several years before its calculation in the West.

Composers now had, at least in theory, complete freedom to modulate to any key without hearing wolf intervals - but at a price. In the present, E_{12} is the most spread tone system.

6.4.2 Meantone zones

Meantones, the single chains of equal tempered fifths, are a very large and significant class of tone systems. Meantones have nonvoid intersections with other classes of tone systems. Mention some relations to other tone systems classes—equal temperaments, well temperaments, and just intonation. These classes are neither proper subsets nor supersets of meantones. For instance, not all equal temperaments are meantones, since some consist of 2 or more closed chains of fifths, equally spaced within the octave; well temperaments are not in this class in general, because they have more than one size of usable fifth; Just Intonations are not in this class because their intervals (other than fifths) are not generated by chains of fifths at all.

Tempered fifths with maximal 10 cent temperature, i.e. fifths with values less than 692 cents and greater than 712 cents, are often considered as unusable wolves. The region of used tempered fifths can be conventionally split into four overlapping zones.

• The *zone of historical meantones* contains the tempered fifths of 1/3-, 2/7-, 1/4-, 1/5-, 1/6- Didymus comma meantones. It is the zone where 4 fifths remains the well temperatures to a major third 5/4, within a 12-tone scale.

• The *well-temperatures* zone contains 1/8-, 1/11-, 1/14- Pythagorean comma meantones.

• Then there is a zone of *Pythagorean tone system* between 700.0 and about 705.8 cents (Comma Schizma is used).

• The fourth zone is a zone, let us call it the *22-TET zone*, which involves tempered fifths between 705.8 and 711.8 cents.

The meantones with fifths less than 700 cents are called to be *positive* (in this case, $c_\sharp < d_\flat$, etc.) and meantones with fifths greater than 700 are called to be *negative* (in this case, $c_\sharp > d_\flat$, etc.). For instance, Praetorius 1/4-comma tone system is a positive menatone while Pythagorean tone system is a negative one.

In Figure 6.14, there are graphed meantone temperatures of 3/2, 5/4, 6/5, and 7/4 depending of the size of the tempered fifth (the lengths of dashes are decreasing in this order). The wolf fifth temperament intersects the horizontal axis in the value 700 cents. Both horizontal and vertical axis are cent (i.e., logarithm) scaled.

If there was no limit to the number of fifths or fourths we could stack to approximate a given ratio we could approximate any ratio as closely as we liked with any size of fifth, but there are practical limits. We somewhat arbitrarily chose a limit of 12 fifths per octave for the Figure 6.14 to avoid ambiguities.

Note that this is not a limit on the number of values in a meantone system. If a tone system contains more than 12 consecutive values in a chain of fifths, then it might contain a better approximation to some ratio than that shown in Figure 6.14.

For example, the choice of 14 consecutive fifths allows an approximation to ratios of 7 in the Pythagorean zone. This greatly exaggerates the usefulness of Pythagorean tone system for harmony involving ratios of 7. In 1/4-comma meantone we get a good approximation to a 7/4 ratio between $B_\flat$ and $G_\sharp$ (an augmented sixth), this is equivalent to stacking 10 fifths, $F/B_\flat$, C/F, G/C, D/G, A/D, E/A, B/E, $F_\sharp/B$, $C_\sharp/F_\sharp$, $G_\sharp/C_\sharp$.

This is not as useful an approximation as the one we get in the zone around 22-tone Equal Temperament where only 2 stacked fourths (-2 fifths) are needed. If our scale were limited to 12 con-

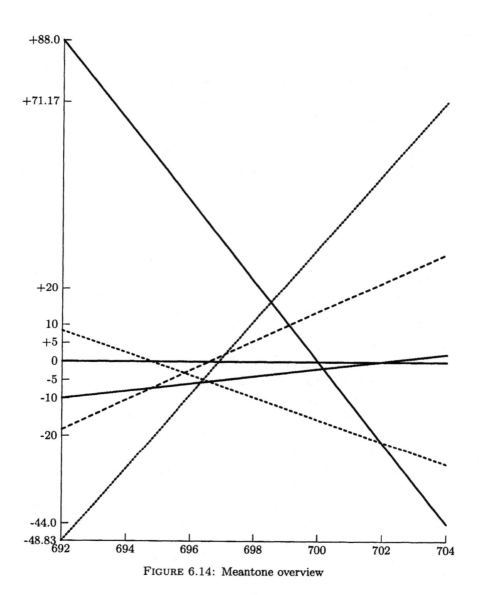

FIGURE 6.14: Meantone overview

secutive values from the chain, then in the meantone zone we would have only two 7/4's while in the 22-tone Equal Temperament zone we would have ten of them. In the Pythagorean zone we would have none.

In the historical meantone zone we get a good approximation to the just major third 5/4 by stacking only 4 fifths, e.g., G/C, D/G, A/D, E/A. In the Pythagorean zone 8 fourths is best. In 12-tone Equal Temperament these two approximations are equivalent. In the 22-tone Equal Temperament zone 9 fifths is best. In 17-tone Equal Temperament 8 fourths is the same as 9 fifths. This is why we have used E_{12} and E_{17} as the boundaries of the Pythagorean zone.

In the following examples of mean tones systems, there are underlined the rational values but 2 .

Example 46

EQUAL(1/12-PYTH)-12; 12-tone Equal Temperament, 1/12-Pythagorean comma meantone (Werckmeister, Salinas, Stevin, etc.);
100, 200, 300, 400, 500, 600, 700, 800, 900, 1 000, 1 100, 1 200

MEAN(3/10-DIDY)-12; 3/10-comma meantone;
68.522, 191.006, 259.528, 382.012, 504.497, 573.019, 695.503, 764.025, 886.509, 125/72, 1 077.516, 1 200.000

MEAN(3/11-DIDY)-12; 3/11-comma meantone;
72.628, 192.179, 264.807, 384.359, 125/96, 576.538, 696.090, 768.717, 888.269, 960.897, 1 080.448, 1 200.000

MEAN(4/17-DIDY)-19; 4/17-comma meantone, least square error of 5/4 and 3/2;
78.263, 156.526, 193.789, 272.052, 350.315, 387.579, 465.842, 503.105, 581.368, 659.631, 696.895, 775.158, 853.420, 890.684, 968.947,
1 875/1 024, 1 084.474, 1 162.736, 1 200.000

MEAN(2/7-DIDY)-19; 19-tone 2/7-comma meantone;
25/24, 120.948, 191.621, 262.293, 312.569, 383.241, 453.914, 504.190, 574.862, 625.138, 695.810, 766.483, 816.759, 887.431, 958.103, 1 008.379, 1 079.052, 1 149.724, 1 200.000

MEAN(2/7-DIDY)-31; 31=tone 2/7-comma meantone;
20.397, 25/24, 120.948, 625/576 191.621, 241.897, 262.293, 312.569, 332.966, 383.241, 433.517, 453.914, 504.190, 524.586, 574.862, 625.138, 645.535, 695.810,

716.207, 766.483, 816.759, 837.155, 887.431, 937.707, 958.103, 1 008.379, 1 028.776, 1 079.052, 48/25, 1 149.724, 1 200.000

MEAN(2/9-DIDY)-12; 2/9-comma meantone;
80.231, 194.352, 75/64, 388.703, 502.824, 583.055, 697.176, 777.407, 891.528, 1 005.648, 1 085.879, 1 200.000

MEAN(2/9-DIDY)-19; 19-tone 2/9-comma meantone;
80.231, 114.121, 194.352, 75/64, 308.473, 388.703, 468.934, 502.824, 583.055, 616.945, 697.176, 777.407, 811.297, 891.528, 971.758, 1 005.648, 1 085.879, 1 166.110, 1 200.000

MEAN(1/5-DIDY)-12; 1/5-comma meantone;
83.577, 195.308, 307.039, 390.615, 502.347, 585.923, 697.654, 781.230, 892.962, 1 004.693, 15/8, 2

MEAN(1/5-DIDY)-19; 19-tone 1/5-comma meantone;
83.576, 16/15, 195.307, 278.884, 307.039, 390.615, 474.191, 502.346, 585.922, 614.078, 697.654, 781.230, 809.385, 892.961, 225/128, 1 004.693, 15/8, 1 171.845, 1 200.000

MEAN(1/5-DIDY)-43; 1/5-comma meantone;
28.155, 55.421, 83.576, 16/15, 138.997, 167.152, 195.307, 256/225 250.729, 278.884, 307.039, 334.305, 362.460, 390.615, 418.770, 446.036, 474.191, 502.346, 530.501, 557.767, 585.922, 614.078, 759 375/524 288, 669.499, 697.654, 725.809, 50 625/32 768, 781.230, 809.385, 836.651, 3 375/2 048, 892.961, 921.116, 948.382, 225/128, 1 004.693, 1 032.848, 1 060.114, 15/8, 1 116.424, 1 143.690, 1 171.845, 1 200.000

MEAN(1/4-DIDY)-12; 1/4-comma meantone (Praetorius, Aaron, Pereja, etc.);
76.049, 193.157, 310.265, 5/4, 503.421, 579.470, 696.578, 25/16, 889.735, 1 006.843, 1 082.892, 2

MEAN(1/4-DIDY)-19; 19-tone 1/4-comma meantone;
76.049, 117.108, 193.157, 269.206, 310.265, 5/4, 462.363, 503.422, 579.471, 620.529, 696.578, 25/16, 8/5, 889.735, 965.784, 1 006.843, 1 082.892, 125/64, 2

MEAN(1/4)-27; 27-tone 1/4-comma meantone;
76.049, 117.108, 152.098, 193.157, 234.216, 269.206, 310.265, 5/4, 32/25, 462.363, 503.422, 579.471, 620.529, 655.520, 696.578, 737.637, 25/16, 8/5, 848.676, 889.735, 930.794, 965.784, 1 006.843, 1 082.892, 1 123.951, 125/64, 2

MEAN(1/7-SCHIZMA)-12; 1/7-schisma meantone;

91.621, 203.352, 294.972, 386.593, 498.324, 589.945, 701.676, 793.296, 884.917, 996.648, 15/8, 2

MEAN(35/32)-19;
Meantone with septimal neutral second;
77.570, 35/32, 193.591, 271.161, 348.730, 387.183, 464.752, 503.204, 580.774, 658.344, 696.796, 774.365, 851.935, 890.387, 967.957, 1 045.526, 1 083.978, 1 161.548, 2

MEAN(7/6)-41; Meantone with septimal minor third;
35.950, 62.453, 88.957, 124.907, 151.410, 177.914, 213.864, 240.367, 7/6, 293.374, 329.324, 355.828, 382.331, 418.281, 444.785, 471.288, 497.791, 49/36, 560.245, 586.748, 622.698, 649.202, 675.705, 711.655, 738.159, 764.662, 791.166, 827.116, 853.619, 880.122, 916.073, 942.576, 969.079, 995.583, 1 031.533, 1 058.036, 1 084.540, 1 120.490, 1 146.993, 1 173.497, 2

MEAN(21/16)-41;
Meantone with septimal narrow fourth;
34.820, 61.791, 88.762, 123.582, 150.553, 177.524, 212.344, 239.315, 266.286, 293.257, 328.077, 355.048, 382.019, 416.839, 443.810, 21/16, 497.752, 532.572, 559.543, 586.514, 621.334, 648.305, 675.277, 710.096, 737.068, 764.039, 791.010, 825.830, 852.801, 879.772, 914.592, 441/256, 868.534, 995.505, 1 030.325, 1 057.296, 1 084.267, 1 119.087, 1 146.058, 1 173.029, 2

MEAN(7/5)-12; Meantone with septimal diminished fifth;
79.598, 194.171, 273.769, 388.342, 502.915, 7/5, 697.085, 776.683, 891.256, 970.854, 1 085.427, 2

MEAN(7/5)-29; Meantone with septimal diminished fifth;
50.451, 85.427, 135.878, 170.854, 221.305, 256.280, 291.256, 341.707, 376.683, 427.134, 462.110, 497.085, 547.536, 7/5, 632.963, 667.939, 718.390, 753.366, 788.341, 838.793, 873.768, 924.219, 959.195, 994.171, 1 044.622, 1 079.598, 1 130.049, 49/25, 2

MEAN(35/32)-41; Meantone with septimal neutral second;
40.866, 65.335, 89.805, 130.671, 35/32, 179.609, 220.475, 244.944, 269.414, 293.883, 334.749, 359.218, 383.687, 424.554, 449.023, 473.492, 497.961, 538.827, 563.296, 587.765, 628.632, 653.101, 677.570, 718.436, 742.905, 767.375, 791.844, 832.710, 857.179, 881.648, 922.514, 946.984, 971.453, 995.922, 1 036.788, 1 061.257, 1 085.726, 1 126.593, 1 151.062, 1 175.531, 2

MEAN(1/7-DIDY)-12; 1/7-comma meantone;
135/128, 197.765, 303.352, 395.531, 501.117, 593.296, 698.883, 791.061, 896.648,

1 002.235, 1 094.413, 2

MEAN(1/6-DIDY)-19; 19-tone 1/6-comma meantone;
88.594, 108.147, 196.741, 285.336, 304.888, 393.482, 482.077, 501.629, 45/32, 64/45, 698.371, 786.965, 806.518, 895.112, 983.706, 1 003.259, 1 091.853, 2 025/1 024, 2

MEAN(1/3-DIDY)-12; 1/3-comma meantone;
63.504, 189.573, 6/5, 379.145, 505.214, 25/18, 694.787, 758.290, 5/3, 1 010.428, 1 073.932, 2

MEAN(1/3-DIDY)-19; 1/3-comma meantone;
63.504, 126.069, 189.572, 125/108, 6/5, 379.145, 442.649, 505.214, 25/18, 36/25, 694.786, 758.290, 820.855, 5/3, 947.862, 1 010.428, 1 073.931, 1 136.496, 2

RICCATI(3/14-DIDY)-12; 3/14-comma meantone (G. Riccati, 1762);
81.426, 194.693, 307.960, 389.386, 502.654, 584.079, 697.347, 778.772, 892.040, 1 005.307, 1 086.733, 1 200.000

SABAT/GARIBALDI(1/9-SCHIZMA)-12;
1/9-schisma meantone Sabat–Garibaldi's;
112.165, 203.476, 6/5, 406.952, 498.262, 610.428, 701.738, 813.903, 905.214, 1 017.379, 1 108.690, 2

SALINAS(1/6-DIDY)-12;
1/6-comma meantone (tritonic temperament of Salinas);
88.594, 196.741, 304.888, 393.482, 501.629, 45/32, 698.371, 786.965, 895.112, 1 003.259, 1 091.853, 2

SMITH(5/18-DIDY)-12; 5/18-comma meantone Smith);
71.867, 191.962, 312.057, 383.924, 504.019, 575.886, 695.981, 767.848, 887.943, 1 008.038, 1 079.905, 1 200.000

HELM(1/8-SCHIZMA)-12; 1/8-schisma meantone Helmholtz;
91.446, 203.422, 294.868, 5/4, 498.289, 589.735, 701.711, 793.157, 884.603, 996.578, 1 088.025, 2

ZARLINO(2/7-DIDY)-12; 2/7-comma meantone, Zarlino;
25/24, 191.621, 312.569, 383.242, 504.190, 574.862, 695.811, 816.759, 887.431, 1 008.380, 1 079.052, 1 200.000

6.4.3   Further examples of mean-tone systems

Example 47 *BAMBOO-23; LING-LUN-12.*

Example 48

ARABIC-17;
Arabic 17-tone Pythagorean mode, Safi al-Din;
256/243, 65 536/59 049, 9/8, 32/27, 8 192/6 561, 81/64, 4/3,
1 024/729, 262 144/177 147, 3/2, 128/81, 32 768/19 683, 27/16, 16/9, 4 096/2 187,
1 048 576/531 441, 2

AEOLIC-7;
Ancient Greek Aeolic, also tritriadic scale of the 54:64:81 triad;
9/8, 32/27, 4/3, 3/2, 128/81, 16/9, 2

AL-DIN-35; Safi al-Din's complete lute scale on 5 strings 4/3 apart;
256/243, 65 536/59 049, 9/8, 32/27, 8 192/6 561, 81/64, 4/3, 1 024/729,
262 144/177 147, 3/2, 128/81, 32 768/19 683, 27/16, 16/9, 4 096/2 187,
1 048 576/531 441, 2 , 512/243, 131 072/59 049, 9/4, 64/27, 16 384/6 561,
4 194 304/1 594 323, 8/3, 2 048/729, 524 288/177 147, 3/1, 256/81, 65 536/19 683,
16 777 216/4 782 969, 32/9, 8 192/2 187, 2 097 152/531 441, 4/1, 1 024/243

AL-DIN-19; Arabic scale by Safi al-Din;
256/243, 65 536/59 049, 9/8, 32/27, 8 192/6 561, 81/64, 2 097 152/1 594 323, 4/3,
1 024/729, 262 144/177 147, 3/2, 128/81, 32 768/19 683, 27/16, 8 388 608/4 782 969,
16/9, 4 096/2 187, 1 048 576/531 441, 2

ERATOS/DIAT-7; Dorian mode of Eratosthenes's Diatonic, Pytha-
gorean, 2 tetrachord periods;
256/243, 32/27, 4/3, 3/2, 128/81, 16/9, 2

KORNERUP-19;
Kornerup's temperament with fifth of (15 -$\sqrt{5}$) / 22 octaves;
73.501, 118.928, 192.429, 265.930, 311.357, 384.858, 458.359, 503.786, 577.287,
622.713, 696.215, 769.716, 815.142, 888.643, 962.145, 1 007.571, 1 081.072,
1 154.574, 1 200.000

LUCY-7; Diatonic Lucy's scale, $L = s\phi, \phi = (1 + \sqrt{5})/2$, octave =
$5L + 2 = 1\,200$, fourth = $2L + s \approx 405$, fifth = $3L + s \approx 696$;
190.986, 381.972, 504.507, 695.493, 886.479, 1 077.465, 1 200.000

LUCY-19; Lucy's 19-tone scale;
68.451, 122.535, 190.986, 245.070, 313.521, 381.972, 436.056, 504.507, 572.958,
627.042, 695.493, 763.944, 818.028, 886.479, 940.563, 1 009.014, 1 077.465,
1 131.549, 1 200.000

LUCY-21; Charles Lucy's scale;
68.750, 121.875, 190.625, 245.313, 259.375, 314.063, 381.250, 435.938, 504.688, 573.438, 626.563, 704.688, 750.000, 764.063, 818.750, 885.938, 940.625, 1 009.375, 1 078.125, 1131.250, 1 200.000

LUCY-31; Lucy Tuning from A;
54.084, 68.451, 122.535, 136.903, 190.986, 245.070, 259.438, 313.521, 367.605, 381.972, 436.0656, 450.424, 504.507, 558.591, 572.958, 627.042, 641.042, 695.493, 749.577, 763.944, 818.028, 872.112, 886.479, 940.563, 954.931, 1 009.014, 1 063.098, 1 077.465, 1 131.549, 1 145.917, 1 200.000

MEAN(1/9-HARRISON)-12; 1/9-Harrison's comma meantone;
74.233, 192.638, 7/6, 385.276, 503.681, 577.914, 696.319, 770.552, 888.957, 963.190, 1081.595, 2

MEAN(1/11-HARRISON)-12; 1/11-Harrison's comma meantone;
81.406, 194.687, 276.093, 389.375, 21/16, 584.062, 697.344, 778.750, 892.031, 973.437, 1 086.719, 2

MEAN(1/4-HARISSON)-12; 1/4-Harrison's comma meantone;
78.178, 193.765, 271.943, 387.530, 503.117, 581.296, 696.883, 775.061, 890.648, 7/4, 1 084.413, 2

MEAN(7/4)-41; Meantone with harmonic seventh;
35.425, 62.146, 88.866, 124.292, 151.012, 177.733, 213.158, 239.879, 266.599, 293.320, 328.745, 355.466, 382.186, 417.612, 444.332, 471.053, 497.773, 533.199, 559.919, 586.640, 622.065, 648.786, 675.506, 710.931, 49/32, 764.373, 791.093, 826.518, 853.239, 879.960, 915.385, 942.105, 7/4, 995.547, 1 030.972, 1 057.692, 1 084.413, 1 119.838, 1 146.559, 1 173.279, 2

MEAN(LST357)-19; 19 of meantone, least square error in 3/2, 5/4 and 7/4;
78.190, 115.578, 193.769, 271.959, 309.347, 387.537, 465.728, 503.116, 581.306, 618.694, 696.884, 775.075, 812.463, 890.653, 968.844, 1006.231, 1084.422, 1162.612, 1 200.000

MEAN(π, NO.1)-12; π-based meantone with Harrison's major third by Erv Wilson;
88.733, 204.507, 293.240, 381.972, 497.747, 586.479, 702.254, 790.986, 879.718, 995.493, 1 084.225, 1 200.000

MEAN(π, NO.2)-12; π-based meantone by Erv Wilson analogous to 22-TET;

163.756, 218.216, 381.972, 436.432, 600.188, 654.648, 709.108, 872.864, 927.324, 1 091.080, 1 145.540, 1 200.000

METAMEAN-12; Erv Wilson's Meta-Meantone tuning;
69.413, 191.261, 260.674, 382.522, 504.370, 573.783, 695.630, 765.043, 886.891, 956.304, 1078.152, 1 200.000

SHERWOOD-12; Sherwood's improved meantone temperament;
114.420, 194.501, 308.921, 389.002, 503.422, 583.503, 697.923, 812.342, 892.424, 1 006.843, 1 086.925, 1 201.344

6.5 Anatomy of the Pythagorean whole tone

The class of tones $T = \Pi$ is usually restricted to a useful set of tones $\mathbb{T}$ with some further conditions which are important for music. Such conditions are e.g.: (1) the number of qualitative musical granules in the octave (e.g., 12 granules); (2) some harmonic structures; (3) number of keys within the octave on the keyboard; etc.

As a concentrated result of these restrictions are *Pythagorean Systems* $\Pi_{[q_1,q_2]}$ which are subsets of Π and are defined as follows:

$$\Pi_{[q_1,q_2]} = \{2^p 3^q; \; q_1 \leq q \leq q_2, \; p,q,q_1,q_2 \in \mathbb{Z}\}.$$

If for 12 granules, qualitative musical degrees, there are used 12, 17, 22, or more different values, then clearly these *Pythagorean Systems are many valued coding systems of information in music*, e.g., there are considered Pythagorean Systems Π_{12}, Π_{17}, Π_{22}, Π_{27}, Π_{31}, c.f. [125], [126], [129], [133]; the index denotes the number of values per octave. Here are values of Π_{17} and Π_{31}:

$\Pi_{17} = \Pi_{[-6;10]} = \langle[1/1], [256/243, 2\,187/2\,048], [9/8], [32/27, 19\,683/16\,384], [81/64], [4/3], [1\,024/729, 729/512], [3/2], [128/81, 6\,561/4\,096], [27/16], [16/9, 59\,049/32\,768], [243/128]\rangle$,

$\Pi_{31} = \Pi_{[-13;17]} = \langle[1/1, 531\,441/524\,288], [256/243, 2\,187/2\,048], [1\,162\,261\,467/1\,073\,741\,824, 9/8, 4\,782\,969/4\,194\,304], [32/27, 19\,683/16\,384], [3^{21}/2^{33}, \; 8\,192/6\,561, 81/64], [43\,046\,721/33\,554\,432, 4/3, 177\,147/131\,072], [1\,024/729, 729/512], [387\,420\,489/268\,435\,456, 3/2, 1\,594\,323/1\,048\,576], [128/81, 6\,561/4\,096], [3^{20}/2^{31}, 27/16, 14\,348\,907/8\,388\,608], [16/9, 59\,049/32\,768], [3^{22}/2^{34}, 4\,096/2\,187, 243/128], [129\,140\,163/67\,108\,864]\rangle$.

The values belonging to Garbuzov 12 granules are clustered with
[...]. For example, in the cluster $[32/27, 19\,683/16\,384]$ are two values for the minor third. In Π_{31}, the values $1/1, 531\,441/524\,288$ and
$129\,140\,163/67\,108\,864$ belong to one cluster. The using of individual
values chosen from a cluster depends on the musical context.

The perfect fifth $3/2$ and octave $2/1$ can be expressed as X^4Y^3
and X^7Y^5 respectively, where $X = 256/243 = 2^8/3^5$ are the minor
and $Y = 9/8 : X = 3^7/2^11$ the major Pythagorean semitones, c.f.
Section 6.1. In this section, Pythagorean System is studied when
couples of semitones need not be considered in rational numbers; we
deal with semitone couples (X, Y) in algebraic numbers expressed
with dth roots, $d = 1, 2, 3, \ldots$.

The semitones *limma* and *apotome* are unique rational numbers
which enables to express Pythagorean Tuning Π_{31} (and also Π_{12}, Π_{17},
Π_{22}, Π_{28}; i.e. *PYTH-12, PYTH-17, PYTH-22, PYTH-28, PYTH-31*)
for expression of Π_{17} via the *geometric net in rational numbers* with
some natural conditions, c.f. [133]. More precisely, in the form of the
geometric net

$$\langle \Gamma_i = X^{m_i}Y^{n_i}; \ i \in \mathbb{Z} \rangle \tag{6.2}$$

where X, Y are positive rationals, and $(m_i, n_i) \in I \subset \mathbb{Z}^2$ (possibly
not unique for every $i \in \mathbb{Z}$). I is a lattice such that

$$m_i + n_i = i, \ m_i \leq m_{i+1}, n_i \leq n_{i+1}, \ i \in \mathbb{Z}. \tag{6.3}$$

The characteristic equations for Pythagorean System (c.f. Preface,
Definition 1, (S3)) are

$$X^{m_{12}}Y^{n_{12}} = 2, \tag{6.4}$$

$$X^{m_7}Y^{n_7} = 3/2, \tag{6.5}$$

The condition (6.4) shows that an octave contains 12 granules.
The condition (6.5) asserts that the seventh granule contains the
perfect fifth (3/2).

Twelve qualitative music degrees, the 12 granulation of the octave is a characteristic attribute of the European tone systems and
culture. Each granule may contain one or more tone system values.
For instance, the geometric nets $\Pi_{17}, \Pi_{22}, \Pi_{28}, \Pi_{31}$ have 17, 22, 28,

31 values for 12 granules within the octave $[1, 2)$, respectively. There are different I for $\Pi_{12}, \Pi_{17}, \Pi_{22}, \Pi_{28}, \Pi_{31}$. Examples of index sets I for geometric nets we can see also in Figure 4.3. Since I are lattices, the index sets are directed both forward and backward (the two side geometric nets).

Pythagorean Systems $\Pi_{12}, \Pi_{17}, \Pi_{22}, \Pi_{28}, \Pi_{31}$ are geometric nets, the solutions of the equation system (6.2), (6.3), (6.4), (6.5) with rational (X, Y). Therefore, there arises a question: *does there exist any geometric net, the solution of the equation system* (6.2), (6.3), (6.4), (6.5) *with irrational* (X, Y).

6.5.1 The second level: algebraic numbers

Observe, [126], that Π_i, $i = 2, 3, 5, 7, 12, 17, 22, 28, 31$, can be obtained as the solution of the following Problem A and using the enlargement algorithm based on the fifth $X^{m_7} Y^{n_7} = 3/2$ and the octave $X^{m_{12}} Y^{n_{12}} = 2$, i.e.:

$$\Pi_{[k_1, k_2]} = \{(3/2)^k \pmod 2);\ k \in [k_1, k_2],\ k_1 < k_2,\ k_1, k_2 \in \mathbb{Z}\}.$$

For instance, $k = 1, 2, 3, 4, 5$ for P_5.
Problem A
To find X, Y rational numbers such that (6.4), (6.5) for some m_i, n_i nonnegative integer numbers such that

$$m_i + n_i = i,\ i = 0, 7, 12,\ 0 \leq m_7 \leq m_{12}, 0 \leq n_7 \leq n_{12}. \qquad (6.6)$$

Problem B. To find all triplets (m_{12}, m_7, d) in nonnegative integers such that

$$7m_{12} - 12m_7 = d, 0 \leq 7 - m_7 \leq 12 - m_{12}, 0 \leq m_7 \leq m_{12}.$$

The solution of Problem A is given by the solution of Problem B. Indeed, for $d = 0$, we have $m_{12} = 12, m_7 = 7$, i.e. E_{12} Equal Temperament $(X = Y = \sqrt[12]{2})$ which does not contain any perfect fifth $3/2$, (analogously for $d = 12$). If $d \neq 0$ and (m_{12}, m_7, d) is a solution of Problem B, then the solution of (6.4), (6.5) we obtain as follows:

$$X = 2^{E_{2,X}} 3^{E_{3,X}}, \quad Y = 2^{E_{2,Y}} 3^{E_{3,Y}}, \qquad (6.7)$$

where $E_{2,X} = (19 - m_7 - m_{12})/d$, $E_{3,X} = (m_{12} - 12)/d$, $E_{2,Y} = (-m_{12} - m_7)/d$, $E_{3,Y} = m_{12}/d$. The numbers X, Y are rational solutions of Problem A if and only if $d = 1$ (or $d = -1$, the symmetrical case). For $d = 1$, the values are $X = 256/243$, $Y = 3^7/2^{11}$.

Now it is possible to extend Pythagorean System (in any form: Π; Π_2, ..., Π_{31}, etc.) to algebraic numbers: to solve Problem A1.

Problem A1

To find X, Y in algebraic numbers such that (6.4), (6.5) for some m_i, n_i nonnegative integer numbers such that

$$m_i + n_i = i, \ i = 0, 7, 12,$$

$$0 \le m_7 \le m_{12}, 0 \le n_7 \le n_{12}.$$

Theorem 33 *For every $d \in \mathbb{Z}$, there exists unique or no solution (X_d, Y_d) of Problem A1 such that X_d, Y_d are expressed via d-th roots, multiplication, and/or dividing of numbers 2 and 3.*

Proof. The solutions of the Diophantine equation

$$7m_{12} - 12m_7 = d, \ 0 \le m_7 \le m_{12}, \tag{6.8}$$

where $d, m_7, m_{12} = 0, 1, 2, 3, 4 \ldots$ are collected in Table 6.4. By (6.3), we obtained values n_{12}, n_7. Take only the solutions such that $n_{12} \ge n_7$. Then the solutions of Problem B are collected in Table 6.5. By (6.7) we obtain the solutions of Problem A1 ($d < 0$ are the symmetrical cases). $\qquad\qquad\square$

Theorem 34 *Denote by $\mathcal{G}_d$ the class T of all tones generated by the solution (X_d, Y_d) of Problem A1, d in Table 6.5. Then*

$$\begin{aligned}
\mathcal{G}_1 &= \Pi = \{2^\alpha 3^\beta; \ \alpha, \beta \in \mathbb{Z}\}, \\
\mathcal{G}_2 &= \Pi \cup (\sqrt{2}) \cdot \Pi \\
\mathcal{G}_3 &= \Pi \cup (\sqrt[3]{4}) \cdot \Pi \cup (3\sqrt[3]{2}) \cdot \Pi
\end{aligned}$$

$$\cdots$$

$$\mathcal{G}_d = \bigcup_{i=1}^{d} \beta_{i,d} \cdot \Pi$$

$$\cdots$$

$$\mathcal{G}_{35} = \bigcup_{i=1}^{35} \beta_{i,35} \cdot \Pi,$$

where $\beta_{1,d}, \ldots, \beta_{d,d}$ are pairwise incommensurable algebraic numbers for every d.

TABLE 6.4: Solutions of the Diophantine equation (6.8)

m_7	d	
0	7, 14,21,28,35,	42,49,56,63,70,77
1	2,9,16,23,30,	37,44,51,58,65
2	4, 11,18,25,	32,39,46,53
3	6,13,20,	27,34,41
4	1,8,15	22, 29
5	3,10,	17
6	5	

TABLE 6.5: Solutions of Problem B

d	1	2	3	4	5	6	7	8	9	10	11	
m_{12}	7	2	9	4	11	6	1	8	3	10	5	
n_{12}	5	10	3	8	1	6	11	4	9	2	7	
m_7	4	1	5	2	6	3	0	4	1	5	2	
n_7	3	6	2	5	1	4	7	3	6	2	5	

d	13	14	15	16	18	20	21	23	25	28	30	35
m_{12}	7	2	9	4	6	8	3	5	7	4	6	5
n_{12}	5	10	3	8	6	4	9	7	5	8	6	7
m_7	3	0	4	1	2	3	0	1	2	0	1	0
n_7	4	7	3	6	5	4	7	6	5	7	6	7

TABLE 6.6: The square root: shifted black and white keys

C	$X_2^0 Y_2^0$	1	0.00	$U^0 K^0$	○
$C\sharp$	$X_2^0 Y_2^1$	$3\sqrt{8}$	101.96	$U^1 K^{1/2}$	●
D	$X_2^0 Y_2^2$	$9/8$	203.91	$U^2 K^1$	○
$E\flat$	$X_2^0 Y_2^3$	$27\sqrt{512}$	305.86	$U^3 K^{3/2}$	●
E	$X_2^0 Y_2^4$	$81/64$	407.82	$U^4 K^2$	○
F	$X_2^1 Y_2^4$	$4/3$	498.04	$U^5 K^2$	○
$F\sharp$	$X_2^1 Y_2^5$	$\sqrt{2}$	600.00	$U^6 K^{5/2}$	●
G	$X_2^1 Y_2^6$	$3/2$	701.96	$U^7 K^3$	○
$G\sharp$	$X_2^1 Y_2^7$	$9\sqrt{32}$	803.91	$U^8 K^{7/2}$	●
A	$X_2^1 Y_2^8$	$27/16$	905.86	$U^9 K^4$	○
$B\flat$	$X_2^1 Y_2^9$	$81\sqrt{2\,048}$	1 007.82	$U^{10} K^{9/2}$	●
B	$X_2^1 Y_2^{10}$	$243/128$	1 109.78	$U^{11} K^5$	○
C'	$X_2^2 Y_2^{10}$	2	1 200.00	$U^{12} K^5$	

Proof.   Consider the set

$$\Pi_{\mathbb{Z}} = \{(3/2)^k \ (\text{mod } 2); \ k \in \mathbb{Z}\}$$
$$= \{(X^{m_7} Y^{n_7})^k \ (\text{mod } X^{m_{12}} Y^{n_{12}}); \ k \in \mathbb{Z}\},$$

where $X > 0$, $Y > 0$ are suitable numbers.

Clearly, $\Pi_{\mathbb{Z}} = \Pi$ and $\Pi_{\mathbb{Z}} \supset \Pi_{[k_1, k_2]}$ for every $k_1, k_2 \in \mathbb{Z}$, $k_1 < k_2$.

Let $\alpha \in \mathbb{R}$ and $R \subset \mathbb{R}$ be a set. Denote by $\alpha R = \{v \in \mathbb{R}; \ v = \alpha r, r \in R\}$.

The assertion $\mathcal{G}_1 = \Pi$ is obvious for $d = 1$ when $X_1 = X = 256/243$ and $Y_1 = Y = (9/8) : X_1$.

For $d = 2$, $X_2 = 256/243$ and $Y_2 = 3\sqrt{8}$. We have (for illustration, c.f. Figure 6.15 and Table 6.6, where Π_7 is marked with ○ and $(\sqrt{2}) \cdot \Pi_5$ with ●, $U = 256/243$):

$$\Pi_{\mathbb{Z}} = \Pi = \bigcup_{p \in \mathbb{Z}} \bigcup_{q \in \mathbb{Z}} X_2^p Y_2^{2q}.$$

Then

$$\mathcal{G}_2 = \Pi \cup (Y_2) \cdot \Pi = \Pi \cup (\sqrt{2}) \cdot \Pi.$$

For $d = 3$, $X_3 = \sqrt[3]{32}/3$ and $Y_3 = 27/\sqrt[3]{16384}$. We have (for illustration, c.f. Figure 6.16 and Table 6.7, where Π_5 is marked with

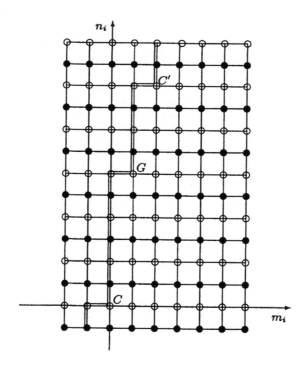

FIGURE 6.15: Generalized Pythagorean System $\mathcal{G}_d$, $d=2$

TABLE 6.7: The cubic root: 3 pentatonics

C	$X_3^0 Y_3^0$	1	0.00	$U^0 K^0$	○
$D_\flat$	$X_3^1 Y_3^0$	$\sqrt[3]{32}/3$	98.04	$U^1 K^{1/3}$	●
$C_\sharp$	$X_3^0 Y_3^1$	$27/\sqrt[3]{16384}$	105.86	$U^1 K^{2/3}$	■
D	$X_3^1 Y_3^1$	$9/8$	203.91	$U^2 K^1$	○
$E_\flat$	$X_3^2 Y_3^1$	$3/\sqrt[3]{16}$	301.96	$U^3 K^{4/3}$	●
$D_\sharp$	$X_3^1 Y_3^2$	$243/\sqrt[3]{8388608}$	309.78	$U^3 K^{5/3}$	■
E	$X_3^2 Y_3^2$	$81/64$	407.82	$U^4 K^2$	○
F	$X_3^3 Y_3^2$	$27/\sqrt[3]{8\,192}$	505.86	$U^5 K^{7/3}$	●
$F_\sharp$	$X_3^4 Y_3^2$	$9/\sqrt[3]{256}$	603.91	$U^6 K^{8/3}$	■
G	$X_3^5 Y_3^2$	$3/2$	701.96	$U^7 K^3$	○
$A_\flat$	$X_3^6 Y_3^2$	$\sqrt[3]{4}$	800.00	$U^8 K^{10/3}$	●
$G_\sharp$	$X_3^5 Y_3^3$	$81/\sqrt[3]{931072}$	807.82	$U^8 K^{11/3}$	■
A	$X_3^6 Y_3^3$	$27/16$	905.86	$U^9 K^4$	○
$B_\flat$	$X_3^7 Y_3^3$	$9/\sqrt[3]{128}$	1\,003.91	$U^{10} K^{13/3}$	●
B	$X_3^8 Y_3^3$	$3/\sqrt[3]{4}$	1\,101.96	$U^{11} K^{14/3}$	■
C'	$X_3^9 Y_3^3$	2	1\,200.00	$U^{12} K^5$	

○, $(\sqrt[3]{32}/3) \cdot \Pi_5$ with ●, and $(3/\sqrt[3]{4}) \cdot \Pi_5$ with ■, $U = 256/243$):

$$\Pi_Z = \Pi = \bigcup_{p \in Z} \bigcup_{q \in Z} X_3^p Y_3^p \cdot X_3^{3q}.$$

Then

$$\mathcal{G}_3 = \Pi \cup (X_3) \cdot \Pi \cup (X_3^2) \cdot \Pi = \Pi \cup (\sqrt[3]{4})\Pi \cup (3\sqrt[3]{2}) \cdot \Pi.$$

Analogously for every $d = 4, 5, 6, 7, 8, 9, 10, 11, 13, 14, 15, 16,$ 18, 20, 21, 23, 25, 28, 30, and 35. □

The solution of Problem A, c.f. [131], in the explicit form is the following:
$(X_1, Y_1) = (2^8/3^5, 3^7/2^{11}) = (256/243, 2\,187/2\,048),$
$(X_2, Y_2) = (\sqrt[2]{2^{16}/3^{10}}, \sqrt[2]{3^2/2^3}) = (256/243, 3/\sqrt[2]{8}),$
$(X_3, Y_3) = (\sqrt[3]{2^5/3^3}, \sqrt[3]{3^9/2^{14}}) = (\sqrt[3]{32}/3, 27/\sqrt[3]{16\,384}),$
$(X_4, Y_4) = (\sqrt[4]{2^{13}/3^8}, \sqrt[4]{3^4/2^6}) = (\sqrt[4]{8\,192}, 3/\sqrt[2]{8}),$
$(X_5, Y_5) = (\sqrt[5]{2^2/3^1}, \sqrt[5]{3^{11}/2^{17}}) = (\sqrt[5]{4/3}, \sqrt[5]{177\,147/131\,072})',$
$(X_6, Y_6) = (\sqrt[6]{2^{10}/3^6}, \sqrt[6]{3^6/2^9}) = (\sqrt[3]{32}/3, 3/\sqrt[2]{8}),$
$(X_7, Y_7) = (\sqrt[7]{2^{18}/3^{11}}, \sqrt[7]{3^1/2^1}) = (\sqrt[7]{262\,144/177\,147}, \sqrt[7]{3/2}),$

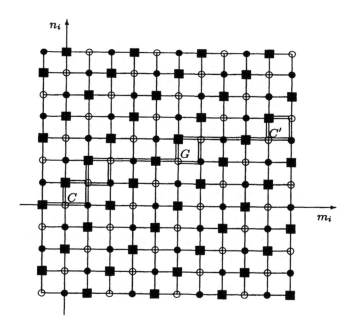

FIGURE 6.16: Generalized Pythagorean System $\mathcal{G}_d$, $d=3$

$(X_8, Y_8) = (\sqrt[8]{2^7/3^4}, \sqrt[8]{3^8/2^{12}}) = (\sqrt[8]{128/81}, 3/\sqrt[2]{8}),$

$(X_9, Y_9) = (\sqrt[9]{2^{15}/3^9}, \sqrt[9]{3^3/2^4}) = (\sqrt[3]{32}/3, \sqrt[9]{27/16}),$

$(X_{10}, Y_{10}) = (\sqrt[10]{2^4/3^2}, \sqrt[10]{3^{10}/2^{12}}) = (\sqrt[5]{4/3}, 3/\sqrt[5]{64}),$

$(X_{11}, Y_{11}) = (\sqrt[11]{2^{12}/3^7}, \sqrt[11]{3^5/2^7}) = (\sqrt[11]{4\,096/2\,187}, \sqrt[11]{243/128}),$

$(X_{13}, Y_{13}) = (\sqrt[13]{2^9/3^5}, \sqrt[13]{3^7/2^{10}}) = (\sqrt[13]{512/243}, \sqrt[13]{2\,187/1\,024}),$

$(X_{14}, Y_{14}) = (\sqrt[14]{2^{17}/3^{10}}, \sqrt[14]{3^2/2^2}) = (\sqrt[14]{131\,072/59\,049}, \sqrt[7]{3/2}),$

$(X_{15}, Y_{15}) = (\sqrt[15]{2^7/3^3}, \sqrt[15]{3^9/2^{13}}) = (\sqrt[15]{128/27}, \sqrt[15]{19687/8\,192}),$

$(X_{16}, Y_{16}) = (\sqrt[16]{2^{14}/3^8}, \sqrt[16]{3^4/2^5}) = (\sqrt[8]{128/3}, \sqrt[16]{81/32}),$

$(X_{18}, Y_{18}) = (\sqrt[18]{2^{11}/3^6}, \sqrt[18]{3^6/2^8}) = (\sqrt[18]{2\,048/729}, \sqrt[9]{27/16}),$

$(X_{20}, Y_{20}) = (\sqrt[20]{2^8/3^4}, \sqrt[20]{3^8/2^{11}}) = (\sqrt[5]{4/3}, \sqrt[20]{6\,561/2\,048}),$

$(X_{21}, Y_{21}) = (\sqrt[21]{2^{16}/3^9}, \sqrt[21]{3^3/2^3}) = (\sqrt[21]{65\,536/19\,683}, \sqrt[7]{3/2}),$

$(X_{23}, Y_{23}) = (\sqrt[23]{2^{13}/3^7}, \sqrt[23]{3^5/2^6}) = (\sqrt[23]{8\,192/2\,187}, \sqrt[23]{243/64}),$

$(X_{25}, Y_{25}) = (\sqrt[25]{2^{10}/3^5}, \sqrt[25]{3^7/2^9}) = (\sqrt[5]{4/3}, \sqrt[25]{2\,187/512}),$

$(X_{28}, Y_{28}) = (\sqrt[28]{2^{15}/3^8}, \sqrt[28]{3^4/2^4}) = (\sqrt[28]{32\,768/6\,561}, \sqrt[7]{3/2}),$

$(X_{30}, Y_{30}) = (\sqrt[30]{2^{12}/3^6}, \sqrt[30]{3^6/2^7}) = (\sqrt[5]{4/3}, \sqrt[30]{729/128}),$

$(X_{35}, Y_{35}) = (\sqrt[35]{2^{14}/3^7}, \sqrt[35]{3^5/2^5}) = (\sqrt[5]{4/3}, \sqrt[7]{3/2}).$

6.5.2 Semitone metric space

Now we show that Pythagorean tone systems generalized to rational numbers in the previous section are closer in some sense to Equal Temperament the greater the positive integer d is.

Theorem 35 *For* $\mathcal{K} = 53\,441\,524/524\,288 = X_1/Y_1$ *(Pythagorean comma, comat), denote by* $\rho(X, Y) = |\log_{\mathcal{K}}(X) - \log_{\mathcal{K}}(Y)|$, $X, Y \in \mathbb{R}^+$. *Let* (X_{d_1}, Y_{d_1}), (X_{d_2}, Y_{d_2}) *are solutions of* Problem A1 *corresponding to* $d_1 < d_2$, $d_1, d_2 = 1, 2, \ldots$. *Then* $\rho(X_1, Y_1) > \rho(X_2, Y_2)$.

Proof.   It can be verified directly that $(\mathbb{R}^+, \rho)$ is a metric space.

If (X_d, Y_d) is a couple of numbers which is a solution of Problem A1 corresponding to d, $d = 0, 1, 2, 3, \ldots$, then by (6.7),

$$
\begin{aligned}
|\log_{\mathcal{K}}(Y_d) - &\ \log_{\mathcal{K}}(X_d)| \\
&= \log_{\mathcal{K}} 2^{\frac{-m_{12}-m_7}{d}} 3^{\frac{m_{12}}{d}} - \log_{\mathcal{K}} 2^{\frac{19-m_{12}-m_7}{d}} 3^{\frac{m_{12}-12}{d}} \\
&= \tfrac{12}{d} \log_{\mathcal{K}} 3 - \tfrac{19}{d} \log_{\mathcal{K}} 2 \\
&= \tfrac{1}{d} \log_{\mathcal{K}} \tfrac{3^{12}}{2^{19}} = \tfrac{1}{d} \log_{\mathcal{K}} \mathcal{K} = \tfrac{1}{d}.
\end{aligned}
$$

If $d_1 < d_2$, then

$$\rho(X_{d_1}, Y_{d_1}) = \frac{1}{d_1} > \frac{1}{d_2} = \rho(X_{d_2}, Y_{d_2}). \qquad \square$$

Corollary 36 *If $Y_d > X_d$, then $Y_d/X_d = \sqrt[d]{K}$.*

Definition 16 Denote by $\mathcal{X}$ the set of all numbers $X \in \mathbb{R}^+$ such that (X, Y) is a solution of Problem A1 for some integer m_7, m_{12}, d and $Y \in (0, \infty)$. We will call the metric space $(\mathcal{X}, \rho)$ *the metric space of semitones generating the perfect fifths.*

6.5.3 Searching in transcendental numbers

Problem T. To solve Problem A in real numbers.
  Thus is, to find all couples (X, Y) in real numbers such that (6.4), (6.5) for some m_i, n_i nonnegative integer numbers.

Theorem 37 *Let $m_{12}, n_{12}, m_7, n_7 \in \mathbb{Z}$ be numbers such that (6.4), (6.5) for some real numbers $X > 0$, $Y > 0$. Denote by*

$$d = \begin{vmatrix} m_{12} & n_{12} \\ m_7 & n_7 \end{vmatrix}.$$

Then X, Y are algebraic numbers and

$$X = \sqrt[d]{\frac{2^{n_7+n_{12}}}{3^{n_{12}}}}, Y = \sqrt[d]{\frac{3^{m_{12}}}{2^{m_{12}+m_7}}}.$$

Proof. The trivial cases $d = 0$ and $m_7 m_{12} = 0$ contradict (6.4) or (6.5). Without loss of generality suppose $d \neq 0$ and $m_7 m_{12} \neq 0$.
  Consider the following bijection $\ell : Z \mapsto \lambda$, where

$$Z = 2^{8-19\lambda} 3^{-5+12\lambda}, \ \lambda \in (-\infty, +\infty), Z \in (0, +\infty). \qquad (6.9)$$

  Denote by λ_X, λ_Y the corresponding values of λ to X, Y in bijection ℓ, respectively. So,

$$X = 2^{8-19\lambda_X} 3^{-5+12\lambda_X}, \qquad (6.10)$$

$$Y = 2^{8-19\lambda_Y} 3^{-5+12\lambda_Y}.$$

By (6.4) and (6.5),

$$\begin{array}{rcl} X^{m_7} \cdot (2^{8-19\lambda_Y} 3^{-5+12\lambda_Y})^{7-m_7} & = & 3/2, \\ X^{m_{12}} \cdot (2^{8-19\lambda_Y} 3^{-5+12\lambda_Y})^{12-m_{12}} & = & 2. \end{array} \qquad (6.11)$$

By (6.11),

$$X = \frac{2^{1/m_{12}}}{(2^{8-19\lambda_Y} 3^{-5+12\lambda_Y})^{n_{12}/m_{12}}} = \frac{(3/2)^{1/m_7}}{(2^{8-19\lambda_Y} 3^{-5+12\lambda_Y})^{n_7/m_7}}.$$

This implies

$$\frac{1}{m_{12}} \log_{\mathcal{K}} 2 - \frac{n_{12}}{m_{12}} R = \frac{1}{m_7} \log_{\mathcal{K}}(3/2) - \frac{n_7}{m_7} R,$$

where $R = \log_K Y$ for some $K \neq 1$, $K > 0$. We have:

$$\frac{1}{m_{12}} \log_K 2 - \frac{1}{m_7} \log_K(3/2) = R\left(\frac{n_{12}}{m_{12}} - \frac{n_7}{m_7}\right) = -R\frac{d}{m_{12} m_7}.$$

Consequently,

$$\begin{aligned} R & = \frac{-m_7 \log_K 2 + m_{12} \log_K(3/2)}{d} \\ & = (8 - 19\lambda_Y) \log_K 2 + (-5 + 12\lambda_Y) \log_K 3. \end{aligned}$$

Now,

$$\lambda_Y(-19 \log_K 2 + 12 \log_K 3) = \frac{\dfrac{-m_7 \log_K 2 + m_{12} \log_K(3/2)}{d}}{-8 \log_K 2 + 5 \log_K 3}.$$

Put $K = 3^{12}/2^{19} = 531\,441/524\,288 = \mathcal{K}$ (i.e., the *Pythagorean comma*) and $U = 2^8/3^5 = 256/243$ (i.e., the *minor Pythagorean semitone*). Then $-19 \log_{\mathcal{K}} 2 + 12 \log_{\mathcal{K}} 3 = \log_{\mathcal{K}} \mathcal{K} = 1$ and $8 \log_{\mathcal{K}} 2 + 5 \log_{\mathcal{K}} 3 = \log_{\mathcal{K}} U$ and

$$\lambda_Y = \frac{\begin{vmatrix} m_{12} & \log_{\mathcal{K}} 2 \\ m_7 & \log_{\mathcal{K}} 3/2 \end{vmatrix}}{d} - \log_{\mathcal{K}} U.$$

Symmetrically,

$$
\lambda_X = \frac{\begin{vmatrix} n_{12} & \log_{\mathcal{K}} 2 \\ n_7 & \log_{\mathcal{K}} 3/2 \end{vmatrix}}{\begin{vmatrix} \log_{\mathcal{K}} n_{12} & m_{12} \\ \log_{\mathcal{K}} n_7 & m_7 \end{vmatrix}} - \log_{\mathcal{K}} U = \frac{\begin{vmatrix} \log_{\mathcal{K}} 2 & n_{12} \\ \log_{\mathcal{K}} 3/2 & n_7 \end{vmatrix}}{d} - \log_{\mathcal{K}} U.
$$

(6.10) implies (is equivalent to) $\lambda_X = \log_{\mathcal{K}} \frac{X}{U}$. Then $\log_{\mathcal{K}} X = \lambda_X + \log_{\mathcal{K}} U$. Analogously for Y. We have:

$$
X = \mathcal{K}^{\frac{\begin{vmatrix} \log_{\mathcal{K}} 2 & n_{12} \\ \log_{\mathcal{K}} 3/2 & n_7 \end{vmatrix}}{d}}, \quad Y = \mathcal{K}^{\frac{\begin{vmatrix} m_{12} & \log_{\mathcal{K}} 2 \\ m_7 & \log_{\mathcal{K}} 3/2 \end{vmatrix}}{d}}.
$$

Finally,

$$
X = \mathcal{K}^{\frac{\begin{vmatrix} \log_{\mathcal{K}} 2 & n_{12} \\ \log_{\mathcal{K}} 3/2 & n_7 \end{vmatrix}}{d}} = \mathcal{K}^{\log_{\mathcal{K}} \frac{2^{n_7/d}}{(3/2)^{n_{12}/d}}} = \sqrt[d]{\frac{2^{n_7 + n_{12}}}{3^{n_{12}}}}.
$$

Analogously,

$$
Y = \sqrt[d]{\frac{3^{m_{12}}}{2^{m_{12} + m_7}}}.
$$

So, both $X = X(d)$ and $Y = Y(d)$ are algebraic numbers. $\quad\square$

Corollary 38 *The solution of* Problem T *coincides with the solution of* Problem A.

6.5.4 Adaptive Pythagorean system

According to the Garbuzov theory, c.f.[19], tempo/rhythm in western culture is structured also into 12 main zones. After the Garbuzov's theory about pitch zones, the faster the tempo/rhythm, the closer the tone system to equal temperament. It is natural to ask about a correlation between tempo/rhythm and Pythagorean tone system when semitones are considered in real numbers. We suggest via Table 6.8 such a relation, where $X = 256/243$ and $\mathcal{K}$ is the Pythagorean comma. Note that to a given scale there exists (possibly not

TABLE 6.8: Adaptive Pythagogrean system

tempo/rhythm	M 1/4	minor semitone	major semitone
grave	40–45	$X_1 = X$	$Y_1 = X \cdot K$
largo	46–51	$X_2 = X$	$Y_2 = X \cdot K^{1/2}$
lento	52–55	$X_3 = X \cdot K^{2/3}$	$Y_3 = X \cdot K^{1/3}$
adagio	56–62	$X_4 = X \cdot K^{2/4}$	$Y_4 = X \cdot K^{1/4}$
larghetto	63–71	$X_5 = X \cdot K^{2/5}$	$Y_5 = X \cdot K^{1/5}$
andante	72–79	$X_6 = X \cdot K^{3/6}$	$Y_6 = X \cdot K^{1/6}$
andantino	80–91	$X_7 = X \cdot K^{3/7}$	$Y_7 = X \cdot K^{1/7}$
moderato	92–107	$X_8 = X \cdot K^{4/8}$	$Y_8 = X \cdot K^{1/8}$
allegretto	108–131	$X_9 = X \cdot K^{4/9}$	$Y_9 = X \cdot K^{1/9}$
allegro	132–159	$X_{10} = X \cdot K^{5/10}$	$Y_{10} = X \cdot K^{1/10}$
vivace	160-183	$X_{11} = X \cdot K^{5/11}$	$Y_{11} = X \cdot K^{1/11}$
presto	184–207	$X_{12} = \sqrt[12]{2}$	$Y_{12} = \sqrt[12]{2}$

unique) a related timbre, c.f. [60]. In Table 6.8, the tempo values are according to the Maentzel metronome scale (M=1/4). The tempo descriptions (i.e., grave, largo, etc.) are linguistic names of fuzzy sets and interval zones (the supports of the fuzzy sets) after the Maentzel metronome are made artificially disjoint. In fact, they are, of course, not overlapping. Note also, that tempo/rhythm values are of exponential nature (like pitch values). The tone system described with this table represents a synergetic join of the two trend of "modern" tone systems, c.f. Section 1.2, for the "Pythagorean Gothic harmony", c.f. [196].

Remarks and ideas for exploring

4 For fifths of equal temperaments, c.f. Subsection 5.3.3.

5 For $d = 2$, c.f. Table 6.6 and Figure 6.15. The set Π_7 is the Pythagorean heptatonic F, C, G, D, A, E, B (the "white keys"). The set $(\sqrt{2}) \cdot \Pi_5$, is a Pythagorean pentatonic $F\sharp$, $C\sharp$, $G\sharp$, $D\sharp$, $A\sharp$ (the "black keys").

6 For $d = 3$, c.f. Table 6.7 and Figure 6.16, we have many valued tone systems. The sets Π_5 (i.e. $\{C, G, D, A, E\}$), $(3/\sqrt[3]{4}) \cdot \Pi_5$, (i.e. $\{B, F\sharp, C\sharp, G\sharp, D\sharp\}$), and $(\sqrt[3]{32}/3) \cdot P_5$ (i.e. $\{D\flat, A\flat, E\flat, B\flat, F\}$) are three Pythagorean pentatonics.

7 The sensitivity of human perception apparat is 5–6 cents. The psycho-acoustical boundary (to what the listener or player tends to) is 2–3 cents. This is the reason why there is no practical sense to use geometric nets for $d > 12$. Indeed, we hear no difference between E_{12} and values of the constructed geometric net (we do not suppose too many values within 12 qualitative degrees). From the acoustical point of view, the interesting are cases of small values of $d = 1,2,3,\ldots,$ 11 for which $Y_d/X_d \approx 23.46, 11.73, 7.82, 5.86, 4.69, 3.91, 3.35,$ $\ldots,$ 2.13 cents, respectively. From the viewpoint of modulations, the ordering is reverse: $d = 11, 10, \ldots 3, 2, 1$. So, we have to choose an appropriate compromise of d depending on the kind of modulations within a given musical composition (set of compositions).

8 Consider the "anatomy" of the whole Praetorius tone, i.e. find all real $X > 0$, $Y > 0$ such that $X^{m_{12}} Y^{n_{12}} = 2$, $X^{m_4} Y^{n_4} = 5/4$, such that $m_{12} + n_{12} = 12$, $m_4 + n_4 = 4$ and $0 \le m_4 \le m_{12}, 0 \le n_4 \le n_{12}$.

Which role does comma of Didymus play here? Are there rational couples (X, Y)?

9 Sections 6.3.3 and 6.3.4 are only a very short introduction into the problematics of keyboards. For more detailed explanation about historical keyboards and als, c.f.

Chapter 7

Well tempered systems

7.1 Well temperament periods

7.1.1 Historical well temperaments

In fact, the acoustical disadvantages of E_{12} started a new development activity in tone systems, the so-called well temperaments. The problem of well temperaments is to permit all relevant intervals in a given style to be playable in any transposition. All keys are usable, but the sizes of the intervals between various degrees of the scale are not exactly the same in each key. This is thought of as an advantage, since it lends a distinctive character to the different keys (the idea similar to Indian ragas).

As historical well temperaments we denote the subclass of the class of all well temperaments such that every scale has 12-tones per octave. Historical well temperaments generalize the subclass of meantones having a sequence of 11 equally tempered fifth and one single wolf fifth. Maybe, the simplest and satisfactorily general definition of the class of all historical well temperaments can be formulated as follows:

Definition 17 Let $S = \{C, C_\sharp, D, E_\flat, E, F, F_\sharp, G, G_\sharp, A, B_\flat, B, C'\}$ be an increasing sequence of reals such that $C = 1$, $C' = 2$. Denote by

$$\Theta_0, \Theta_1, \ldots, \Theta_{11} = \begin{array}{l} G/C, 2D/G, A/D, 2E/A, B/E, 2F_\sharp/B, \\ 2C_\sharp/F_\sharp, G_\sharp/C_\sharp, 2E_\flat/G_\sharp, B_\flat/E_\flat, 2F/B_\flat, 2C/F, \end{array}$$

respectively. Let $\varepsilon > 0$. We say that S is a ε-*well tempered historical temperament* if

$$\frac{\Pi_{i=0}^{11}\Theta_i}{(3/2)^{12}} = \mathcal{K},$$

where $\mathcal{K} = 3^{12}/2^{19} = 531441/524258$ (Pythagorean comma), and

$$\left| \log_\mathcal{K} \frac{\Theta_i}{3/2} \right| \le \varepsilon, \quad \forall i = 0, 1, \ldots, 11.$$

Usually, $\varepsilon \le 1/2$, c.f. *KIRNBERGER(NO.2)-12*. We see also that historical well temperaments may have more than one tempered fifth; consequently, many values of semitones (and other intervals). For illumination, in Figure 7.1 and Figure 7.2, there are some historical well temperaments where numbers in the figures are logarithms of fifth temperatures, i.e. $\log_\mathcal{K} \frac{\Theta_i}{3/2}$, $\mathcal{K}$ is Pythagorean comma, $i = 0, 1, \ldots, 11$.

Example 49 A well-tempered tone system adopted for organs in the time of Bach is known as Werkmeister III (*WERCK(NO.3)-12* in our collection of tone systems). In this tone system, the fifths G/C, D/G, A/D and $F_\sharp/B_\flat$ are each tempered by 1/4 of the Pythagorean comma (approximately 5.87 cents). In Table 7.1, there are schemes of spread Pythagorean comma of some historic well temperaments. In the tables of this subsection, τ denotes the temperature in cents from the 12-tone Equal Temperament when A is the common point of the both tone systems (the pitch standard, 440 Hz physically). In *WERC(NO.3)-12*, $F_\sharp$ (and, of course A) are the same as in 12-tone Equal Temperament, and the others are all "sharper." The major thirds are stretched by 2/12, 5/12, 8/12, or 11/12 of the Pythagorean comma, and minor thirds flattened by 2/12, 5/12, 8/12, or 11/12 of it.

Example 50 Another popular well-tempered tone system is the Vallotti tone system, historically accurate for music of Mozart's time.

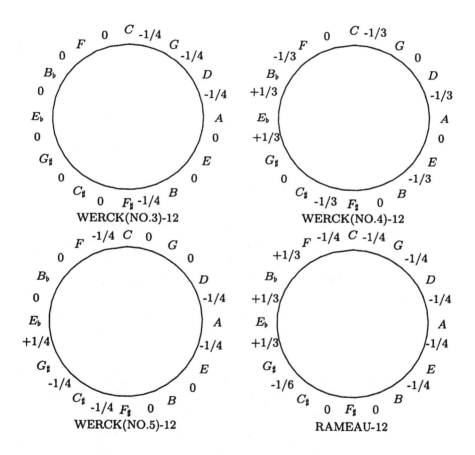

FIGURE 7.1: Historic well temperaments (Werckmeister, Rameau)

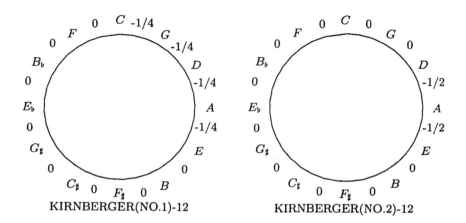

FIGURE 7.2: Historic well temperaments (Kirnberger II)

In Valotti tone system (*VALLOTTI-12* in our notation system), c.f. Table 7.1 the fifths C/F, G/C, D/G, A/D, E/A, B/E are each tempered by 1/6 of the Pythagorean comma (approx. 3.91 cents). The tempered fifths are then 698.04 cents and the tempered fourths are 501.96 cents. The result is that major thirds are stretched by 3/12, 5/12, 7/12, 9/12, or 11/12 of the Pythagorean comma, and minor thirds flattened by 5/12, 7/12, 9/12, or 11/12 of it.

Example 51 A pair of tone systems devised by Charles Fisk and Harald Vogel, *FISK/VOGEL(NO.1)-12* and *FISK/VOGEL(NO.2)-12*, for the Fisk organ in Memorial Church at Stanford University, which can be switched by a lever above the music desk from meantone to well-tempered; the lever activates an alternate set of tracker mechanisms for the sharp keys, so that there are 17 pipes per octave. Stanford's Fisk organ uses a modified meantone system which shares the natural keys with the well-tempered tone system. This means that the six intervals C/F, G/C, D/G, A/D, E/A, B/E are identical in both systems; they are flattened by 1/5 of the Pythagorean comma (approx. 4.69 cents). The five intervals B_b/F, E_b/B_b and $C_\sharp/G_\sharp$, $F_\sharp/C_\sharp$, $B/F_\sharp$ are flattened by 1/4 of the syntonic comma (approx.

	cents	τ		cents	τ
C	0.00	11.73	C	0.00	5.87
$C\sharp$	90.22	1.96	$C\sharp$	94.13	0.00
D	192.18	3.91	D	196.09	1.96
$D\sharp$	294.13	5.87	$D\sharp$	298.04	3.91
E	390.22	1.96	E	392.18	-1.96
F	498.04	9.78	F	501.96	7.82
$F\sharp$	588.27	0.00	$F\sharp$	592.18	-1.96
G	696.09	7.82	G	698.04	3.91
$G\sharp$	792.18	3.91	$G\sharp$	796.09	1.96
A	888.27	0.00	A	894.13	0.00
$A\sharp$	996.09	7.82	$A\sharp$	1 000.00	5.87
B	1092.18	3.91	B	1090.22	-3.91
C'	1 200.00	11.73	C'	1 200.00	5.87

TABLE 7.1: Werkmeister III and Valloti

5.38 cents). So natural fifths are tuned to 697.26 cents and "other" fifths are tuned to 696.58 cents, except the wolf which must be seven octaves minus six natural fifths minus five accidental fifths (approx. 733.53 cents). Here the wolf is shown inverted as the fourth of 466.47 cents, between $G\sharp$ and $E_\flat$. All natural fourths are 502.74 cents, and other fourths are 503.42 cents, c.f. Table 7.2.

These should be compared with quarter-comma meantone; the differences are hardly ever more than a couple of cents. The nifty thing is that by switching only the sharps, you get the well-tempered tone system described next, c.f. Table 7.2. Stanford's Fisk organ uses a well-tempered tone system which shares the natural keys with the modified meantone system. This means that the six intervals

$$C/F, \quad G/C, \quad D/G, \quad A/D, \quad E/A, \quad B/E$$

are identical in both systems; they are flattened by 1/5 of the Pythagorean comma. $C\sharp/G\sharp$ is similarly flattened in the well-tempered tone system. The three intervals $E_\flat/B_\flat$, $G\sharp/E_\flat$ and $B/F\sharp$ are pure; the two intervals $B_\flat/F$ and $F\sharp/C\sharp$ are sharpened (widened) by one-fifth Pythagorean comma. So natural fifths are tuned to 697.26 cents, and wide fifths are tuned to 706.65 cents, compared with pure fifths of 701.96 cents. All natural fourths are 502.74 cents, and wide fourths are 493.35 cents, compared with pure fourths of 498.04 cents.

	cents	τ		cents	τ
C	0.00	8.21	C	0.00	8.21
$C\sharp$	79.47	-12.32	$C\sharp$	94.92	3.13
D	194.53	2.74	D	194.53	2.74
$D\sharp$	309.58	17.79	$D\sharp$	294.13	2.35
E	389.05	-2.74	E	389.05	-2.74
F	502.74	10.95	F	502.74	10.95
$F\sharp$	582.89	-8.90	$F\sharp$	588.27	-3.52
G	697.26	5.47	G	697.26	5.47
$G\sharp$	776.05	-15.74	$G\sharp$	792.18	0.39
A	891.79	0.00	A	891.79	0.00
$A\sharp$	1006.16	14.37	$A\sharp$	996.09	4.30
B	1086.31	-5.47	B	1086.31	-5.47
C'	1 200.00	8.21	C'	1 200.00	8.21

TABLE 7.2: Fisk–Vogel I and Fisk–Vogel II

Example 52 It has been suggested that J. S. Bach wrote "The Well-Tempered Clavier" for 12-tone Equal Temperament but this is erroneous. Research by the American musicologist J. Barnes in the 1970s shows that what Bach probably used was a variation on meantone temperament devised independently by F. Vallotti and Th. Young. It is almost certain that Bach knew of the existence of equal temperament but would have never used it himself as it would have been impractical to tune a clavichord this way since its pitch alters depending on how hard the keys are struck; in extreme circumstances, the pitch can vary by up to a minor 3rd.

YOUNG-12; Young's Well-Tempered Piano;

567/512, 9/8, 147/128, 21/16, 1 323/1 024, 189/128, 3/2, 49/32, 7/4, 441/256, 63/32, 2

BACH/BARNES-12; Barnes–Bach, variation of YOUNG-12, likely meant for Das Wohltemperierte Klavier;

256/243, 196.198, 32/27, 392.072, 4/3, 592.200, 698.025, 128/81, 894.223, 16/9, 1 090.245, 2

BARNES-12; J. Barnes' temperament (1979);

94.135, 196.090, 298.045, 392.180, 501.955, 592.180, 698.045, 796.090, 894.135, 1 000.000, 1 094.135, 1 200.000

In the following example we bring some other typical tone systems of the historical period of well temperaments.

Example 53

ANONYMUS-12; Anonymus: Pro clavichordiis faciendis, Erlangen 15th century;
256/243, 4 096/3645, 32/27, 5/4, 4/3, 1 024/729, 3/2, 128/81, 2 048/1 215, 16/9, 15/8, 2

CAUS(NO.1)-12; A 12-tone just scale by the 17th century inventor and philosopher S. de Caus;
126/121, 10/9, 75/64, 5/4, 4/3, 25/18, 3/2, 25/16, 5/3, 16/9, 15/8, 2

CAUS(NO.2)-12; De Caus (a mode of Ellis's DUODENE);
25/24, 10/9, 75/64, 5/4, 4/3, 25/18, 3/2, 25/16, 5/3, 16/9, 15/8, 2

CHOQUEL-12; Choquel/Barbour/Marpurg?;
25/24, 9/8, 6/5, 5/4, 4/3, 45/32, 3/2, 25/16, 5/3, 20/11, 15/8, 2

ERLANGEN-12; Revised Erlangen;
135/128, 9/8, 32/27, 5/4, 4/3, 45/32, 3/2, 405/256, 27/16, 16/9, 15/8, 2

EULER-12; Euler (a mode of Ellis's duodene), genus [33355];
25/24, 9/8, 5/64, 5/4, 4/3, 45/32, 3/2, 25/16, 5/3, 225/128, 15/8, 2

FOGLIANO(NO.1)-12; Fogliano I;
25/24, 10/9, 6/5, 5/4, 4/3, 25/18, 3/2, 25/16, 5/3, 16/9, 15/8, 2

FOGLIANO(NO.2)-12; Fogliano II, also Mandelbaum;
25/24, 9/8, 6/5, 5/4, 4/3, 25/18, 3/2, 25/16, 5/3, 9/5, 15/8, 2

KEPLER(NO.1)-12; Kepler I;
135/128, 9/8, 6/5, 5/4, 4/3, 45/32, 3/2, 405/256, 27/16, 9/5, 15/8, 2

KEPLER(NO.2)-12; Kepler II;
135/128, 9/8, 6/5, 5/4, 4/3, 45/32, 3/2, 8/5, 27/16, 9/5, 15/8, 2

KIRNBERGER(NO.1)-12; Kirnberger I;
256/243, 9/8, 32/27, 5/4, 4/3, 45/32, 3/2, 128/81, 895.112, 16/9, 15/8, 2

KIRNBERGER(NO.2)-12; Kirnberger's scale;
256/243 193.849, 32/27, 386.314, 4/3, 45/32, 696.663, 128/81, 889.650, 16/9, 15/8, 2

LAMBERT-12; Lambert's temperament (1774);
93.576, 197.207, 297.486, 394.414, 501.396, 591.621, 698.604, 795.531, 895.811, 999.441, 1 093.018, 1 200.000

MERSEN(LUTE/NO.1)-12; Mersenne lute I;
16/15, 10/9, 6/5, 5/4, 4/3, 64/45, 3/2, 8/5, 5/3, 9/5, 15/8, 2

MERSEN(LUTE/NO.2)-12; Mersenne lute II;
16/15, 9/8, 6/5, 5/4, 4/3, 64/45, 3/2, 8/5 5/3, 9/5 15/8, 2

MERSEN(SPINET/NO.1)-12; Mersenne spinet I;
16/15, 10/9, 6/5, 5/4, 4/3, 64/45, 3/2, 8/5, 5/3, 16/9, 15/8, 2

MERSEN(SPINET/NO.2)-12; Mersenne spinet II;
25/24, 9/8, 75/64, 5/4, 4/3, 25/18, 3/2, 25/16, 5/3, 16/9, 15/8, 2

NEIDHARDT-12; Neidhardt temperament (1724);
96.090, 196.090, 298.045, 394.135, 500.000, 596.090, 698.045, 796.090, 894.135, 1 000.000, 1 096.090, 1 200.000

RAMEAU-12; Rameau scale (1725);
86.803, 193.157, 297.801, 386.314, 503.421, 584.848, 696.579, 788.758, 889.735, 1 006.843, 1 082.892, 1 200.000

RAMEAU(NS)-12; C.f. Rameau, book Nouveau Systeme, 1726;
92.472, 193.157, 302.053, 5/4, 503.422, 587.682, 696.578, 797.263, 889.735, 1 006.843, 1 082.892, 2

RAMIS-12; Ramis's Monochord;
135/128, 10/9, 32/27, 5/4, 4/3, 45/32, 3/2, 8/5, 5/3, 16/9, 15/8, 2

ROMIEU-12; Romieu;
25/24, 9/8, 6/5, 5/4, 4/3, 45/32, 3/2, 25/16, 5/3, 16/9, 15/8, 2

ROUSSEAU-12; Rousseau;
25/24, 9/8, 6/5, 5/4, 4/3, 25/18, 3/2, 8/5, 5/3, 125/72, 15/8, 2

VALLOTI/YOUNG(NO.1)-12; Vallotti & Young I (Vallotti version);
94.135, 196.090, 298.045, 392.180, 501.955, 592.180, 698.045, 796.090, 894.135, 1 000.000, 1 090.225, 1 200.000

VALLOTI/YOUNG(NO.2)-12; Vallotti & Young II (Young version);
256/243, 196.090, 32/27, 392.180, 4/3, 1 024/729, 698.045, 128/81, 894.135, 16/9, 1 090.225, 2

WERC(NO.1)-20; A. Werckmeister I (Just Intonation);
25/24, 16/15, 10/9, 9/8, 75/64, 6/5, 5/4, 4/3, 25/18, 45/32, 3/2, 25/16, 8/5, 5/3, 27/16, 225/128, 16/9, 9/5, 15/8, 2

WERC(NO.3)-12;
A. Werckmeister's temperament III (the most famous one, 1681);

256/243, 192.180, 32/27, 390.225, 4/3, 1 024/729, 696.090, 128/81, 888.270, 16/9, 1 092.180, 2

WERC(NO.4)-12;
A. Werckmeister's temperament IV;
82.406, 196.090, 32/27, 392.180, 4/3, 1 024/729 694.135, 784.361, 890.225, 1 003.910, 4 096/2 187, 2

WERC(NO.5)-12;
A. Werckmeister's temperament V;
96.091, 9/8, 300.000, 396.090, 503.910, 600.000, 3/2, 128/81, 900.000, 1 001.955, 1 098.045, 2

WERC(NO.6)-12; A. Werckmeister's "septenarius" tone system VI;
98/93, 49/44, 196/165, 49/39, 4/3, 196/139, 196/131, 49/31, 196/117, 98/55, 49/26, 2

WERC(NO.6/DUPONT)-12; A. Werckmeister's VI, Dupont version (1935);
256/243, 187.153, 297.486, 394.414, 4/3, 594.973, 698.604, 792.180, 892.459, 999.441, 1 096.369, 2

7.1.2 More than 12

From a time perspective, the so-called *historical well-temperaments* are a family of tone systems spanning the period of about 1680–1885. A well-temperament, in a broadest sense is any temperament that permits all relevant intervals in a given musical style playable in any transposition (a temperament "avoiding wolves"). By this test, E_{12} (was known before 1620) is a special and uncharacteristic case of well-temperament.

More typically, a well-temperament involves tempering different fifths by differing amounts, thus producing a variety of sizes and "colors" for thirds and other derivative intervals. This shading and 12 tones per octave is the distinguishing feature of well-temperaments in their classic form (the first, *historical* period of well temperaments).

More than 12 tones per octave, this is a typical attribute for the second period of the well tempered tone systems. Therefore, let us refer to this period as period *more than 12*. The key persons are Huygens, Petzval, Hába, Sumec, Wilson, Helmholtz, Wilson, Ellis, etc. This family of tone system is spanning the time period of about 1850–1960 and certainly ends with the mass extension of personal computers.

The most imposing person of the period *more than 12* is Ch. Huygens (1629–1695). E_{12} was propagated in the 2nd half of the 17th century and fully advocated maintaining the triad with the pure third. The result of Huygens' investigations, arising from dissatisfaction with the well-tempered tone system E_{12}, was the discovery of a "Nouveau Cycle Harmonique," which did not close after 12, but after 31 steps. He called the smallest interval in this 31-part system the diese, the fifth part of a whole tone. He showed that the division of the octave into 31 parts had all the advantages of 12-tone well-tempered tone system, particularly unlimited possibilities of modulation, without the triad being marred in an unacceptable way through the third that was tempered to far too great an extent. Huygens stated himself that in his 31-tone system each diese contains all consonances in a rising and falling tone and in all cases in an equally pure way. A distinction is made in his system between the large (major) and small (minor) semitone, being respectively diatonic and chromatic. The large semitone containing 3/5 and the small semitone 2/5 of the 5-part whole tone. An important part in his 31-tone cycle is the presence of the harmonic seventh, the seventh in the series of natural tones. The sign with the seventh tone points to the lowering of 1/5 interval, one diese. The physicist A. D. Fokker took note of the view of Huygens on the theory of tone systems at the time 1940–1945.

The innocent period of tone systems ends with the appearance of the book [26] of H. Helmholtz in 1863 (reedition in 1954). In this book Helmholtz described all knowledge of his time about tone systems (including acoustics, anatomy of the human ear, etc.). Helmholtz belonged to the so-called "empirical stream" of mathematics in his time. The concrete contents (of tone system, in this case) were strong arguments again "apriorism" (another stream in mathematics of this time). But, in general, this position is labile since it explains the mathematical knowledge not searching its specificity. Helmholtz, *de facto*, identified mathematics with theoretical physics.

J. M. Petzval published two extensive mathematical papers about tone systems in the early 1860s. His results about tone systems were then republished in an integrated version in the book [12] by Petzval's pupil L. Eményi. Petzval was the first who thoroughly mathematically studied tone systems based on tempered fifths.

Recall that the main reason for all the history with discrete tone systems in Europe was the burgeoning development of the keyboard. Because there is a physical limitation on the number of keys which can be used to play tones, some means had to be found to permit the tonalities demanded by developments in polyphonic music to work within this limitation. In fact, this way keyboard becomes the dominant force in Western composition and composers seek to explore new tonal palettes.

While the 12-tone Equal Temperament E_{12} has its disadvantages, it has led the way for the full development of harmonic music, and the rich variety of musical styles which have grown up in the last one hundred and fifty years. It took nearly two centuries for E_{12} to find universal acceptance by the musical world; the first pianos to be tuned this way were produced by Broadwoods in the middle of the 19th century and by the beginning of the 20th, virtually all pianos were tuned this way. So, we see a very paradoxical great come back of E_{12} to music.

The advantages of E_{12} led to its non-critical use and to consequences. Musicians (and mathematicians ...) in Europe stopped to think about tone system. They often false identify tones (physical or perceptual objects) and notes (coding of tones). So can happen that music supposing other than E_{12} (using wolf tones; music bases mainly on melody; contrapunct based essentially on the pure major and minor thirds and sixths, timbre and exploiting the building acoustics; music outside Europe; etc.) is interpreted on instruments tuned in E_{12}.

One fact to note is that the figures used to calculate E_{12} are based on the theoretical values. In practice, even equally tempered instruments, such as the piano, sound flat in their upper octaves when they are tuned in strict accordance with the equal tempered scale. Piano-tuners employ a trick called, "brightening the treble" or "stretch tuning," which means that the top one and a half to two octaves are sharpened slightly; the low bass octaves are also lowered in a similar fashion. So, surprising and paradoxically, every clavier cannot be principally tuned into E_{12}: When tuning clavier into E_{12} "via computer," the upper and lower octaves sound not in E_{12}. On the other hand, when tuning clavier into E_{12} "via ear," it is not E_{12} physically. Moreover, these tunings are depending of individual psychic. These are 3-4 octaves from the totally 7 ones. So, any compromise should be made using an uncertainty theory. Recall that E_{12} itself is a psychoacoustic compromise.

The victory march of E_{12} through the world is a perfect mockery of those mathematical ignoramuses who ever claim that their music has no common ground with mathematics, that it is based only on their absolute pitch hearing or genial musical feeling, cosmic soul, and various other snob declarations. It is because all numbers used (excluding octaves) are irrational. So, was it true Pythagoras (although he did not think it so...) that "music should be judged intellectually better through numbers than with the sensitivity of human ear"?

A revival of interest in tone systems among musicians and in the industry in the late 20th century we can observe in connection with the development of the so-called computer music and production of the electronic musical instruments with computer control. At present, there are no real technological boundaries for practical use of arbitrary tone systems (including the so-called pure tunings). The "golden mine" for the tone systems study is an archive of scales [192] collected by M. Op de Coul which contains about 3000 items. Here can be found also an extensive miscellaneous bibliography.

In the following example, there are well temperatures of Fokker, Helmholtz, Öttingen, Opelt, Wilson, and Wurschmidt.

Example 54

FOKKER(6-STAR/MIX)-16; Harmonic 6-star, groups A, B and C mixed, from Fokker;
21/20, 16/15, 28/25, 8/7, 6/5, 32/25, 4/3, 48/35, 7/5, 3/2, 8/5, 7/4, 64/35, 28/15, 48/25, 2

FOKKER(H)-19; Fokker(H) 19-tone scale, 1968;
25/24, 16/15, 10/9, 75/64, 6/5, 5/4, 32/25, 4/3, 25/18, 36/25, 3/2, 25/16, 8/5, 5/3, 128/75, 9/5, 15/8, 48/25, 2

FOKKER(K)-19; Fokker(K) 19-tone scale, 1968;
25/24, 27/25, 10/9, 125/108, 6/5, 5/4, 162/125, 4/3, 25/18, 36/25, 3/2, 125/81, 8/5, 5/3, 216/125, 9/5, 50/27, 48/25, 2

FOKKER(L)-19; Fokker(L) 7-limit 19-tone scale, 1969;
28/27, 175/162, 125/112, 144/125, 6/5, 56/45, 35/27, 75/56, 25/18, 36/25, 112/75, 54/35, 45/28, 5/3, 125/72, 224/125, 324/175, 27/14, 2

FOKKER-19; Fokker's 19-tone scale;
25/24, 16/15, 9/8, 75/64, 6/5, 5/4, 32/25, 4/3, 45/32, 36/25, 3/2, 25/16, 8/5, 5/3, 128/75, 9/5, 15/8, 48/25, 2

FOKKER-31; Fokker's 31-tone just system;
64/63, 135/128, 15/14, 35/32, 9/8, 8/7, 7/6, 135/112, 315/256, 5/4, 9/7, 21/16, 4/3, 175/128, 45/32, 10/7, 35/24, 3/2, 32/21, 4/9, 45/28, 105/64, 5/3, 12/7, 7/4, 16/9, 945/512, 15/8, 40/21, 63/32, 2

FOKKER(A)-31; Fokker's 31-tone first alternate septimal scale;
36/35, 25/24, 15/14, 35/32, 9/8, 8/7, 7/6, 25/21, 315/256, 5/4, 9/7, 21/16, 4/3, 175/128, 45/32, 10/7, 35/24, 3/2, 32/21, 63/40, 45/28, 105/64, 5/3, 12/7, 7/4, 9/5, 175/96, 15/8, 40/21, 63/32, 2

FOKKER(B)-31; Fokker's 31-tone second alternate septimal scale;
49/48, 21/20, 15/14, 35/32, 9/8, 8/7, 7/6, 6/5, 315/256, 5/4, 9/7, 21/16, 4/3, 175/128, 45/32, 10/7, 35/24, 3/2, 32/21, 25/16, 45/28, 105/64, 5/3, 12/7, 7/4, 25/14, 90/49, 15/8, 40/21, 63/32, 2

FOKKER-53; Fokker's 53-tone system, degree 37 has alternatives;
126/125, 525/512, 25/24, 21/20, 16/15, 27/25, 35/32, 10/9, 9/8, 8/7, 147/128, 7/6, 189/160, 6/5, 243/200, 315/256, 5/4, 63/50, 32/25, 125/96, 21/16, 4/3, 27/20, 175/128, 441/320, 7/5, 10/7, 36/25, 35/24, 189/128, 3/2, 32/21, 49/32, 384/245, 63/40,8/5, 81/50, 105/64, 5/3, 42/25, 12/7, 441/256, 7/4, 16/9, 9/5, 175/96, 147/80, 15/8, 40/21, 48/25, 35/18, 63/32, 2

FOKKER(AV)-31; Fokker's suggestion for a shrinked octave by averaging approximations;
38.652, 77.303, 115.955, 154.606, 193.258, 231.910, 270.561, 309.213, 347.865, 386.516, 425.168, 463.819, 502.471, 541.123, 579.774, 618.426, 657.077, 695.729, 734.381, 773.032, 811.684, 850.335, 888.987, 927.639, 966.290, 1 004.942, 1 043.594, 1 082.245, 1 120.897, 1 159.548, 1 198.200

OETTINGEN(NO.1)-53; von Oettingen's Orthotonophonium scale;
81/80, 128/125, 25/24, 135/128, 16/15, 27/25, 1 125/1 024, 10/9, 9/8, 729/640,

144/125, 75/64, 1 215/1 024, 6/5, 243/200, 10 125/8 192, 5/4, 81/64, 32/25,
162/125, 675/512, 4/3, 27/20, 512/375, 25/18, 45/32, 729/512, 36/25, 375/256,
6 075/4 096, 3/2, 243/160, 192/125, 25/16, 405/256, 8/5, 81/50, 3 375/2 048,
5/3, 27/16, 128/75, 216/125, 225/128, 16/9, 9/5, 729/400, 30 375/16 384, 15/8,
243/128, 48/25, 125/64, 2025/1 024, 2

OETTINGEN(NO.2)-53;
von Oettingen's Orthotonophonium scale with central 1/1;
81/80, 128/125, 25/24, 135/128, 16/15, 27/25, 1 125/1 024, 10/9, 9/8, 256/225,
144/125, 75/64, 32/27, 6/5, 4 096/3 375, 100/81, 5/4, 81/64, 32/25, 125/96,
675/512, 4/3, 27/20, 512/375, 25/18, 45/32, 64/45, 36/25, 375/256, 40/27, 3/2,
1 024/675, 192/125, 25/16, 128/81, 8/5, 81/50, 3 375/2 048, 5/3, 27/16, 128/75,
125/72, 225/128, 16/9, 9/5, 2 048/1 125, 50/27, 15/8, 256/135 48/25, 125/64,
160/81, 2

OPELT-19; Opelt 19-tone;
25/24, 27/25, 9/8, 75/64, 6/5, 5/4, 32/25, 4/3, 25/18, 36/25, 3/2, 25/16, 8/5,
5/3, 125/72, 9/5, 15/8, 48/25, 2

WIL(NO.1)-19; Wilson 19-tone, 1976;
25/24, 16/15, 9/8, 75/64, 6/5, 5/4, 32/25, 4/3, 45/32, 36/25, 3/2, 25/16, 8/5,
5/3, 225/128, 9/5, 15/8, 48/25, 2

WIL(NO.2)-19; Wilson 19-tone, 1975;
28/27, 16/15, 9/8, 7/6, 6/5, 5/4, 35/27, 4/3, 112/81, 64/45, 3/2, 14/9, 8/5, 2
7/16, 7/4, 9/5, 15/8, 35/18, 2

WIL(NO.3)-19; Wilson 19-tone;
21/20, 35/32, 9/8, 7/6, 6/5, 5/4, 21/16, 4/3, 7/5, 35/24, 3/2, 63/40, 105/64,
27/16, 7/4 9/5, 15/8, 63/32, 2

WIL(11-LIM)-19;
Wilson 11-limit 19-tone scale, 1977;
28/27, 35/33, 49/44, 7/6, 105/88, 56/45, 14/11, 4/3, 7/5, 63/44, 3/2, 14/9,
35/22, 147/88, 7/4, 315/176, 28/15, 21/11, 2

WIL(5-LIM)-22; Wilson's 22-tone 5-limit scale;
25/24, 16/15, 10/9, 9/8, 75/64, 6/5, 5/4, 32/25, 4/3, 27/20, 45/32, 36/25, 3/2,
25/16, 8/5, 5/3, 27/16, 225/128, 9/5, 15/8, 48/25, 2

WIL(7-LIM/NO.1)-22;
Wilson's 22-tone 7-limit "marimba" scale;
28/27, 16/15, 10/9, 9/8, 7/6, 6/5, 5/4, 35/27, 4/3, 27/20, 45/32, 35/24, 3/2,
14/9, 8/5, 5/3, 27/16, 7/4, 9/5, 15/8, 35/18, 2

WIL(7-LIM/NO.2)-22; Wilson 7-limit scale;
126/125, 21/20, 35/32, 9/8, 7/6, 6/5, 5/4, 63/50, 21/16, 27/20, 7/5, 36/25, 3/2, 25/16, 63/40, 5/3 42/25, 7/4, 9/5, 15/8, 189/100, 2

WIL(7-LIM/NO.3)-22; Wilson 7-limit scale;
128/125, 16/15, 10/9, 9/8, 32/27, 6/5, 5/4, 32/25, 4/3, 27/20, 64/45, 36/25, 3/2, 25/16, 8/5, 5/3, 128/75, 16/9, 9/5, 15/8, 48/25, 2

WIL(DIA/NO.1)-22; Wilson Diaphonic cycle, tetrachordal form;
36/35, 18/17, 12/11, 9/8, 36/31, 6/5 36/29, 9/7, 4/3, 18/13 27/19, 54/37, 3/2, 54/35, 27/17 18/11, 27/16, 54/31, 9/5, 54/29, 27/14, 2

WIL(DIA/NO.2)-22; Wilson Diaphonic cycle, conjunctive form;
39/38, 39/37, 13/12, 39/35, 39/34, 13/11, 39/32, 39/31, 13/10, 39/29, 39/28, 13/9, 52/35, 26/17, 52/33, 13/8, 52/31, 26/15, 52/29, 13/7, 52/27, 2

WIL(DIA/NO.3)-22; Wilson Diaphonic cycle on 3/2;
39/38, 39/37, 13/12, 39/35, 39/34, 13/11, 39/32, 39/31, 13/10, 39/29, 39/28, 13/9, 3/2, 54/35, 27/17, 18/11, 27/16, 54/31, 9/5, 54/29, 27/14, 2

WIL(DIA/NO.4)-22; Wilson Diaphonic cycle on 4/3;
36/35, 18/17, 12/11, 9/8, 36/31, 6/5, 36/29, 9/7, 4/3, 26/19, 54/37, 13/9, 52/35, 26/17, 52/33, 13/8, 52/31, 26/15, 52/29, 13/7, 52/27, 2

WIL(DUO)-22; Wilson 'duovigene';
28/27, 16/15, 35/32, 9/8, 7/6, 6/5, 5/4, 35/27, 4/3, 112/81, 45/32, 35/24, 3/2, 14/9, 8/5 5/3, 27/16, 7/4, 9/5, 15/8, 35/18, 2

WIL(FACET)-22; Wilson study in "conjunct facets," Hexany based;
28/27, 21/20, 10/9, 9/8, 7/6, 6/5, 5/4, 35/27, 4/3, 27/20, 7/5, 40/27, 3/2, 14/9, 63/40, 5/3, 140/81, 7/4, 9/5, 28/15, 35/18, 2

WIL(L1)-22; Wilson 11-limit scale;
33/32, 21/20, 35/32, 9/8, 7/6, 77/64, 5/4, 165/128, 21/16, 11/8, 7/5, 231/160, 3/2, 99/64, 77/48, 33/20, 55/32, 7/4, 231/128, 15/8, 77/40, 2

WIL(L2)-22; Wilson 11-limit scale;
49/48, 77/72, 11/10, 9/8, 7/6, 77/64, 5/4, 77/60, 4/3, 11/8, 77/54, 35/24, 3/2, 11/7, 77/48, 5/3, 77/45, 7/4, 11/6, 15/8, 77/40, 2

WIL(L3)-22; Wilson 11-limit scale;
33/32, 21/20, 35/32, 9/8, 7/6, 6/5, 5/4, 14/11, 21/16, 11/8, 7/5, 35/24, 3/2, 14/9, 8/5, 105/64, 27/16, 7/4, 9/5, 15/8, 21/11, 2

WIL(L4)-22; Wilson 11-limit scale;
49/48, 21/20, 10/9, 8/7, 7/6, 6/5, 5/4, 35/27, 4/3, 49/36, 7/5, 35/24, 3/2, 14/9,

8/5, 5/3, 12/7, 7/4, 9/5, 28/15, 35/18, 2

WIL(L5)-22; Wilson 11-limit scale;
49/48, 77/72, 12/11, 8/7, 7/6, 6/5, 5/4, 14/11 4/3, 49/36, 7/5, 35/24, 3/2, 14/9
8/5, 5/3, 12/7, 7/4, 11/6, 28/15, 35/18, 2

WIL(31)-19; A septimal interpretation of 19 out of 31 tones, after Wilson, XH7+8;
25/24, 16/15, 9/8, 7/6, 6/5, 5/4, 9/7, 4/3, 7/5, 10/7, 3/2, 14/9, 8/5, 5/3, 7/4,
16/9, 15/8, 27/14, 2

WURSCHMIDT-12; Wuerschmidt's normalised 12-tone system;
135/128, 9/8, 6/5, 81/64, 27/20, 45/32, 3/2, 405/256, 27/16, 9/5, 15/8, 2

WURSCHMIDT(NO.1)-19; Wuerschmidt I 19-tone scale;
25/24, 16/15, 9/8, 75/64, 6/5, 5/4, 32/25, 4/3, 25/18, 36/25, 3/2, 25/16, 8/5,
5/3, 128/75, 16/9, 15/8, 48/25, 2

WURSCHMIDT(NO.2)-19; Wuerschmidt II 19-tone scale;
25/24, 27/25, 9/8, 75/64, 6/5, 5/4, 32/25, 4/3, 25/18, 36/25, 3/2, 25/16, 8/5,
5/3, 128/75, 16/9, 50/27, 48/25, 2

WURSCHMIDT(NO.1)-31; Wuerschmidt's 31-tone system;
128/125, 25/24, 16/15, 1 125/1 024, 9/8, 144/125, 75/64, 6/5, 625/512, 5/4,
32/25, 125/96, 4/3, 512/375, 25/18, 36/25, 375/256, 3/2, 192/125, 25/16, 8/5,
1 024/625, 5/3, 128/75, 125/72, 16/9, 2 048/1 125, 15/8, 48/25, 125/64, 2

WURSCHMIDT(NO.2)-31; Wuerschmidt's 31-tone system with alternative tritone;
128/125, 25/24, 16/15, 1 125/1 024, 9/8, 144/125, 75/64, 6/5, 625/512, 5/4,
32/25, 125/96, 4/3, 512/375, 25/18, 64/45, 375/256, 3/2, 192/125, 25/16, 8/5,
1 024/625, 5/3, 128/75, 125/72, 16/9, 2 048/1 125, 15/8, 48/25, 125/64, 2

WURSCHMIDT-53; Wuerschmidt's 53-tone system;
81/80, 128/125, 25/24, 135/128, 16/15, 27/25, 1125/1 024, 10/9, 9/8, 256/225,
144/125, 75/64, 32/27, 6/5, 625/512, 768/625, 5/4, 81/64, 32/25, 125/96, 675/512,
4/3, 27/20, 512/375, 25/18, 45/32, 64/45, 36/25, 375/256, 40/27, 3/2, 1 024/675,
192/125, 25/16, 128/81, 8/5, 625/384, 1 024/625, 5/3, 27/16, 128/75, 125/72,
225/128, 16/9 9/5, 2 048/1 125, 50/27, 15/8, 256/135, 48/25 125/64, 160/81, 2

In the following example, there are miscellaneous well temperatures.

Example 55

DARVISH-30; After Shaykh Darvish, Syria, from [11], *vol.5, p.29;*
800/779, 256/243, 2 187/2 048, 35 073/32 000, 9/8, 500/433, 32/27, 19 683/16 384,
315 657/256 000, 8 192/6 561, 81/64, 1 299/1 000, 4/3, 16 000/11 691, 1 024/729,
729/512, 11 691/8 000, 3/2, 2 000/1 299, 128/81, 6 561/4 096, 105 219/64 000,
27/16, 433/250, 16/9, 64 000/35 073, 4 096/2 187 243/128, 3 897/2 000, 2

CHALFOUN-24; [11], *vol.5, p. 40. After Alexandre Chalfoun;*
1 000/971, 10 000/9429, 2 500/2 289, 9/8, 125/108, 25/21, 1 250/1 021, 500/397,
5 000/3 859, 4/3, 2 000/1 457, 400/283, 10 000/6 869, 3/2, 1 250/809, 2 000/1 257,
10 000/6 103, 27/16, 10 000/5 757, 5 000/2 797, 10 000/5 437, 5 000/2 643,
10 000/5 141, 2

ELLIS-24; c.f. [26], *p. 421, 24-tones of JI for 1 manual harmonium;*
81/80, 25/24, 135/128, 9/8, 729/640, 75/64, 1 215/1 024, 5/4, 81/64, 4/3, 27/20,
45/32, 729/512, 3/2, 243/160, 25/16, 405/256, 5/3, 27/16, 225/128, 3 645/2 048,
15/8, 243/128, 2

MUSAQAH-24; From [11], *vol.5, p.34,*
after Mih-a'il Mùsaqah, 1899, a Lebanese scholar;
48, 97, 146, 195, 244, 294, 344, 394, 445, 495, 546, 597, 648, 699, 750, 801, 852,
902, 953, 1 003, 1 053, 1 103, 1 152, 1 200

SECOR-17; George Secor's well temperament with 5 pure 11/7 and
3 near just 11/6;
66.7425, 144.855, 214.440, 278.340, 353.610, 428.880, 492.780, 562.365, 640.4775,
707.220, 771.120, 849.2325, 921.660, 985.560, 1 057.9875, 1 136.100, 1 200.000

RVF(NO.1)-19; $n \cdot 695.5 \pm m \cdot 1\,200 \pm p \cdot 0.25$, A/D=695, interval
range is 49.5–75.5 cents;
68.250, 118.750, 190.750, 261.000, 312.000, 381.000, 454.750 (427.000), 504.250,
572.250,621.750, 695.500, 764.500, 815.500, 885.750, 957.750, 1 008.250, 1 076.500,
1 152.000 (1 124.500), 1 200.000

RVF(NO.2)-19; $n \cdot 696.2 \pm m \cdot 1\,200 \pm p \cdot 0.607$; 695, interval range
is 31–90 cents, $A_\sharp/C = 7/4$;
72.900, 109.800, 191.800, 269.600, 307.700, 382.400, 468.700 (408.400), 503.200,
575.500, 610.000, 696.200, 770.900, 809.100, 886.800, 968.800, 1 005.700, 1 078.600,
1 169.100 (1 109.500), 1 200.000

RVF(NO.3)-19; $n \cdot 696.3 \pm m \cdot 1\,200 \pm p \cdot 0.082$; 694.737, interval
range is 25-97 cents, $B_\sharp/E_\sharp = 3/2$;

73.600, 106.300, 191.900, 271.900, 306.200, 382.200, 473.400 (400.400), 502.900, 575.600, 605.100, 696.300, 772.300, 806.600, 886.600, 972.200, 1 004.900, 1 078.500, 1 175.300 (1 103.200), 1 200.000

7.1.3 A revival of interest in the 20th century

For an overview of the topic, including historical aspects, c.f. [2], [3], [50], [60].

In the late 20th century we can observe a revival of interest in tone systems among musicians and in the industry in connection with the development of the so-called computer music and production of the electronic musical instruments with computer control. At present, there are no real technological boundaries for practical use of arbitrary tone systems within the human aural perceptive abilities and sensitivity. In particular, this concerns the Just Intonation (pure tunings), equal temperaments (freedom in transpositions), and tone systems with the uncertain pitch (including noises).

Since the main motivation for the development of tone systems in Europe was a physical limitation on the number of keys of the keyboard which can be used to play tones, some means had to be found to permit the tonalities demanded by developments in polyphonic music to work within this limitation. In fact, this way the keyboard became the dominant force in Western composition and composers seek it out to explore new tonal palettes.

At present, choices of various discrete tone systems by composers are based on the conviction that the process of limitation is necessary to give the musician the material with which he can work. The point, amply discussed in aesthetics, need not be labored here.

7.1.4 Temperature and mistuning

In this Chapter, we will understand under tone system S a discrete subset of the set of all fuzzy numbers $\mathbb{F}$, where

$$\mathbb{F} = \{F : \mathbb{R}^+ \to [0,1]; \ \exists f \in \mathbb{R}^+, \ F(f) = 1\},$$

i.e. *tone system* $S \subset \mathbb{F}$ is a discrete set of fuzzy numbers F which "approximate in the psychological sense" musical intervals ω. Denote the class of all tone systems by $\mathfrak{S}_{\mathbb{F}}$.

Moreover, we suppose that the kernel of a fuzzy number F, i.e. $\ker(F) = \{\omega \in \mathbb{R}^+; F(\omega) = 1\}$ is nonempty and need not be a unique value (e.g., when $C_\sharp \neq D_\flat$ in music). Denote by $\ker(S) = \{\omega \in \mathbb{R}^+; F(\omega) = 1, F \in S\} = \bigcup_{F \in S} \ker(F)$ and call the *kernel of the tone system* S.

The order on $\mathbb{F}$ we define as follows: for $F_a, F_b \in \mathbb{F}$,

$$F_a \leq_\mathbb{F} F_b \Leftrightarrow \forall \omega_a \in \ker(F_a), \forall \omega_b \in \ker(F_b); \ \omega_a \leq \omega_b.$$

In general, the order $\leq_\mathbb{F}$ is not linear since elements $F \in \mathbb{F}$ with non-disjoint kernels need not be comparable.

Let $K > 1$ be a real number called *comma*. Mathematically, the concrete comma value does not play any principal role in our theory. Usually, it is bounded with the size of the fuzzy number support. Analogously to $\varepsilon > 0$ in calculus, which has to be close to 0, it is reasonable to deal with commas $K > 1$ which are close to 1 in some sense. The examples of commas with historical names: Comma of Didymus $\mathcal{D} = 81/80$, Pythagorean comma $\mathcal{K} = 531\,441/524\,288$, c.f. also List of intervals. For more about commas and they structure, c.f. [137].

Let $K > 1$ be a comma. A tone system $\Sigma \in \mathfrak{S}_\mathbb{F}$ is called the *K-temperament* of the tone system $S \in \mathfrak{S}_\mathbb{F}$, if for every $\omega \in \ker(S)$ there exists $\tau_\omega \in \mathcal{R}^+$, $1/K \leq \tau_\omega \leq K$, such that $\phi = \tau_\omega \cdot \omega \in \ker(\Sigma)$, where $\ker(\Sigma) = \{\phi \in \mathbb{R}^+; \Phi(\phi) = 1, \Phi \in \Sigma\}$. Denote by $\Theta_{\omega,\tau_\omega} = \omega \cdot \tau_\omega$.

Let $K > 1$. The kernel $\ker(\Sigma)$ is said the set of *K-tempered values* of S. The set $\Gamma = \{\tau_\omega; \omega \in \ker(S), \Theta_{f,\tau_\omega} \in \ker(\Sigma), 1/K \leq \tau_\omega \leq K\}$ is called the *K-temperature* of S and the set $\Xi = \{\mu_\omega; \mu_\omega = \tau_\omega - 1, \tau_\omega \in \Gamma\}$ is called the *K-mistuning* of S.

For $\omega \in \ker(S)$ (one single element), we will simply say that τ_ω, $1/K \leq \tau_\omega \leq K$, is a *$K$-temperature* of ω and μ_ω is a *K-mistuning* of ω.

Clearly, if Σ is a K-temperament of S, then S is a K-temperament of Σ. The relation "to be K-temperament" is reflexive, symmetric, but not transitive (because of the condition $1/K \leq \tau_\omega \leq K$).

Everywhere in the ongoing text in this Chapter we will suppose the given fixed comma $K > 1$ and will say simply about temperament instead of K-temperament, similarly: mistuning, tempered value, etc. For concrete tone systems, we will put everywhere $K = \mathcal{D} = 81/80$ (comma of Didymus).

It is technologically reasonable to give S as a theoretical crisp set (e.g., the 12-tone Equal Temperament E_{12} is given as a set of positive real numbers). However this never really occurs. For this reason and also for the sake of symmetry, we consider both tone systems S and its temperament Σ as sets of fuzzy numbers.

7.1.5   Three types of temperaments – examples

To a given tone system $S \in \mathfrak{S}_{\mathbb{F}}$, there are distinguished three types of temperaments: the psychoacoustic, musical, and variable ones. We may consider also the psychoacoustic, musical, and variable sides of one temperament S. According to the classification of uncertainty types in [33] (i.e. quantifying the real world object), the uncertainty of the psychoacoustic temperament is *fuzziness*, of the musical temperament is *strife* and of the variable temperament is *nonspecificity*.

In practice, temperatures depends on the aural sensitivity of a considered individual human and is of psychoacoustic nature. Especially, a temperature τ_ω^* (similarly, mistuning) is *virtual* if a temperature (resp. mistuning) of the musical interval $\tau_\omega^* = \Theta_{f,\tau_\omega}^* / \omega$ is not "hearable." Dealing with *psychoacoustic* temperaments (or, the psychoacoustic side of a temperament), we are interested, e.g., in the supports of fuzzy numbers $\Phi \in \Sigma$.

Example 56
(a) The Petzval's conventional suggestion about virtual temperature, c.f. [12]:
$$239/240 < \tau^* < 240/239.$$
(b) Average sensitivity of the human ear is about 10 cents, tuners do 5-6 cents, the boundary is 2-3 cents. The aural sensitivity depends on many factors, e.g., on the age or health of an individual, c.f. [3].

Music is not an acoustic consideration of musical intervals. In each musical composition there are consonant passages and passages with dissonances (tensions) and this together makes music interesting. So, every musical composition uses and requires its own structure of used intervals, the relations among possible consonances and dissonances from the harmonic and melodic musical context. This side of temperament, the structure of played tones, considered in a concrete situation and time depends on the individual human musical education, given musical style, construction of musical instruments, building acoustics, concrete musical composition, etc. Dealing with the *musical* temperaments (or, the musical sides of a temperament), we take notice of the function shape of fuzzy numbers $\Phi \in \Sigma$.

Example 57 In Example 64, the musical mistuning

$$\mu_{3/2}{}^{(m')} = \mu_{6/5}{}^{(m')} = \mu_{5/4}{}^{(m')} = \mu_{7/4}{}^{(m')}.$$

In Example 65, the musical mistuning

$$\mu_{3/2}^{(m'')} = \mu_{5/4}^{(m'')}/5 = \mu_{6/5}^{(m'')}/5 = \mu_{7/4}^{(m'')}/5.$$

In Example 28, the weights of $\tau_{3/2}$, $\tau_{5/4}$, $\tau_{7/4}$ are $1/9$, $1/25$, $1/49$, respectively. In Example 29, the weights of $\tau_{3/2}, \tau_{5/4}, \tau_{7/4}$ are the same as in Example 28, but the temperament depends proportionally on the number of keys per octave.

Temperature can be mentioned also as a union of equivalent (equally probable) variants, alternatives, or possibilities of a tone system. Dealing with the *variable* temperament (or, the variable side of a temperament) we are interested, e.g., in cardinality of $\ker(\Sigma)$.

Example 58 Tone systems consisting of values of 12 major and 12 minor scales (not specifying here this notions) are defined ambiguously and also the cardinality of these systems may vary. Observe that the sets involving the values of 12 major and minor scales in Lemma 45 and Theorem 43, and in Theorem 48 may have different numerical expressions and also cardinalities.

7.2 Harmonic mean based measures

Given a set $\mathbb{F}$ of all fuzzy numbers, let $\mathfrak{S}_{\mathbb{F}}$ be an algebra of tone systems $S \subset \mathbb{F}$, such that $\emptyset \in \mathfrak{S}_{\mathbb{F}}$ and the operations $\cap, \cup$ we define on $\mathfrak{S}_{\mathbb{F}}$ as follows: let $S_a, S_b \in \mathfrak{S}_{\mathbb{F}}$, then

$$S_a \cup S_b = \{F \in \mathbb{F}; \; F = F_a \vee F_b, F_a \in S_a, F_b \in S_b\} \in \mathfrak{S}_{\mathbb{F}},$$

$$S_a \cap S_b = \{F \in \mathbb{F}; \; F = F_a \wedge F_b, F_a \in S_a, F_b \in S_b\} \in \mathfrak{S}_{\mathbb{F}},$$

where for every $t \in \mathbb{R}^+$,

$$(F_a \vee F_b)(t) = \max(F_a(t), F_b(t)),$$

and

$$(F_a \wedge F_b)(t) = \begin{cases} \min(F_a(t), F_b(t)), & \ker(F_a) \cap \ker(F_b) \neq \emptyset, \\ \emptyset, & \ker(F_a) \cap \ker(F_b) = \emptyset. \end{cases}$$

Let $S_1, S_2 \in \mathfrak{S}_{\mathbb{F}}$. We say that $S_1 \subset S_2$ if $S_1 \cap S_2 = S_1$.

An *uncertainty measure* $\sigma : \mathfrak{S}_{\mathbb{F}} \to [0, \infty]$ (the adjective "uncertainty" we will omit) is a non negative extended real valued set function with the properties:

(1) $\sigma(\emptyset) = 0$;

(2) if $S_1, S_2 \in \mathfrak{S}_{\mathbb{F}}$, $S_1 \subset S_2$, then $\sigma(S_1) \leq \sigma(S_2)$.

In [33] we can find also the review of the known uncertainty-based measures. Our construction of uncertainty measures is based on distance measuring between two sets. In Example 28 and Example 29, there are measured distances between two crisp sets: $L = \{5/4, 3/2, 7/4\}$ and E_N, the N-tone Equal Temperament, $N \in \mathbb{N}$. In Subsection 7.2.1, Equation (7.2), there is used an asymmetric set distance between E_{12} (crisp) and S (fuzzy). A bimeasure construction based on a distance between two uncertain sets we can find in Subsection 7.2.2, Equation (7.11).

Traditional quantities comparing musical temperatures are based on *harmonic* means of temperatures. We construct uncertainty measures using (generalized) harmonic mean in their construction as following.

Definition 18 Let A be a finite set of positive numbers. Let $p \in \mathbb{N}$. Let $\mathfrak{S}_{\mathbb{F}}$ be the class of all tone systems. A set function $\sigma_p : \mathfrak{S}_{\mathbb{F}} \to$

$[0, \infty]$ is called the temperament measure, if $\sigma_p(\emptyset) = 0$ and if $S \neq \emptyset$, then

$$\sigma_p(S) = \frac{1}{\sum_{\phi \in A} w_\phi \mu_\phi^p(S)},$$

where $S \in \mathfrak{S}_F$ and $w_\phi \geq 0$, $\sum_{\phi \in A} w_a = 1$, are weights of mistunings, i.e.

$$\mu_\phi(S) = \min_{\omega \in \ker(S)} \left| 1 - \frac{\omega}{\phi} \right|, \phi \in A.$$

Since we will deal only with temperament measures in this chapter, we will often omit the adjective "temperament" for the sake of simplicity.

The proof of the following lemma is easy and omitted.

Lemma 39 *Let $S \in \mathfrak{S}_F$. Let $p \in \mathcal{N}$. Let σ_p' and σ_p'' are two temperament measures on $S \in \mathfrak{S}_F$. Put*

$$\sigma_p^*(S) = (\sigma_p' \oplus \sigma_p'')(S) = \frac{1}{w_1/\sigma_p'(S) + w_2/\sigma_p''(S)}, S \in \mathfrak{S}_F, \quad (7.1)$$

where $w_1 + w_2 = 1$, $w_1 > 0, w_2 > 0$. Then σ_p^* is a temperament measure as well.

7.2.1 Possibility of free transpositions

The system E_{12} is considered by some musicians as the definitive ideal with respect to free transpositions in music. The possibility of free transpositions within a tone system S can be expressed then as a small distance between sets E_{12} and $S \in \mathfrak{S}_F$ (the denominator in the formula (7.2)).

Definition 19 *Let $\kappa > 0$. Let $p \in \mathbb{N}$. We say that the tone system S with the octave equivalence enables $(\kappa, E_{12}, \sigma_p')$-free transpositions if*

$$\sigma_p'(S) = \frac{1}{\sum_{\phi \in E_{12} \cap [1;2)} w_\phi \mu_\phi^p(S)} > \kappa, \quad (7.2)$$

where $w_\phi > 0$, $\sum_{\phi=1}^{12} w_\phi = 1$, $\mu_\phi(S) = \min_{\omega \in \ker(S) \cap [1,2)} \left| 1 - \frac{\omega}{\phi} \right|, \phi \in E_{12}$.

7.2.2 Basic musical intervals sound as pure

Similarly to the construction of the measure σ'_p, we may understand what means if we say that basic acoustic intervals "sound as pure" in a tone system S.

Definition 20 Let $\kappa > 0$. Let $p \in \mathbb{N}$. Let $L = \{6/5, 5/4, 3/2, 7/4\}$. We say that *intervals* 6/5, 5/4, 3/2, 7/4 sound (κ, L, σ''_p)-*pure* in the tone system $S \in \mathfrak{S}_\mathbb{F}$ if

$$\sigma''_p(S) = \frac{1}{\sum_{\phi \in L} w_\phi \mu_\phi^p(S)} > \kappa, \tag{7.3}$$

where $w_\phi > 0$, $\sum_{\phi=1}^4 w_\phi = 1$, $\mu_\phi(S) = \min_{\omega \in \ker(S)} \left|1 - \frac{\omega}{\phi}\right|$, $\phi \in L$.

7.2.3 We cannot avoid uncertainty

Roughly speaking (the definition will be given in Section 7.3.4), well tempered tone systems are those tone systems which enable, at the same time, both free transpositions and basic musical intervals sound as pure.

Let us open the question about the suitability of the chosen constructions of measures σ'_p and σ''_p in Definitions 19 and 20. For instance, instead of mistuning we may consider temperature with values expressed in cents or consider special weights, c.f. Section 5.3.2.

The sets E_{12}, L, and $\ker(S)$ are crisp and uncertainty is present in the formulas (7.2), (7.1), and (7.11), because of ambiguity given by $\ker(S)$ (possible more kernel values for one fuzzy number). Of course, we may construct also measures σ'_p, σ''_p which are "more soft," including in addition fuzziness-type uncertainty in their constructions. But, we can also avoid uncertainty in all in these formulas by dealing with unambiguous and crisp sets.

However, there is a moment which should be underlined in connection with the uncertainty of tone systems: since $L \cap E_N = \emptyset$ for every $N \in \mathbb{N}$ (the uncomensurability of rationals and irrationals), the value

$$\frac{1}{\sigma'_p(S)} + \frac{1}{\sigma''_p(S)} > 0$$

strictly for every crisp tone system $S \in \mathfrak{S}_{\mathbb{F}}$. Therefore, in crisp tone systems it is impossible to freely transpose and at the same time to sound the basic intervals as pure. We *must* make any appropriate compromise. Therefore, it is "more honest" to deal with tone systems as objects including uncertainty from the beginning (as we do in this chapter). Thus, considering well tempered tone systems, we observe the principal using and sense of uncertainty-based information.

For our purposes, it is reasonable to consider the following quantity:

$$(\sigma'_p \oplus \sigma''_p)(S) = \frac{1}{w_1/\sigma'_p(S) + w_2/\sigma''_p(S)}, S \in \mathfrak{S}_{\mathbb{F}}, \qquad (7.4)$$

where $p \in \mathcal{N}$ and $w_1 + w_2 = 1$, $w_1 > 0, w_2 > 0$. Tone system S is then the "better" well tempered the bigger is the value of $(\sigma'_p \oplus \sigma''_p)(S)$. However, this value is ever finite (we can normalize it with respect to 1).

We can see the corollary of Lemma 39, that the function

$$\sigma^*_p = \sigma'_p \oplus \sigma''_p \qquad (7.5)$$

is a temperament measure on $S \in \mathbb{S}$, where σ'_p and σ''_p are defined as in Definitions 19 and 20, respectively.

7.3 Well tempered tone systems

7.3.1 Symmetry and octave equivalence

The notion of the tone system symmetry corresponds to the notion of even function in real analysis. For our purposes, it is enough.

Definition 21 A tone system $S \in \mathfrak{S}_{\mathbb{F}}$ is said to be *symmetric* if

$$\omega \in \ker(S) \Rightarrow 1/\omega \in \ker(S).$$

The octave (ratio 2/1) plays a role of "unit" in the theory of tone systems (as number 1 in the group of all real numbers with the operation of addition).

Definition 22 We say that the relation of *octave equivalence* $\equiv$ is given on a tone system $S \in \mathfrak{S}_{\mathbb{F}}$, if for every $\omega_1, \omega_2 \in \ker(S)$,

$$\omega_1 \equiv \omega_2 \Leftrightarrow (\exists z \in \mathbb{Z}; \; \omega_1 = \omega_2 \cdot 2^z).$$

For $\omega \in \ker(S)$, denote by $[\omega]$ the octave equivalence class such that $\omega \in [\omega]$.

Lemma 40 *Let* $S \in\in \mathfrak{S}_{\mathbb{F}}$, $\Sigma \in \mathfrak{S}_{\mathbb{F}}$ *be two symmetrical tone systems with octave equivalence. Then*
    (a) *if* $\omega \in \ker(S)$, *then* $2/\omega \in \ker(S)$;
    (b) *if* τ_ω *is a temperature, then* $\tau_\omega = \tau_{2\omega} = 1/\tau_{2/\omega}$.

Proof.
    (a) If $k \in \ker(S)$, then $1/\omega \in \ker(S)$ according to the symmetry. By the octave equivalence, $2^z/\omega \in \ker(S)$, $z \in \mathbb{Z}$.
    (b) If $\omega \in \ker(S)$, $\omega \cdot \tau_\omega = \phi \in \ker(\Sigma)$, then $2^z\omega \cdot \tau_{2^z\omega} = 2^z\phi \in \ker(\Sigma)$. Thus, $\tau_\omega = \tau_{2^z\omega}$, $z \in \mathbb{Z}$.
    We have: $(1/\omega)/(\tau_{1/\omega}) = (1/\phi)$. Thus, $\tau_\omega = 1/\tau_{1/\omega} = 1/\tau_{2^z/\omega}$, $z \in \mathbb{Z}$. $\qquad\square$

7.3.2 The tempered fifth approximations

The classical tuning algorithms based on the so-called *spiral of fifths* uses the fifth as "yard stick." The idea is as follows. If we consider the tone system $S \in \mathfrak{S}_{\mathbb{F}}$ with $\ker(S) = \{\Theta_{3/2,\tau_{3/2}}^z \cdot 2^c; \; c, z \in \mathbb{Z}\}$, then for some $\Theta_{3/2,\tau_{3/2}}$ the set $\ker(S) \cap [1, 2]$ is finite (e.g., for $\Theta_{3/2,\tau_{3/2}} = \sqrt[12]{2^7}$, we have E_{12}) or for other ones it will be infinite and dense in $[1, 2]$ (Kronecker's lemma, c.f. [35]); e.g., for $\Theta_{3/2,\tau_{3/2}} = 3/2$, the Pythagorean Tuning. The tone systems of the first kind we will call to be *cyclic* and the second ones *open*, respectively. More precisely:

Definition 23 A tone system $S \in \mathfrak{S}_{\mathbb{F}}$ with $\ker(S) = \{\Theta_{3/2,\tau_{3/2}}^z \cdot 2^c; \; c, z \in \mathbb{Z}\}$ is *cyclic* if there exist natural numbers n, N such that

$$2^n = \Theta_{3/2,\tau_{3/2}}^N. \tag{7.6}$$

It is obvious that every equal tempered tone system is cyclic (for some $z_0 \in \mathbb{Z}$, put $\Theta_{3/2,\tau_{3/2}} = 2^{z_0/N}$. Then $2^{z_0} = \Theta_{3/2,\tau_{3/2}}^N$). There are cyclic tone systems which are not equally tempered, c.f. Subsection 7.4.1 and Subsection 7.4.2.

Lemma 41 *A cyclic tone system* $S \in \mathfrak{S}_{\mathbb{F}}$ *with* $\ker(S) = \{\Theta_{3/2,\tau_{3/2}}^z$ $2^c;\ c, z \in \mathbb{Z}\}$ *is equal tempered if and only if the numbers* n *and* N *in* Definition 23 *are relatively prime.*

Proof. Our assertion is equivalent to the following:

Let N, n be two relatively prime natural numbers.

Let $k_i = i \cdot n (\mathrm{mod}\ N)$ for every $i = 0, 1, \ldots, N - 1$.

If $\{p_0, p_1, \ldots, p_{N-1}\}$ is a permutation of the set $\{k_0, k_1, \ldots, k_{N-1}\}$ such that $k_i \leq k_{i+1}$ for every $i = 0, 1, \ldots, N - 2$, then $p_{i+1} - p_i = 1$.

Indeed, suppose that $i, j \in \{0, 1, \ldots, N - 1\}$ and $k_i = k_j$. Then $i \cdot n = j \cdot n\,(\mathrm{mod}\ N)$. Therefore $n \cdot (i - j) = 0\,(\mathrm{mod}\ N)$. Since N, n are relatively prime, we obtain $i - j = 0\,(\mathrm{mod}\ N)$. We conclude $i = j$. That means that the set $\{p_0, p_1, \ldots, p_{N-1}\}$ is equal to the set $\{0, 1, \ldots, N - 1\}$. $\qquad\square$

7.3.3 Basic law of tempering

The equation

$$\tau_{6/5} \cdot \tau_{5/4} = \tau_{3/2}, \tag{7.7}$$

is called the *basic law of tempering*. This equation holds in most of the historical well tempered tone systems.

Considering the tone systems constructed as the pure fifth approximations, we obtain the following helpful lemma.

Lemma 42 *Let* $S \in \mathfrak{S}_{\mathbb{F}}$ *and* $\ker(S) = \{(3/2)^z \cdot 2^c;\ c, z \in \mathbb{Z}\}$. *Let* $n, m \in \mathbb{Z}$ *be such that* $(3/2)^n 2^\alpha = \tau_{6/5} \cdot 6/5$, $(3/2)^m 2^\beta = \tau_{5/4} \cdot 5/4$ *(powers of the fifth temperatures of the major and minor thirds) for some* $\alpha \in \mathbb{Z}$ *and* $\beta \in \mathbb{Z}$. *Let* $\tau_{6/5} \cdot \tau_{5/4} = \tau_{3/2}$. *Then* $n + m = 1$.

The proof is trivial and we omit it.

Since there is a possibility of many good approximations of 6/5, 5/4, 3/2 in the same tone system $S \in \mathfrak{S}_{\mathbb{F}}$, we will understand that the basic law of tempering is satisfied if there exist trinities of elements of S with temperatures $\tau_{6/5}, \tau_{5/4}, \tau_{3/2}$ such that (7.7).

7.3.4   A formalization of well tempered tone systems

We integrate the essential ideas of various constructions of *well tempered tone systems* based on the tempered fifths in

Definition 24   Let $K > 1$. Let $\tau_{3/2} \in \mathbb{R}$, $1/K < \tau_{3/2} < K$. Let $\mathbb{Z}_1 \subset \mathbb{Z}$. Let $\kappa > 0$. Let $p \in \mathbb{N}$. Let σ_p^* be defined by (7.5). We say that a tone system $S \in \mathfrak{S}_\mathbb{F}$, is a *well tempered* (more precisely, $(K, \kappa, \sigma_p^*, \tau_{3/2}, \mathbb{Z}_1)$-*well tempered tone system*) if

(1)
$$\ker(S) = \bigcup_{z \in \mathbb{Z}_1} [\Theta_{3/2,\tau_{3/2}}^z];$$

(2)
$$\omega \in \ker(S) \Rightarrow 1/\omega \in \ker(S);$$

(3)
$$\exists \Theta_{6/5,\tau_{6/5}} \in \ker(S), \ \exists \Theta_{5/4,\tau_{5/4}} \in \ker(S); \ \tau_{6/5} \cdot \tau_{5/4} = \tau_{3/2};$$

(4)
$$\sigma_p^*(S) > \kappa.$$

In what follows, we will suppose $p = 2$ for temperament measures.

7.4   The Petzval's tone systems

Consider tone systems $S \in \mathfrak{S}_\mathbb{F}$ with $\ker(S) = \bigcup_{z \in \mathbb{Z}_1 \subset \mathbb{Z}} [\Theta_{3/2,\tau_{3/2}}^z]$, c.f. Definition 24, 1. (which are, of course, not equally tempered in general). Besides the fifth virtual mistuning $\tau_{3/2}$ we ask also for appropriate mistuning of $\{3/2, 5/4, 6/5, 7/4\}$.

Taking $\Theta_{3/2,\tau_{3/2}} = 3/2$ (exactly, $\tau_{3/2} = 1$), we have the following approximating sequence $\{K_n^*\}_{n=1}^\infty$ of the tempered major thirds such that $\lim_{n \to \infty} K_n^* = 5/4$ and with the increasing absolute value of the power of $\Theta_{3/2,\tau_{3/2}}$ (we do not describe the obvious algorithm how to obtain this sequence):

$$K_n^* = \Theta_{3/2,\tau_{3/2}}^4, \Theta_{3/2,\tau_{3/2}}^{-8}, \Theta_{3/2,\tau_{3/2}}^{45}, \Theta_{3/2,\tau_{3/2}}^{-314}, \Theta_{3/2,\tau_{3/2}}^{351}, \ldots$$
$$\approx 1.265\,625, 1.248\,6, 1.251\,205, 1.249\,832, 1.249\,96, \ldots.$$

The analogous approximating sequence $\{F_n^*\}_{n=1}^\infty$ of the tempered minor thirds is as follows:

$$F_n^* = \Theta_{3/2,\tau_{3/2}}^{-3}, \Theta_{3/2,\tau_{3/2}}^9, \Theta_{3/2,\tau_{3/2}}^{-44}, \Theta_{3/2,\tau_{3/2}}^{315}, \Theta_{3/2,\tau_{3/2}}^{-350}, \ldots$$
$$\approx 1.185\,185, 1.201\,355, 1.198\,849, 1.200\,128, 1.200\,075, \ldots.$$

The analogous approximating sequence $\{G_n^*\}_{n=1}^\infty$ of the tempered natural sevenths is as follows:

$$G_n^* = \Theta_{3/2,\tau_{3/2}}^{10}, \Theta_{3/2,\tau_{3/2}}^{-14}, \Theta_{3/2,\tau_{3/2}}^{39}, \Theta_{3/2,\tau_{3/2}}^{-67}, \Theta_{3/2,\tau_{3/2}}^{239} \ldots$$
$$\approx 1.77\,778, 1.802\,03, 1.753\,84, 1.757\,52, 1.750\,18, 1.740\,40, \ldots.$$

Now, let $\Theta_{3/2,\tau_{3/2}}$ be a *tempered* fifth and consider sequences

$$\{K_n\}_{n=1}^\infty, \{F_n\}_{n=1}^\infty, \{G_n\}_{n=1}^\infty,$$

where

$$K_n = \Theta_{3/2,\tau_{3/2}}^4, \Theta_{3/2,\tau_{3/2}}^{-8}, \Theta_{3/2,\tau_{3/2}}^{45}, \Theta_{3/2,\tau_{3/2}}^{-314}, \Theta_{3/2,\tau_{3/2}}^{351}, \ldots,$$
$$F_n = \Theta_{3/2,\tau_{3/2}}^{-3}, \Theta_{3/2,\tau_{3/2}}^9, \Theta_{3/2,\tau_{3/2}}^{-44}, \Theta_{3/2,\tau_{3/2}}^{315}, \Theta_{3/2,\tau_{3/2}}^{-350}, \ldots,$$
$$G_n = \Theta_{3/2,\tau_{3/2}}^{10}, \Theta_{3/2,\tau_{3/2}}^{-14}, \Theta_{3/2,\tau_{3/2}}^{39}, \Theta_{3/2,\tau_{3/2}}^{-67}, \Theta_{3/2,\tau_{3/2}}^{239}, \ldots.$$

Definition 25 The well tempered tone system S_n, $n = 1, 2, 3, \ldots$, is said to be the *Petzval's tone system of the n-th type* if

$$\{K_n, F_n, G_n\} \subset \ker(S_n).$$

Especially,

$$\{\Theta_{3/2,\tau_{3/2}}^{-3}, \Theta_{3/2,\tau_{3/2}}^4, \Theta_{3/2,\tau_{3/2}}^{10}\} \subset \ker(S_1)$$

and

$$\{\Theta_{3/2,\tau_{3/2}}^9, \Theta_{3/2,\tau_{3/2}}^{-8}, \Theta_{3/2,\tau_{3/2}}^{-14}\} \subset \ker(S_2)$$

characterize the *Petzval's tone systems of the first type* and *Petzval's tone systems of the second type*, respectively. Systems S_n, $n \geq 3$, are not used in practice.

7.4.1   The Petzval's Tone Systems I

Properties   For the Petzval's tone systems of the first Type we have the following expressions:

$$\Theta^{-3}_{3/2,\tau_{3/2}} = \tau_{3/2}{}^{-3} \cdot \frac{2^5}{3^3}, \; \Theta^{4}_{3/2,\tau_{3/2}} = \tau_{3/2}{}^{4} \cdot \frac{3^4}{2^6}, \; \Theta^{10}_{3/2,\tau_{3/2}} = \tau_{3/2}{}^{10} \cdot \frac{3^{10}}{2^{15}}.$$

These equations imply:

$$\tau_{6/5} \cdot \frac{6}{5} = \tau_{3/2}{}^{-3} \cdot \frac{2^5}{3^3}, \; \tau_{5/4} \cdot \frac{5}{4} = \tau_{3/2}{}^{4} \cdot \frac{3^4}{2^6}, \; \tau_{7/4} \cdot \frac{7}{4} = \tau_{3/2}{}^{10} \cdot \frac{3^{10}}{2^{15}}.$$

Thus,

$$\begin{aligned} \tau_{6/5} &= \tau_{3/2}{}^{-3} \cdot \tfrac{80}{81}, \\ \tau_{5/4} &= \tau_{3/2}{}^{4} \cdot \tfrac{81}{80}, \\ \tau_{7/4} &= \tau_{3/2}{}^{10} \cdot \tfrac{59\,049}{57344}. \end{aligned} \qquad (7.8)$$

The consequences from the formulas (7.8) we collect into the following

Theorem 43 *For every Petzval tone system $S_1 \in \mathfrak{S}_{\mathbb{F}}$ of the first type,*

(1) $\tau_{6/5} \cdot \tau_{5/4} = \tau_{3/2}$;

(2) *the major scale (based on the value $\Theta^i_{3/2,\tau_{3/2}}$):*

$$\begin{aligned} D_i^{(\text{major})} = \; &\{\Theta^i_{3/2,\tau_{3/2}}, \Theta^{i+2}_{3/2,\tau_{3/2}}, \Theta^{i+4}_{3/2,\tau_{3/2}}, \Theta^{i-1}_{3/2,\tau_{3/2}}, \\ &\Theta^{i+1}_{3/2,\tau_{3/2}}, \Theta^{i+3}_{3/2,\tau_{3/2}}, \Theta^{i+5}_{3/2,\tau_{3/2}}, \Theta^{i}_{3/2,\tau_{3/2}}\}; \end{aligned}$$

(3) *the minor scale(based on the value $\Theta^i_{3/2,\tau_{3/2}}$):*

$$\begin{aligned} D_i^{(\text{minor})} = \; &\{\Theta^i_{3/2,\tau_{3/2}}, \Theta^{i+2}_{3/2,\tau_{3/2}}, \Theta^{i-3}_{3/2,\tau_{3/2}}, \Theta^{i-1}_{3/2,\tau_{3/2}}, \\ &\Theta^{i+1}_{3/2,\tau_{3/2}}, \Theta^{i-4}_{3/2,\tau_{3/2}}, \Theta^{i-2}_{3/2,\tau_{3/2}}, \Theta^{i}_{3/2,\tau_{3/2}}\}; \end{aligned}$$

(4) *there is only one whole tone interval, $W_1 = \tau_{3/2}{}^{2} \cdot 9/8$;*

(5) *there are two semitones:* $X = \tau_{3/2}^{-5} \cdot 2^8/3^5$ *(the minor semitone) and* $Y = \tau_{3/2}^{7} \cdot 3^7/2^{11}$ *(the major semitone), and* $W_1 = X \cdot Y$;

(6) $X = Y$ *is equivalent to* $\tau_{3/2} = \sqrt[12]{2^{19}/3^{12}}$ *(the temperature of the* E_{12} *fifth);*

(7) *the major and minor scales based on* $\Theta^i_{3/2,\tau_{3/2}}$ *and* $\Theta^{i+3}_{3/2,\tau_{3/2}}$ *are related: if we take the basic value* $\Theta^i_{3/2,\tau_{3/2}}$ *for the major scale and* $\Theta^{i+3}_{3/2,\tau_{3/2}}$ *for minor scale, then both these scales consist of the same numbers;*

(8) *the major scale is a union of two equal tetrachords:*

$$\{\Theta^{i+0}_{3/2,\tau_{3/2}}, \Theta^{i+2}_{3/2,\tau_{3/2}}, \Theta^{i+4}_{3/2,\tau_{3/2}}, \Theta^{i-1}_{3/2,\tau_{3/2}}\}$$

and

$$\{\Theta^{(i+0)+1}_{3/2,\tau_{3/2}}, \Theta^{(i+2)+1}_{3/2,\tau_{3/2}}, \Theta^{(i+4)+1}_{3/2,\tau_{3/2}}, \Theta^{(i-1)+1}_{3/2,\tau_{3/2}}\}.$$

Example 59 The Pythagorean tone system. $\tau_{3/2} = 1$, $\tau_{5/4} = 81/80$, $\tau_{6/5} = 80/81$, $\tau_{7/4} = 64/63$.

Example 60 A rational approximation of E_{12}. $\tau_{3/2} = 1 - 1/886$, $\tau_{5/4} = 1 + 1/126$, $\tau_{6/5} = 1 - 1/111$, $\tau_{7/4} = 1 + 1/5$.

Example 61 An open tone system with the exact minor third. $\tau_{3/2} = \tau_{5/4} = \sqrt[3]{80/81}$, $\tau_{6/5} = 1$.

Example 62 $N = 19, x = 11$, the Opelt's system, c.f. [26], the approximation of the previous tone system, a cyclic system. $\tau_{3/2} \approx 239/240$, $\tau_{5/4} \approx 234/235$, $\tau_{6/5} \approx 11561/11560$, $\tau_{7/4} \approx 80/81$.

Images in the complex plane For $X > 1$ and $Y > 1$ (the minor and major semitones), denote by

$$\mathbb{P}_{X,Y} = \{X^\alpha Y^\beta;\ \alpha, \beta \in \mathbb{Z}\}.$$

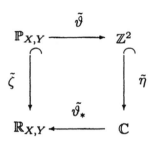

FIGURE 7.3: The embedding $\tilde{\zeta}$

Consider the map $\tilde{\vartheta}$: $\mathbb{P}_{X,Y} \rightarrow \mathbb{Z}^2, \tilde{\vartheta}(X^\alpha Y^\beta) = (\alpha, \beta)$, c.f. also Subsection 6.1.2. Then

$$\tilde{\vartheta}((X^\alpha Y^\beta)(X^\gamma Y^\delta)) = (\alpha + \gamma, \beta + \delta),$$

$$\tilde{\vartheta}((X^\alpha Y^\beta)^\gamma) = (\gamma\alpha, \gamma\beta), \quad \alpha, \beta, \gamma, \delta \in \mathbb{Z}.$$

Embed $\mathbb{Z}^2$ identically into the complex plane $\mathbb{C}$. Denote this injection by $\tilde{\eta}$. Denote by $\tilde{\vartheta}_*^{-1}$ the extended map $\tilde{\vartheta}_*^{-1} : \mathbb{C} \rightarrow \mathbb{R}_{X,Y}$, where $\mathbb{R}_{X,Y} = \{X^\alpha Y^\beta; \ \alpha, \beta \in \mathbb{R}\}$. So we have the following commutative diagram, c.f. Figure 7.3, which defines the embedding $\tilde{\zeta}$: $\mathbb{P}_{X,Y} \rightarrow \mathbb{R}_{X,Y}$.

In Figure 7.4, we see the images of major and minor scales of tone systems of the first type in the map ϑ.

Cyclic tone systems of the first type

Lemma 44 *Let D_p be the union of $p+1$ major and $p+1$ minor scales over the set:*

$$\{\Theta_{3/2, \tau_{3/2}}^0, \Theta_{3/2, \tau_{3/2}}^1, \dots, \Theta_{3/2, \tau_{3/2}}^p\}. \tag{7.9}$$

Let S be the tone system of the first type. Then the set D_p consists of $(p+10)$ values.

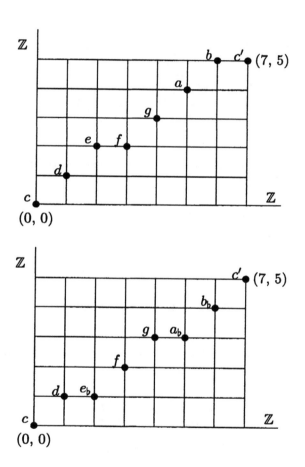

FIGURE 7.4: Major and minor scales of Tone systems I in $\tilde{\vartheta}$

Proof. For the construction of $(p+1)$, major and $(p+1)$ minor scales over the set (7.9) ("the spiral of $p+1$ generalized fifths"), there is used the following sequence of $(p+10)$ values:

$$D_p = \bigcup_{i=0}^{p}(D_i^{(\text{major})} \cup D_i^{(\text{minor})})$$
$$= \{\Theta_{3/2,\tau_{3/2}}^{-4}, \Theta_{3/2,\tau_{3/2}}^{-3}, \Theta_{3/2,\tau_{3/2}}^{-2}, \Theta_{3/2,\tau_{3/2}}^{-1},$$
$$\overbrace{\Theta_{3/2,\tau_{3/2}}^{0}, \Theta_{3/2,\tau_{3/2}}^{1}, \cdots, \Theta_{3/2,\tau_{3/2}}^{p}}^{p+1},$$
$$\Theta_{3/2,\tau_{3/2}}^{p+1}, \Theta_{3/2,\tau_{3/2}}^{p+2}, \Theta_{3/2,\tau_{3/2}}^{p+3}, \Theta_{3/2,\tau_{3/2}}^{p+4}, \Theta_{3/2,\tau_{3/2}}^{p+5}\}.$$

The main advantage of cyclic tone systems of the first type consists of the following:

Lemma 45 *Let S_1 be a Petzval's cyclic tone system of the first type. Let $D_p \subset S_1$ be the union of p major and p minor scales over the set*

$$\Theta_{3/2,\tau_{3/2}}^{0}, \Theta_{3/2,\tau_{3/2}}^{1}, \cdots, \Theta_{3/2,\tau_{3/2}}^{p}.$$

Then the set D_p consists only of $(p+1)$ values.

Proof. Since the tone system is cyclic, there exist n, m such that (7.6). Put $p = m - 1$. Then

$$2^n \Theta_{3/2,\tau_{3/2}}^{-4} = \Theta_{3/2,\tau_{3/2}}^{p-3},$$
$$2^n \Theta_{3/2,\tau_{3/2}}^{-3} = \Theta_{3/2,\tau_{3/2}}^{p-2},$$
$$2^n \Theta_{3/2,\tau_{3/2}}^{-2} = \Theta_{3/2,\tau_{3/2}}^{p-1},$$
$$2^n \Theta_{3/2,\tau_{3/2}}^{-1} = \Theta_{3/2,\tau_{3/2}}^{p},$$
$$2^n \Theta_{3/2,\tau_{3/2}}^{0} = \Theta_{3/2,\tau_{3/2}}^{p+1},$$
$$2^n \Theta_{3/2,\tau_{3/2}}^{1} = \Theta_{3/2,\tau_{3/2}}^{p+2},$$
$$2^n \Theta_{3/2,\tau_{3/2}}^{2} = \Theta_{3/2,\tau_{3/2}}^{p+3},$$
$$2^n \Theta_{3/2,\tau_{3/2}}^{3} = \Theta_{3/2,\tau_{3/2}}^{p+4},$$
$$2^n \Theta_{3/2,\tau_{3/2}}^{4} = \Theta_{3/2,\tau_{3/2}}^{p+5}.$$

We have:

$$
\begin{aligned}
D_p &= \{\Theta_{3/2,\tau_{3/2}}^{-4}, \Theta_{3/2,\tau_{3/2}}^{-3}, \cdots \Theta_{3/2,\tau_{3/2}}^{p+3}, \Theta_{3/2,\tau_{3/2}}^{p+4}, \Theta_{3/2,\tau_{3/2}}^{p+5}\} \\
&\equiv \{\Theta_{3/2,\tau_{3/2}}^{0}, \Theta_{3/2,\tau_{3/2}}^{1}, \cdots, \Theta_{3/2,\tau_{3/2}}^{p}\}
\end{aligned}
,
$$

where $\equiv$ denotes the octave equivalence. □

Let the octave be decomposed into $p+1$ smaller intervals, segments (semitones in the case of 12 degree tone systems). These segments are of two types – minor and major. Let the minor segment have m equal elementary intervals and the major segment have $m+n$, $m > n$ equal elementary intervals. Then the major or minor scales have $5(m+2n) + 2(m+n) = p+1$, i.e.

$$7m + 12n = p + 1$$

elementary intervals.

In the following theorem there are equal tempered tone systems of the first type. The proof is trivial and omitted.

Theorem 46 *If $m = 0$, then*

$$p + 1 = 12,, \ 24, 36, 48, 60, 72, \ldots, 612, \ldots.$$

So, we obtained specially the 12-tone equal temperature, 24-tone (1/4-tone) equal temperature, 36-tone (1/6-tone) equal temperature, 72-tone (1/12-tone) equal temperature (considered in the 1920s by Alois Hába, c.f. [23]), 612-tone system was considered by Josef Sumec in 1917.

Theorem 47 *If $m = 1$, then*

$$p + 1 = 19, 31, 43, 55, 67, \ldots.$$

The corresponding temperatures are:

$$0.995\,84, 0.997\,012, \quad 0.997\,530, \quad 0.997\,822, \quad 0.998\,009, \ldots$$

Proof.
$1.082\,4\,\tau_{3/2}{}^{19} = 1$, $1.097\,2\,\tau_{3/2}{}^{31} = 1$, $1.112\,2\,\tau_{3/2}{}^{43} = 1$, $\ldots$. □

These systems are also well known in the literature in the present day. An article which characterizes the very special properties of 12-, 19- and 31-tone scales is [71].

If $m = 2$, then $p + 1 = 38, 50, 62, \ldots$; etc. for $m = 3, 4, \ldots$. However, for $m \geq 2$ the fifth temperatures $\tau_{3/2_m}$ are greater than those in the cases $m = 0$ or 1.

Note that some tone systems for higher m are supersets of those for lower m. For example, the 38-tone system contains the 19-tone system, 62-tone contains the 31-tone system as their proper subset.

7.4.2 The Petzval's Tone Systems II

Properties For the Petzval's tone systems of the second Type we have the following expressions:

$$\Theta_{3/2,\tau_{3/2}}^{9} = \tau_{3/2}{}^{9} \cdot \frac{3^9}{2^{14}}, \; \Theta_{3/2,\tau_{3/2}}^{-8} = \tau_{3/2}{}^{-8} \cdot \frac{2^{13}}{3^8}, \; \Theta_{3/2,\tau_{3/2}}^{-14} = \tau_{3/2}{}^{-11} \cdot \frac{2^{23}}{3^{11}}.$$

These equations imply:

$$\tau_{6/5} \cdot \frac{6}{5} = \tau_{3/2}{}^{9} \cdot \frac{3^9}{2^{14}}, \; \tau_{5/4} \cdot \frac{5}{4} = \tau_{3/2}{}^{-8} \cdot \frac{2^{13}}{3^8}, \; \tau_{7/4} \cdot \frac{7}{4} = \tau_{3/2}{}^{-11} \cdot \frac{2^{23}}{3^{11}}.$$

Thus,

$$
\begin{aligned}
\tau_{6/5} &= \tau_{3/2}{}^{9} \cdot \frac{5 \cdot 3^8}{2^{15}}, \\
\tau_{5/4} &= \tau_{3/2}{}^{-8} \cdot \frac{2^{15}}{5 \cdot 3^8}, \\
\tau_{7/4} &= \tau_{3/2}{}^{-14} \cdot \frac{2^{25}}{7 \cdot 3^{14}}.
\end{aligned}
\tag{7.10}
$$

The consequences from the formulas (7.10) we collect into the following

Theorem 48 *For every Petzval tone system S_2 of the second type,*

(1) $\tau_{5/4} \cdot \tau_{6/5} = \tau_{3/2}$;

(2) *the major scale (based on the value $\Theta_{\tau_{3/2}}^{i}$):*

$$
\begin{aligned}
D_i^{(\text{major})} = \big\{ &\Theta_{3/2,\tau_{3/2}}^{i}, \Theta_{3/2,\tau_{3/2}}^{i+2}, \Theta_{3/2,\tau_{3/2}}^{i-8}, \Theta_{3/2,\tau_{3/2}}^{i-1}, \\
&\Theta_{3/2,\tau_{3/2}}^{i+1}, \Theta_{3/2,\tau_{3/2}}^{i-9}, \Theta_{3/2,\tau_{3/2}}^{i-7}, \Theta_{3/2,\tau_{3/2}}^{i} \big\};
\end{aligned}
$$

(3) *the minor scale (based on the value $\Theta^i_{\tau_{3/2}}$):*

$$D_i^{(\text{minor})} = \left\{ \Theta^i_{3/2,\tau_{3/2}}, \Theta^{i+2}_{3/2,\tau_{3/2}}, \Theta^{i+9}_{3/2,\tau_{3/2}}, \Theta^{i-1}_{3/2,\tau_{3/2}}, \right.$$
$$\left. \Theta^{i+1}_{3/2,\tau_{3/2}}, \Theta^{i+8}_{3/2,\tau_{3/2}}, \Theta^{i+10}_{3/2,\tau_{3/2}}, \Theta^i_{3/2,\tau_{3/2}} \right\} ;$$

(4) *the major and minor scales based on $\Theta^i_{3/2,\tau_{3/2}}$ and $\Theta^{i-9}_{3/2,\tau_{3/2}}$ are related: if we take the basic value $\Theta^i_{3/2,\tau_{3/2}}$ for the major scale and $\Theta^{i-9}_{3/2,\tau_{3/2}}$ for minor scale, then both these scales consist of the same numbers;*

(5) *there are two whole tone intervals: $W_1 = \tau_{3/2}{}^2 \cdot 9/8$ (the major whole tone) and $W_2 = \tau_{3/2}{}^{-10} \cdot 2^{16}/3^{10}$ (the minor whole tone);*

(6) *there are two semitones: $Y = \tau_{3/2}{}^7 \cdot 3^7/2^{11}$ (the major semitone) and $Z = \tau_{3/2}{}^{17} \cdot 2^{27}/3^{17}$ (the "cross" semitone);*

(7) *the following equations holds:*

$$\frac{Y}{Z} = \left(\tau_{3/2}{}^{12} \cdot \frac{3^{12}}{2^{19}} \right) = \left(\frac{W_1}{W_2} \right)^2 , \quad Z = \frac{X^2}{Y}.$$

Images in the complex plane   In Figure 7.5, we see the images of major and minor scales of Tone systems of the second type in the map ϑ.

Cyclic tone systems of the second type, Petzval scale   Let the octave be decomposed into $p + 1$ smaller intervals, segments (semitones in the case of 12 degree tone systems). The smallest segment is the interval between the major and minor whole tones, let it contain m disjoint equal elementary intervals. Let the minor semitone have $n, n > m$, elementary intervals. Then the major semitone has $n + 2m$ elementary intervals. Then the major and minor scales contain totally $3(2n + 3m) + 2(2n + 2m) + 2(n + 2m)$ elementary intervals, i.e.

$$12n + 17m = p + 1.$$

In the following theorem there are equal tempered tone systems of the second type. The proof is trivial and omitted.

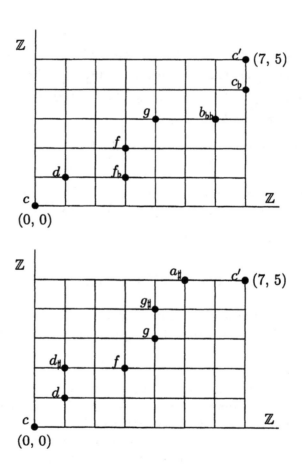

FIGURE 7.5: Major and minor scales of Tone systems II in $\tilde{\vartheta}$

Theorem 49 *If $m = 0$, then*

$$p + 1 = 12, 24, 36, 48, 60, 72, \ldots, 612, \ldots.$$

We see that the equal tempered Petzval's tone systems of the first and second types coincide.

Theorem 50 *If $m = 1$, then*

$$p + 1 = 41, 53, 65, 77, \ldots.$$

The corresponding temperatures are:

$$0.983\,512, \quad 0.999\,961, \quad 0.999\,758, \quad 0.999\,621, \quad \ldots.$$

Example 63 The 53-tone cyclic system, "Petzval scale." Here $m = 1$, the minor semitone ($\sqrt[53]{2^8} \approx 26/25$) contains 3 elementary intervals, the major semitone ($\sqrt[53]{2^9} \approx 1.067577$) has 5 ones, the minor whole tone (≈ 1.110295) has 8, and the major whole tone ($\approx 1.124\,911$) has 9 elementary intervals,

$$\tau_{3/2} = -25\,381/25\,382, \quad \tau_{5/4} = 1229/1230,$$
$$\tau_{6/5} = 1\,293/1\,292, \quad \tau_{7/4} = 364/363.$$

Theorem 51 *If $m = 2$, then*

$$p + 1 = 70, 82, 94, 106, \ldots.$$

We can consider $m = 3, 4, \ldots$. However, these systems seem to bring no new or interesting quality from the acoustic viewpoint.

7.5 Bimeasures and the last square method

What about the interplay between two temperaments of a system S? We need to construct a measure for the distance between two uncertain sets.

Let $\Sigma^{(a)} \in \mathfrak{S}_{\mathbb{F}}$ and $\Sigma^{(m)} \in \mathfrak{S}_{\mathbb{F}}$, be two psychoacoustic and musical temperaments of an (unknown) tone system $S \in \mathfrak{S}_{\mathbb{F}}$ which we would like to be well tempered.

There are questions: given musical temperament or its structure, find concrete S with optimal psychoacoustic temperament in some sense. Specially, to find S with the optimal psychoacoustic temperament when the musical mistuning is arbitrary.

The following uncertainty bimeasure is motivated with the last squares method:

$$\lambda(\Sigma^{(a)}, \Sigma^{(m)}) = \frac{1}{\sum_{\phi \in L} w_\phi \left[\mu_\phi^{(m)} - \mu_\phi^{(a)} \right]^2}, \qquad (7.11)$$

where

$$\mu_\phi^{(m)} = \min_{w \in \ker(\Sigma^{(m)}) \cap [1;2)} \left| 1 - \frac{\omega}{\phi} \right| \cdot \mathrm{sgn}\left(1 - \frac{\omega}{\phi} \right),$$

$$\mu_\phi^{(a)} = \min_{w \in \ker(\Sigma^{(a)}) \cap [1;2)} \left| 1 - \frac{\omega}{\phi} \right| \cdot \mathrm{sgn}\left(1 - \frac{\omega}{\phi} \right),$$

and $\phi \in \{6/5, 5/4, 3/2, 7/4\} = L$, $\sum_{\phi \in L}$, $w_\phi = 1$, $w_\phi > 0$.

Note that Definition 24 works for the case of crisp situation up to this place and the uncertainty context is a direct generalization of the crisp case. In this section we will demonstrate results essentially using the uncertainty based information theory and such that the reduction to crisp case is trivial and pure.

The following Subsection 7.5.1 concerns the Petzval's tone systems of the first type.

7.5.1   Optimal temperatures I

Let us search a tone system $S = \Sigma^{(a)}$ among the Petzval tone systems of the first type. Substitute (7.8) into the function $\lambda(\Sigma^{(a)}, \Sigma^{(m)})$, c.f. (7.11). For the sake of simplicity, let $w_{6/5} = w_{5/4} = w_{3/2} = w_{7/4}$ in Equation (7.11).

The function $\lambda(\Sigma^{(a)}, \Sigma^{(m)})$ is then a function of the variable $\tau_{3/2}$. We evaluate the maximum of the function $\lambda(\Sigma^{(a)}, \Sigma^{(m)})$ applying the well-known method from the mathematical analysis (the first derivative put to zero, the second derivative should be negative in the solution points, etc.). We obtain the following algebraic equation of the 26th degree with respect to psychoacoustic temperature $\tau_{3/2} = \tau_{3/2}^{(a)}$:

$$-2(\tau_{3/2}^{(m)} - \tau_{3/2}) - 8 \cdot \frac{81 \cdot \tau_{3/2}^3}{80} \left(\tau_{5/4}^{(m)} - \frac{81 \cdot \tau_{3/2}^4}{80} \right)$$
$$+6 \cdot \frac{80}{81 \cdot \tau_{3/2}^4} \left(\tau_{6/5}^{(m)} - \frac{80}{81 \cdot \tau_{3/2}^3} \right) + 20 \cdot \tau_{3/2}^9 \left(\tau_{7/4}^{(m)} - \frac{59\,049 \cdot \tau_{3/2}^{10}}{57\,344} \right) = 0.$$

$$(7.12)$$

Since $\mu_{3/2}^{(m)}, \mu_{5/4}^{(m)}, \mu_{6/5}^{(m)}, \mu_{7/4}^{(m)}$ we can choose arbitrary within reasonable boundaries, let $\mu_{3/2}^{(m)} = 0, \mu_{5/4}^{(m)} = 0, \mu_{6/5}^{(m)} = 0, \mu_{7/4}^{(m)} = 0$ (pure intervals). As the result, we obtained the following "ideal" temperature of the searched tone system $S \in \mathfrak{S}_{\mathbb{F}}$:

$$\tau_{3/2} \approx 0.997\,224, \tau_{5/4} \approx 1.001\,305, \tau_{6/5} \approx 0.995\,925, \tau_{7/4} \approx 1.001\,505.$$

In the following two examples we will suppose that we have some *a priori* knowledge about the structure of the set $\{\mu_{3/2}^{(m)}, \mu_{5/4}^{(m)}, \mu_{6/5}^{(m)}, \mu_{7/4}^{(m)}\}$.

Example 64 Huygens–Fokker tone system. Let $\tau_{3/2}^{(m)} = \tau_{5/4}^{(m)} = \tau_{6/5}^{(m)} = \tau_{7/4}^{(m)}$. The equation (7.12) implies $\mu_{3/2} \approx 1/336$ and

$$\tau_{3/2} \approx 335/336, \tau_{5/4} \approx 2013/2012, \tau_{6/5} \approx 287/288, \tau_{7/4} \approx 335/336.$$

The closest cyclic tone system is given with $x/N = 18/31$, i.e. a 31-tone system with the tempered fifth on the 18-th step. This tone system is known as the Huygens–Fokker tone system, c.f. [108].

Example 65 A 43-tone system. Let $\tau_{3/2}^{(m)} = \tau_{5/4}^{(m)}/5 = \tau_{6/5}^{(m)}/5 = \tau_{7/4}^{(m)}/5$. The equation (7.12) implies

$$\mu_{3/2} \approx 1/397, \quad \tau_{3/2} \approx 396/397.$$

The closest cyclic system is for $x/N = 25/43$, i.e. a 43-tone system with the tempered fifth on the 25-th step.

7.5.2   Optimal temperatures II

Not bringing the boring computations of the expression (7.12), for
the case of the Petzval's tone system of the second type with the
minimal mistuning we have the following results:

$$\tau_{3/2} \approx 1.000\,034,\ \tau_{5/4} = 0.998\,6,\ \tau_{6/5} \approx 1.001\,436,\ \tau_{7/4} \approx 1.001\,722.$$

We see that all basic intervals are mistuned a tiny amount, so we
hear them as pure. The Petzvals tone systems of the second type
are excellent. However, there is a cost to be paid with the relatively
large number of steps per octave.

Remarks and ideas for exploring

10 Since superparticular ratios are mentioned by some theorists as an important
subject for "pure" harmony, for a curiosity, there is a list of the superparticular
ratios used in Section 7.4:
2/1, 3/2, 5/4, 6/5, 64/63, 81/80,
235/234, 240/239, 288/287, 336/335, 364/363, 455/454, 886/885,
1 230/1 229, 1 293/1 292, 2 013/2 012, 11 561/11 560, 25 382/25 381.

11 Note that $\Theta_{3/2,\tau_{3/2}}^{10}$ is a better approximation of 7/4 than $\Theta_{3/2,\tau_{3/2}}^{-2}$ when
$\tau_{3/2} < 1$, c.f. Remark 12. Therefore, we start from $\Theta_{3/2,\tau_{3/2}}^{10}$ when constructing
the Petzval's tone systems sequence and omit $\Theta_{3/2,\tau_{3/2}}^{-2}$ at all.

12 We see also that the reasonable are only the tone systems of the 1st type
with $\tau_{3/2} < 1$ (since if $\tau_{3/2} \geq 1$, then $\tau_{5/4} > 81/80$, analogously $\tau_{6/5}$. Also,
$\tau_{7/4} > 59\,049/57\,344$ is a big mistuning of the sevenths). Also, here is the reason
why we choose $\Theta_{3/2,\tau_{3/2}}^{10}$ and not $\Theta_{3/2,\tau_{3/2}}^{-2}$: the corresponding $\tau_{7/4} = 64/(63\tau_{3/2}^2)$
which is not appropriate for $\tau_{3/2} < 1$.

13 The condition $\tau_{3/2} < 1$ in Theorem 48 is superfluous, we can choose $\tau_{3/2} < 1$, $\tau_{3/2} > 1$, or $\tau_{3/2} = 1$. For $\tau_{3/2} = 1$, we have: $\tau_{5/4} = 885/886$, $\tau_{6/5} = 886/885$,
$\Theta_{7/4} = 455/454$. For $\tau_{3/2} < 1$, we obtain tone systems with purer thirds. For
$\tau_{3/2} > 1$, we obtain tone systems with purer sevenths.
There in no tetrachordal structure of the major scale.

14 J. M. Petzval introduced and considered his tone systems in the crisp case.
He used a naive approach (from pre present viewpoint, not using any measure
theory) when measuring tone systems and verified his evaluations with musical
and physical experience; being himself a fine mechanician, he constructed and
played a reed organ equipped with the keyboard with 53 steps per octave. The

corresponding cyclic 53-tone system is now known also as *Petzval scale*, c.f. Example 63. We do not add any adjective like "fuzzy" or "generalized" to the term "Petzval's tone system."

15 From Figure 7.4 it is clear, c.f. [133], that the Petzval's tone system of the first type is a direct generalization of the Pythagorean Tuning structure.

From Figure 7.5 it is clear, c.f. [129], that the Petzval's tone system of the second type is a direct generalization of Just Intonation.

Chapter 8

12 and 10 granulations

8.1 The major scale extensions

8.1.1 Unimodular matrices and diatonic scales

In this section we will deal with a subset of a group of all unimodular 3×3 matrixes (noncommutative group of matrixes A, such that $\det(A) = 1$) derived from geometric nets with 12 granules per octave and 3 quotients. The research is inspired with diatonic scales in music.

Considering the 12-granulation of octave in European music, the ambiguous intervals in the 17-valued Just Intonation Set are the second and the minor seventh and the tritone (the relative frequencies $10/9$, $9/8$, $8/7$; and $7/4$, $16/9$, $18/10$; and $45/32$, $64/45$, respectively). Pythagorean System we can consider also 17- or more valued. Indian tone systems yield another type of ambiguity—it is present in the form of ragas which are restrictions of 22-tone Sruti systems mostly to 7 tone scales. Finally, scales of the Pacific region, Slendro and Pelog, represent fuzzy type of uncertainty because there is no crisp defined pitch of tones. We will study the ambiguity of European tone systems which contain the major scale.

Precisely the question is as follows: find all 5-limit 12-tone systems containing the C-major scale (relative frequencies $1/1$, $9/8$, $5/4$, $4/3$, $3/2$, $5/3$, $15/8$, and 2).

It is clear that each tone system (and scales in particular) can be represented via a geometric net. So, let us search again the answer to the question in the form of geometric net. The results will be then consequences of solutions of Diophantine equations describing the basic acoustic relations among octave, perfect fifth and major (minor) third. The 12-granule system has properties of the Just Intonation Set (it involves octave, perfect fifth, perfect fourth, major third, minor third, major whole tone, minor whole tone, diatonic semitone and chromatic semitone) and also of Pythagorean System. Further, there are applications to superparticular ratios.

The ratios $256/243, 25/24, 16/15$ are known as the *minor Pythagorean, chromatic,* and *diatonic semitone,* respectively. Just Intonation [when we avoid the minor seventh $(7/4)$ and the second $(8/7)$] is constructed on the basis of the chromatic and diatonic semitones. This tone system is often mentioned in Europe as the most natural tone system from many viewpoints (physical, psycho-acoustical, polyphonic, etc.). On the other side, Pythagorean System is based exclusively on the minor Pythagorean semitone, *diesis* $(256/243)$. Thus the Just Intonation Set and Pythagorean System are considered by music theoreticians as two fully incompatible tone systems.

This viewpoint is not correct. In fact, not only the diatonic and chromatic *but also the minor Pythagorean semitone* (together with the diatonic semitone and its complement to the major whole tone) can serve as a basis for the construction of 12-granule diatonic scales, c.f. Table 8.3. Further, we will show in this section why the *gypsy scales* are important tone systems.

We will use the notation of the geometric nets given in Section 4.2.2.

Definition 26 We say that a matrix

$$\mathbb{A} = \begin{bmatrix} \nu_{12,1} & \nu_{12,2} & \nu_{12,3} \\ \nu_{7,1} & \nu_{7,2} & \nu_{7,3} \\ \nu_{4,1} & \nu_{4,2} & \nu_{4,3} \end{bmatrix}$$

of nonnegative integers, is a (12, 7, 4)-*matrix,* if

$$0 \leq \nu_{4,j} \leq \nu_{7,j} \leq \nu_{12,j}, \quad j = 1, 2, 3$$

and

$$\nu_{i,1} + \nu_{i,2} + \nu_{i,3} = i, \quad i = 4, 7, 12.$$

In other words, we say that a matrix $(\nu_{i,j})_{12,7,4}^{1,2,3} \in \mathbb{N}^3 \times \mathbb{N}^3$ is a (12, 7, 4)-*matrix*, c.f. [128], if $0 \leq \nu_{4,\cdot} \leq \nu_{7,\cdot} \leq \nu_{12,\cdot}$ and $|\nu_{i,\cdot}| = i$, $i = 12, 7, 4$.

Definition 27 If to a given (12, 7, 4)-matrix $\mathbb{A}$ there exists an M-granule n-quotient system S, then we say that the tone system S is *generated* by $\mathbb{A}$.

We will find appropriate (12, 7, 4)-matrices and then construct and consider the needed generated 12-granule 3-quotient systems.

Theorem 52 Let $\mathbb{A} = (\nu_{i,j})_{i=12,7,4}^{j=1,2,3} \in \mathbb{N}^3 \times \mathbb{N}^3$ be a matrix such that $\det \mathbb{A} \neq 0$. Then there exists a unique $X \in \mathbb{Q}^3$, such that

$$X^{\nu_{12,\cdot}} = 2, \quad X^{\nu_{7,\cdot}} = 3/2, \quad X^{\nu_{4,\cdot}} = 5/4, \tag{8.1}$$

and the following statements are equivalent:
(i) $X \in \mathbb{Q}^3$, (ii) $\det \mathbb{A} = 1$,
where $\mathbb{Q}$ denotes the set of all rational numbers. The values are as follows:

$$X_1 = {}^{\det \mathbb{A}}\sqrt{2^{D_{2,1}} 3^{D_{3,1}} 5^{D_{5,1}}},$$
$$X_2 = {}^{\det \mathbb{A}}\sqrt{2^{D_{2,2}} 3^{D_{3,2}} 5^{D_{5,2}}},$$
$$X_3 = {}^{\det \mathbb{A}}\sqrt{2^{D_{2,3}} 3^{D_{3,3}} 5^{D_{5,3}}},$$

where

$$D_{2,1} = \begin{vmatrix} +1 & \nu_{12,2} & \nu_{12,3} \\ -1 & \nu_{7,2} & \nu_{7,3} \\ -2 & \nu_{4,2} & \nu_{4,3} \end{vmatrix}, D_{2,2} = \begin{vmatrix} \nu_{12,1} & +1 & \nu_{12,3} \\ \nu_{7,1} & -1 & \nu_{7,3} \\ \nu_{4,1} & -2 & \nu_{4,3} \end{vmatrix}, D_{2,3} = \begin{vmatrix} \nu_{12,1} & \nu_{12,2} & +1 \\ \nu_{7,1} & \nu_{7,2} & -1 \\ \nu_{4,1} & \nu_{4,2} & -2 \end{vmatrix}$$

$$D_{3,1} = \begin{vmatrix} 0 & \nu_{12,2} & \nu_{12,3} \\ 1 & \nu_{7,2} & \nu_{7,3} \\ 0 & \nu_{4,2} & \nu_{4,3} \end{vmatrix}, D_{3,2} = \begin{vmatrix} \nu_{12,1} & 0 & \nu_{12,3} \\ \nu_{7,1} & 1 & \nu_{7,3} \\ \nu_{4,1} & 0 & \nu_{4,3} \end{vmatrix}, D_{3,3} = \begin{vmatrix} \nu_{12,1} & \nu_{12,2} & 0 \\ \nu_{7,1} & \nu_{7,2} & 1 \\ \nu_{4,1} & \nu_{4,2} & 0 \end{vmatrix},$$

$$D_{5,1} = \begin{vmatrix} 0 & \nu_{12,2} & \nu_{12,3} \\ 0 & \nu_{7,2} & \nu_{7,3} \\ 1 & \nu_{4,2} & \nu_{4,3} \end{vmatrix}, D_{5,2} = \begin{vmatrix} \nu_{12,1} & 0 & \nu_{12,3} \\ \nu_{7,1} & 0 & \nu_{7,3} \\ \nu_{4,1} & 1 & \nu_{4,3} \end{vmatrix}, D_{5,3} = \begin{vmatrix} \nu_{12,1} & \nu_{12,2} & 0 \\ \nu_{7,1} & \nu_{7,2} & 0 \\ \nu_{4,1} & \nu_{4,2} & 1 \end{vmatrix}.$$

Proof. (i) $\Rightarrow$ (ii). If $X_1, X_2, X_3 \in \mathbb{Q}$, then there exist a prime number $p \in \mathbb{P}$ and $E_{2,X_1}, \ldots, E_{p,X_3} \in \mathbb{Z}$ such that

$$X_1 = 2^{E_{2,X_1}} 3^{E_{3,X_1}} \ldots p^{E_{p,X_1}},$$
$$X_2 = 2^{E_{2,X_2}} 3^{E_{3,X_2}} \ldots p^{E_{p,X_2}}, \tag{8.2}$$
$$X_3 = 2^{E_{2,X_3}} 3^{E_{3,X_3}} \ldots p^{E_{p,X_3}}.$$

Combining (8.1) and (8.2),

$$
\begin{bmatrix}
E_{2,X_1} & E_{2,X_2} & E_{2,X_3} \\
E_{3,X_1} & E_{3,X_2} & E_{3,X_3} \\
E_{5,X_1} & E_{5,X_2} & E_{5,X_3} \\
E_{7,X_1} & E_{7,X_2} & E_{7,X_3} \\
& \cdots & \\
E_{p,X_1} & E_{p,X_2} & E_{p,X_3}
\end{bmatrix}
\begin{bmatrix}
\nu_{12,1} & \nu_{7,1} & \nu_{4,1} \\
\nu_{12,2} & \nu_{7,2} & \nu_{4,2} \\
\nu_{12,3} & \nu_{7,3} & \nu_{4,3}
\end{bmatrix}
=
$$

$$
=
\begin{bmatrix}
1 & -1 & -2 \\
0 & 1 & 0 \\
0 & 0 & 1 \\
0 & 0 & 0 \\
& \cdots & \\
0 & 0 & 0
\end{bmatrix}.
$$

$(ii) \Rightarrow (i)$ Since $\det A = 1$,

$$
\begin{bmatrix}
E_{2,X_1} & E_{2,X_2} & E_{2,X_3} \\
E_{3,X_1} & E_{3,X_2} & E_{3,X_3} \\
E_{5,X_1} & E_{5,X_2} & E_{5,X_3} \\
E_{7x} & E_{7,X_2} & E_{7,X_3} \\
& \cdots & \\
E_{p,X_1} & E_{p,X_2} & E_{p,X_3}
\end{bmatrix}
=
\begin{bmatrix}
D_{2,X_1} & D_{2,X_2} & D_{2,X_3} \\
D_{3,X_1} & D_{3,X_2} & D_{3,X_3} \\
D_{5,X_1} & D_{5,X_2} & D_{5,X_3} \\
0 & 0 & 0 \\
& \cdots & \\
0 & 0 & 0
\end{bmatrix}. \qquad \square
$$

Theorem 53 *No 12-granulated system with 1- or 2- rational quotients is generated by any $(12,7,4)$-matrix.*

Proof.   The analysis of all $(12, 7, 4)$-matrices with $\det A = 1$ contains no case $D_{2,X_1} = D_{3,X_1} = D_{5,X_1} = 0$. So, $D_{2,X_1}^2 + D_{3,X_1}^2 + D_{5,X_1}^2 > 0$. Analogously, $D_{2,X_2}^2 + D_{3,X_2}^2 + D_{5,X_2}^2 > 0$, $D_{2,X_3}^2 + D_{3,X_3}^2 + D_{5,X_3}^2 > 0$. The assertion for 1-quotient systems is trivial. $\square$

Corollary 54 *Neither the 12-tone Equal Temperament nor Pythagorean System are generated by any $(12,7,4)$-matrix.*

Theorem 55 *Let A be a $(12,7,4)$-matrix. Then*

$$
D_{2,X_1}^2 + D_{2,X_2}^2 + D_{2,X_3}^2 > 0,
$$
$$
D_{3,X_1}^2 + D_{3,X_2}^2 + D_{3,X_3}^2 > 0,
$$
$$
D_{5,X_1}^2 + D_{5,X_2}^2 + D_{5,X_3}^2 > 0,
$$

where D_{2,X_1}, D_{2,X_2}, D_{2,X_3}, D_{3,X_1}, D_{3,X_2}, D_{3,X_3}, D_{5,X_1}, D_{5,X_2}, D_{5,X_3} are as in Theorem 52.

Proof. At first, note that a $(12, 7, 4)$-matrix exists, e.g., the matrix $\mathbb{A}_3$ in Theorem 59. If $D_{5,X_1}^2 + D_{5,X_2}^2 + D_{5,X_3}^2 = 0$, then there exist no $\nu_{4,1}$, $\nu_{4,2}$, $\nu_{4,3}$ in A such that $X_1^{\nu_{4,1}} X_2^{\nu_{4,2}} X_3^{\nu_{4,3}} = 5/4$. A contradiction. Similarly for the numbers 2 and 3. □

Theorem 56 *Let A be an unimodular $(12, 7, 4)$-matrix. Then $X_1 \neq X_2$, $X_2 \neq X_3$, $X_3 \neq X_1$.*

Proof. If $1 < X_1 = X_2 = X_3$, then $X_1 \notin \mathbb{Q}$. A contradiction.

Suppose $X_1 = X_2 \neq Z$ (the cases $X_2 = X_3 \neq X_1$, $X_1 = X_3 \neq X_2$ are symmetric). By Definition 26 and Definition 27,

$$X_1^{4-\nu_{4,3}} X_3^{\nu_{4,3}} = 5/4, \quad X_1^{7-\nu_{7,3}} X_3^{\nu_{7,3}} = 3/2, \quad X_1^{12-\nu_{12,3}} X_3^{\nu_{12,3}} = 2. \tag{8.3}$$

By Theorem 52, $X_1 = 2^\alpha 3^\beta 5^\gamma$, $X_3 = 2^\delta 3^\epsilon 5^\theta$, for some α, β, γ, δ, ϵ, $\theta \in \mathbb{Z}$. Then (8.3) implies

$$\begin{aligned}
\alpha(12 - \nu_{12,3}) + \nu_{12,3}\delta &= 1, \\
\beta(12 - \nu_{12,3}) + \nu_{12,3}\epsilon &= 0, \\
\gamma(12 - \nu_{12,3}) + \nu_{12,3}\theta &= 0,
\end{aligned} \tag{8.4}$$

$$\begin{aligned}
\alpha(7 - \nu_{7,3}) + \nu_{7,3}\delta &= -1, \\
\beta(7 - \nu_{7,3}) + \nu_{7,3}\epsilon &= 1, \\
\gamma(7 - \nu_{7,3}) + \nu_{7,3}\theta &= 0,
\end{aligned} \tag{8.5}$$

$$\begin{aligned}
\alpha(4 - \nu_{4,3}) + \nu_{4,3}\delta &= -2, \\
\beta(4 - \nu_{4,3}) + \nu_{4,3}\epsilon &= 0, \\
\gamma(4 - \nu_{4,3}) + \nu_{4,3}\theta &= 1.
\end{aligned} \tag{8.6}$$

If $\epsilon = \beta$, then (8.4) implies $\beta = 1/7 \notin \mathbb{Z}$. If $\gamma = \theta$, then (8.5) implies $\gamma = 1/4 \notin \mathbb{Z}$. So, $\epsilon \neq \beta$, $\theta \neq \gamma$. Then (8.4), and (8.5) imply

$$\nu_{12,3} = \frac{-12\beta}{\epsilon - \beta} = \frac{-12\gamma}{\theta - \gamma}, \nu_{4,3} = \frac{-4\beta}{\epsilon - \beta} = \frac{1 - 4\gamma}{\theta - \gamma}. \tag{8.7}$$

If $\beta \neq 0$, then (8.6) implies

$$\frac{-12\beta}{-4\beta} = \frac{-12\gamma}{1-4\gamma},$$

which implies $0 = 1$. If $\beta = 0$ then (8.6) implies $0 = \gamma = 1/4$. A contradiction. $\square$

Corollary 57 *If S is a 12-granule 3-quotient $(2/1, 3/2, 5/4)$-system with $X_1, X_2, X_3 \in \mathbb{Q}$, then we can redenote (order) X_1, X_2, X_3, such that $1 < X_1 < X_2 < X_3 < 10/9$.*

8.1.2   TDS geometric nets

Now, we restrict the class of all 12-granule 3-quotient geometric nets generated by (12, 7, 4)-matrices to TDS geometric nets.

Definition 28 We say that a 12-granule 3-quotient geometric net $\langle \Gamma_i \rangle$ generated by a (12, 7, 4)-matrix A is a *TDS geometric net* if

$$\begin{aligned}
\nu_{2,\cdot} &= 2\nu_{7,\cdot} - \nu_{12,\cdot}, \\
\nu_{5,\cdot} &= \nu_{12,\cdot} - \nu_{7,\cdot}, \\
\nu_{9,\cdot} &= \nu_{12,\cdot} - \nu_{7,\cdot} + \nu_{4,\cdot}, \\
\nu_{11,\cdot} &= \nu_{7,\cdot} + \nu_{4,\cdot},
\end{aligned}$$

and for every $i \in \mathbb{Z}$, there exists $p \in \mathbb{N}$, $0 \leq p < 12$, and $q \in \mathbb{N}$, such that $\nu_{i,\cdot} = q\nu_{12,\cdot} + \nu_{p,\cdot}$.

Note that the members Γ_i, $i = 1, 3, 6, 8, 10$, mentioned in Definition 28, are determined ambiguously.

Theorem 58 *According to the symmetry, all generators $X \in \mathbb{Q}^3$ for TDS geometric nets*

$$\langle \Gamma_i \rangle = \langle X^{\nu_{i,\cdot}}; \ |\nu_{i,\cdot}| = i, 0 \leq \nu_{0,\cdot} \leq \nu_{1,\cdot} \leq \cdots \leq \nu_{i,\cdot} \leq \ldots \rangle_{\nu_{i,\cdot} \in \mathbb{N}^3}$$

with the subsequences

$$\langle \Gamma_{12l} \rangle = \left\langle 2^l \right\rangle, \quad \langle \Gamma_{12l+7} \rangle = \left\langle 3 \cdot 2^{l-1} \right\rangle, \quad \langle \Gamma_{12l+4} \rangle = \left\langle 5 \cdot 2^{l-2} \right\rangle$$

are the following:

$$(25/24, 135/128, 16/15),$$
$$(256/243, 135/128, 16/15),$$
$$(25/24, 16/15, 27/25).$$

Proof. The analysis of the Diophantine equation

$$\det[(\nu_{i,j})_{i=12,7,4}^{j=1,2,3}] = 1, \quad 0 \le \nu_{4,\cdot} \le \nu_{7,\cdot} \le \nu_{12,\cdot}, \quad |\nu_{i,\cdot}| = i$$

in $\mathbb{N}^3 \times \mathbb{N}^3$ with the additional (not restricting the solution) condition

$$2\nu_{7,\cdot} - \nu_{12,\cdot} \ge 0$$

yields the following matrices (excluding symmetries, i.e., permutations of columns):

$$A_1 = \begin{bmatrix} 2 & 7 & 3 \\ 1 & 4 & 2 \\ 1 & 2 & 1 \end{bmatrix}, \quad A_2 = \begin{bmatrix} 2 & 5 & 5 \\ 1 & 3 & 3 \\ 1 & 1 & 2 \end{bmatrix},$$

$$A_3 = \begin{bmatrix} 5 & 4 & 3 \\ 3 & 2 & 2 \\ 2 & 1 & 1 \end{bmatrix}, \quad A_4 = \begin{bmatrix} 1 & 2 & 9 \\ 1 & 1 & 5 \\ 0 & 1 & 3 \end{bmatrix},$$

$$A_5 = \begin{bmatrix} 1 & 3 & 8 \\ 1 & 2 & 4 \\ 0 & 1 & 3 \end{bmatrix}, \quad A_6 = \begin{bmatrix} 1 & 4 & 7 \\ 1 & 2 & 4 \\ 1 & 1 & 2 \end{bmatrix},$$

$$A_7 = \begin{bmatrix} 1 & 5 & 6 \\ 1 & 3 & 3 \\ 1 & 2 & 1 \end{bmatrix}, \quad A_8 = \begin{bmatrix} 2 & 3 & 7 \\ 1 & 2 & 4 \\ 1 & 0 & 3 \end{bmatrix}.$$

Apply Theorem 52 and find all sequences by the algorithm in Definition 28. Excluding all such sequences $\langle \Gamma_i \rangle$ which do not satisfy the condition $\nu_{0,\cdot} \le \nu_{1,\cdot} \le \cdots \le \nu_{i,\cdot} \le \dots$, we obtain the following three matrices: A_1, A_2, A_3.

In Table 8.1 (where $(X_1, X_2, X_3) = (25/24, 16/15, 27/25)$), Table 8.2 (where $(X_1, X_2, X_4) = (25/24, 16/15, 135/128)$), and Table 8.3 (where $(X_5, X_2, X_4) = (256/243, 16/15, 135/128)$) there are

TABLE 8.1: Class $\mathfrak{G}$

...	...	...	...	...
$X_1^0 X_2^0 X_3^0$	$2^0 3^0 5^0$	1/1	1.000 000	C
$X_1^1 X_2^0 X_3^0$	$2^{-3} 3^{-1} 5^2$	25/24	1.0416 667	$C_\sharp$
$X_1^0 X_2^0 X_3^1$	$2^0 3^3 5^{-2}$	27/25	1.080 000	$D_\flat$
$X_1^1 X_2^0 X_3^1$	$2^{-3} 3^2 5^0$	9/8	1.125 000	D
$X_1^2 X_2^0 X_3^1$	$2^{-6} 3^1 5^2$	75/64	1.171 875	$D_\sharp$
$X_1^1 X_2^1 X_3^1$	$2^1 3^1 5^{-1}$	6/5	1.200 000	$E_\flat$
$X_1^2 X_2^1 X_3^1$	$2^{-2} 3^0 5^1$	5/4	1.250 000	E
$X_1^2 X_2^2 X_3^1$	$2^2 3^{-1} 5^0$	4/3	1.333 333	F
$X_1^3 X_2^2 X_3^1$	$2^{-1} 3^{-2} 5^2$	25/18	1.388 889	$F_\sharp$
$X_1^2 X_2^2 X_3^2$	$2^2 3^2 5^{-2}$	36/25	1.440 000	$G_\flat$
$X_1^3 X_2^2 X_3^2$	$2^{-1} 3^1 5^0$	3/2	1.500 000	G
$X_1^4 X_2^2 X_3^2$	$2^{-4} 3^0 5^2$	25/16	1.562 500	$G_\sharp$
$X_1^3 X_2^3 X_3^2$	$2^3 3^0 5^{-1}$	8/5	1.600 000	$A_\flat$
$X_1^4 X_2^3 X_3^2$	$2^0 3^{-1} 5^1$	5/3	1.666 667	A
$X_1^5 X_2^3 X_3^2$	$2^{-3} 3^{-2} 5^3$	125/72	1.736 111	$A_\sharp$
$X_1^4 X_2^3 X_3^3$	$2^0 3^2 5^{-1}$	9/5	1.800 000	$B_\flat$
$X_1^5 X_2^3 X_3^3$	$2^{-3} 3^1 5^1$	15/8	1.875 000	B
$X_1^5 X_2^4 X_3^3$	$2^1 3^0 5^0$	2/1	2.000 000	C'
...	...	...	...	...

TABLE 8.2: Class $\mathfrak{R}$

...	...	...	...	...
$X_1^0 X_2^0 X_4^0$	$2^0 3^0 5^0$	1/1	1.000 000	C
$X_1^0 X_2^0 X_4^1$	$2^{-7}3^3 5^1$	135/128	1.054 688	$C_\sharp$
$X_1^0 X_2^1 X_4^0$	$2^4 3^{-1} 5^{-1}$	16/15	1.066 667	$D_\flat$
$X_1^0 X_2^1 X_4^1$	$2^{-3}3^2 5^0$	9/8	1.125 000	D
$X_1^1 X_2^1 X_4^1$	$2^{-6}3^1 5^2$	75/64	1.171 875	$D_\sharp$
$X_1^0 X_2^2 X_4^1$	$2^1 3^1 5^{-1}$	6/5	1.200 000	$E_\flat$
$X_1^1 X_2^2 X_4^1$	$2^{-2}3^0 5^1$	5/4	1.250 000	E
$X_1^1 X_2^3 X_4^1$	$2^2 3^{-1} 5^0$	4/3	1.333 333	F
$X_1^1 X_2^3 X_4^2$	$2^{-5}3^2 5^1$	45/32	1.406 250	$F_\sharp$
$X_1^1 X_2^4 X_4^1$	$2^6 3^{-2} 5^{-1}$	64/45	1.422 222	$G_\flat$
$X_1^1 X_2^4 X_4^2$	$2^{-1}3^1 5^0$	3/2	1.500 000	G
$X_1^2 X_2^4 X_4^2$	$2^{-4}3^0 5^2$	25/16	1.562 500	$G_\sharp$
$X_1^1 X_2^5 X_4^2$	$2^3 3^0 5^{-1}$	8/5	1.600 000	$A_\flat$
$X_1^2 X_2^5 X_4^2$	$2^0 3^{-1} 5^1$	5/3	1.666 667	A
$X_1^2 X_2^5 X_4^3$	$2^{-7}3^2 5^2$	225/128	1.757 812	$A_\sharp$
$X_1^2 X_2^6 X_4^2$	$2^4 3^{-2} 5^0$	16/9	1.777 777	$B_\flat$
$X_1^2 X_2^6 X_4^3$	$2^{-3}3^1 5^1$	15/8	1.875 000	B
$X_1^2 X_2^7 X_4^3$	$2^1 3^0 5^0$	2/1	2.000 000	C'

TABLE 8.3: Class $\mathfrak{P}$

...	...	...	...	...
$X_5^0 X_2^0 X_4^0$	$2^0 3^0 5^0$	1/1	1.000 000	C
$X_5^0 X_2^0 X_4^1$	$2^{-7} 3^3 5^1$	135/128	1.054 688	$C_\sharp$
$X_5^0 X_2^1 X_4^0$	$2^4 3^{-1} 5^{-1}$	16/15	1.066 667	$D_\flat$
$X_5^0 X_2^1 X_4^1$	$2^{-3} 3^2 5^0$	9/8	1.125 000	D
$X_5^1 X_2^1 X_4^1$	$2^5 3^{-3} 5^0$	32/27	1.185 185	$D_\sharp$
$X_5^0 X_2^1 X_4^2$	$2^{-10} 3^5 5^1$	1215/1 024	1.186 523	$E_\flat$
$X_5^1 X_2^1 X_4^2$	$2^{-2} 3^0 5^1$	5/4	1.250 000	E
$X_5^1 X_2^2 X_4^2$	$2^2 3^{-1} 5^0$	4/3	1.333 333	F
$X_5^1 X_2^2 X_4^3$	$2^{-5} 3^2 5^1$	45/32	1.406 250	$F_\sharp$
$X_5^1 X_2^3 X_4^2$	$2^6 3^{-2} 5^{-1}$	64/45	1.422 222	$G_\flat$
$X_5^1 X_2^3 X_4^3$	$2^{-1} 3^1 5^0$	3/2	1.500 000	G
$X_5^2 X_2^3 X_4^3$	$2^7 3^{-4} 5^0$	128/81	1.580 247	$G_\sharp$
$X_5^1 X_2^3 X_4^4$	$2^{-8} 3^4 5^1$	405/256	1.582 031	$A_\flat$
$X_5^2 X_2^3 X_4^4$	$2^0 3^{-1} 5^1$	5/3	1.666 667	A
$X_5^2 X_2^3 X_4^5$	$2^{-7} 3^2 5^2$	225/128	1.757 812	$A_\sharp$
$X_5^2 X_2^4 X_4^4$	$2^4 3^{-2} 5^0$	16/9	1.777 778	$B_\flat$
$X_5^2 X_2^4 X_4^5$	$2^{-3} 3^1 5^1$	15/8	1.875 000	B
$X_5^2 X_2^5 X_4^5$	$2^1 3^0 5^0$	2/1	2.000 000	C'

listed values of all TDS-geometrical nets $\langle \Gamma_i \rangle$ (in the sixths column, there is a musical denotation) corresponding to the matrices A_1, A_2, and A_3. $\square$

In the connection with the previous theorem we mention here that the analysis of all $(12, 7, 4)$-matrices A with $\det A = 1$ yields the following surprising statement.

Theorem 59 *According to the symmetry,*

$$A_3 = \begin{bmatrix} 5 & 4 & 3 \\ 3 & 2 & 2 \\ 2 & 1 & 1 \end{bmatrix}$$

is the unique unimodular matrix which is the unique solution of the Diophantine equation $\det[(\nu_{i,j})_{i=12,7,4}^{j=1,2,3}] = 1, 0 < \nu_{4,\cdot} < \nu_{7,\cdot} < \nu_{12,\cdot}, |\nu_{i,\cdot}| = i$.

Corollary 60 *By Corollary 57, let* $1 < X < X_2 < X_3 < 10/9$. *By Theorem 52, for* A_3 *we have:*

$$\begin{bmatrix} D_{2,X_1} & D_{2,X_2} & D_{2,X_3} \\ D_{3,X_1} & D_{3,X_2} & D_{3,X_3} \\ D_{5,X_1} & D_{5,X_2} & D_{5,X_3} \end{bmatrix} = \begin{bmatrix} -3 & 4 & 0 \\ -1 & -1 & 3 \\ 2 & -1 & -2 \end{bmatrix}.$$

8.1.3 Construction of generated tone systems

In this section we will generate 12-granule 3-quotient tone systems.

The values $X, X_2, X_3 \in \mathbb{R}$ in the following theorem need not be necessary rational.

Theorem 61 *Let A be a* $(12, 7, 4)$*-matrix and corresponding* X, X_2, $X_3 \in \mathbb{R}$ *as in Theorem 52.*
Put
$\nu_{2,1}^* = 2\nu_{7,1} - \nu_{12,1}, \nu_{2,2}^* = 2\nu_{7,2} - \nu_{12,2}, \nu_{2,3}^* = 2\nu_{7,3} - \nu_{12,3},$
$\nu_{5,1}^* = \nu_{12,1} - \nu_{7,1}, \nu_{5,2}^* = \nu_{12,2} - \nu_{7,2}, \nu_{5,3}^* = \nu_{12,3} - \nu_{7,3},$
$\nu_{9,1}^* = \nu_{12,1} - \nu_{7,1} + \nu_{4,1}, \nu_{9,2}^* = \nu_{12,2} - \nu_{7,2} + \nu_{4,2}, \nu_{9,3}^* = \nu_{12,3} - \nu_{7,3} + \nu_{4,3},$
$\nu_{11,1}^* = \nu_{7,1} + \nu_{4,1}, \nu_{11,2}^* = \nu_{7,2} + \nu_{4,2}, \nu_{11,3}^* = \nu_{7,3} + \nu_{4,3}.$
Put

$C = X_1^0 X_2^0 X_3^0$, $D = X_1^{\nu_{2,1}^*} X_2^{\nu_{2,2}^*} X_3^{\nu_{2,3}^*}$, $E = X_1^{\nu_{4,1}} X_2^{\nu_{4,2}} X_3^{\nu_{4,3}}$,

$F = X_1^{\nu_{5,1}^*} X_2^{\nu_{5,2}^*} X_3^{\nu_{5,3}^*}$, $G = X_1^{\nu_{7,1}} X_2^{\nu_{7,2}} X_3^{\nu_{7,3}}$, $A = X_1^{\nu_{9,1}^*} X_2^{\nu_{9,2}^*} X_3^{\nu_{9,3}^*}$,

$B = X_1^{\nu_{11,1}^*} X_2^{\nu_{11,2}^*} X_3^{\nu_{11,3}^*}$, $C' = X_1^{\nu_{12,1}} X_2^{\nu_{12,2}} X_3^{\nu_{12,3}}$, $D' = 2d$.

Then

(i) $C : E : G = G : B : D' = F : A : C' = 1 : 5/4 : 3/2$,

(ii) $\nu_{2,1}^* + \nu_{2,2}^* + \nu_{2,3}^* = 2$, $\nu_{5,1}^* + \nu_{5,2}^* + \nu_{5,3}^* = 5$, $\nu_{9,1}^* + \nu_{9,2}^* + \nu_{9,3}^* = 9$,
  $\nu_{11,1}^* + \nu_{11,2}^* + \nu_{11,3}^* = 11$.

Proof.   (i) We have: $1 : (5/4) : (3/2) = C : E : G = 3/2 :$ $((3/2)(5/4)) : (3/2)^2 = G : B : D' = ((2)(2/3)) : ((2)(2/3)(5/4)) : 2$ $= F : A : C'$.

(ii) For 5, $(\nu_{12,1} - \nu_{7,1}) + (\nu_{12,2} - \nu_{7,2}) + (\nu_{12,3} - \nu_{7,3}) = (\nu_{12,1} + \nu_{12,2} + \nu_{12,3}) - (\nu_{7,1} + \nu_{7,2} + \nu_{7,3}) = 12 - 7 = 5$, analogously for 2, 9, 11. $\qquad\square$

Corollary 62 *If*

$$0 = m_0 \leq \nu_{2,1}^* \leq \nu_{4,1} \leq \nu_{5,1}^* \leq \nu_{7,1} \leq \nu_{9,1}^* \leq \nu_{11,1}^* \leq \nu_{12,1},$$
$$0 = n_0 \leq \nu_{2,2}^* \leq \nu_{4,2} \leq \nu_{5,2}^* \leq \nu_{7,2} \leq \nu_{9,2}^* \leq \nu_{11,2}^* \leq \nu_{12,2},$$
$$0 = r_0 \leq \nu_{2,3}^* \leq \nu_{4,3} \leq \nu_{5,3}^* \leq \nu_{7,3} \leq \nu_{9,3}^* \leq \nu_{11,3}^* \leq \nu_{12,3},$$

then numbers D, F, A, B *can be taken as the 3rd, 6th, 10th, and 12th coordinates of the* 12-granule scale vector *and put*

$$\nu_{2,1}^* = \nu_{2,1}, \quad \nu_{2,2}^* = \nu_{2,2}, \quad \nu_{2,3}^* = \nu_{2,3},$$
$$\nu_{5,1}^* = \nu_{5,1}, \quad \nu_{5,2}^* = \nu_{5,2}, \quad \nu_{5,3}^* = \nu_{5,3},$$
$$\nu_{9,1}^* = \nu_{9,1}, \quad \nu_{9,2}^* = \nu_{9,2}, \quad \nu_{9,3}^* = \nu_{9,3},$$
$$\nu_{11,1}^* = \nu_{11,1}, \quad \nu_{11,2}^* = \nu_{11,2}, \quad \nu_{11,3}^* = \nu_{11,3}.$$

Theorem 63 *Let* $(X_1, X_2, X_3) = (25/24, 16/15, 27/25)$. *Let*
$C = X_1^0 X_2^0 X_3^0$, $C_\natural = X_1^1 X_2^0 X_3^0$, $D_\flat = X_1^0 X_2^0 X_3^1$, $D = X_1^1 X_2^0 X_3^1$,
$D_\natural = X_1^2 X_2^0 X_3^1$, $E_\flat = X_1^1 X_2^1 X_3^1$, $E = X_1^2 X_2^1 X_3^1$, $F = X_1^2 X_2^2 X_3^1$,
$F_\natural = X_1^2 X_2^2 X_3^2$, $G_\flat = X_1^3 X_2^2 X_3^1$, $G = X_1^3 X_2^2 X_3^2$, $G_\natural = X_1^4 X_2^2 X_3^2$,
$A_\flat = X_1^3 X_2^3 X_3^2$, $A = X_1^4 X_2^3 X_3^2$, $A_\natural = X_1^5 X_2^3 X_3^2$, $B_\flat = X_1^4 X_2^3 X_3^3$,
$B = X_1^5 X_2^3 X_3^3$, $C' = X_1^5 X_2^4 X_3^3$, $D' = 2d$.

Then the all 12-granule scales generated by $\mathbb{A}_3$ satisfying the condition $C : E : G = G : B : D' = F : A : C' = 1 : 5/4 : 3/2$ are the next:

$$(C, i, D, j, E, F, k, G, l, A, m, B, C'),$$

where $i = C_\sharp, D_\flat;\ j = D_\sharp, E_\flat;\ k = F_\sharp, G_\flat;\ l = G_\sharp, A_\flat;\ m = A_\sharp, B_\flat$.

Analogously for matrices $\mathbb{A}_1$ (c.f. Table 8.2) and $\mathbb{A}_2$ (c.f. Table pypri), respectively.

Proof.   Combine Corollary 57, Theorem 61, and Corollary 62. It is easy to verify, c.f. Table 8.1, that $i = C_\sharp, D_\flat;\ j = D_\sharp, E_\flat;\ k = F_\sharp, G_\flat;$ $l = G_\sharp, A_\flat;\ m = A_\sharp, B_\flat$, are all the possibilities how to complete $\{C, D, E, F, G, A, B, C'\}$ to the 12-granule scales. $\square$

Corollary 64 *The tone system*

$$S_3 = \{C, C_\sharp, D_\flat, D, D_\sharp, E_\flat, E, F, F_\sharp, G_\flat, G, G_\sharp, A_\flat, A, A_\sharp, B_\flat, B, C'\},$$

c.f. Table 8.1, is a 17-valued 12-granule 3-quotient $(2/1, 3/2, 5/4)$-system.

Observe that the structure of S_3 is similar to the 17-valued Pythagorean System (two values for "black keys" on the standard keyboard).

8.1.4   Comment to superparticular ratios

In this section we show that the found systems (and S_3 in particular) are very near also to Just Intonation.

The only pairs of naturals $(N + 1, N)$, for which $(N + 1)$ and $N, N \in \mathbb{N}$, are divisible only by 2, 3, or 5, are $(2, 1)$, $(3, 2)$, $(4, 3)$, $(5, 4)$, $(6, 5)$, $(9, 8)$, $(10, 9)$, $(16, 15)$, $(25, 24)$, $(81, 80)$.

The following *superparticular ratios*, c.f. [122],

$$2/1,\ 3/2,\ 4/3,\ 5/4,\ 6/5,\ 9/8,\ 10/9,\ 16/15,\ 25/24,\ 81/80$$

account for common music intervals (they denominate the relative acoustic frequency or, inversely, the length of the pipe or the string)

TABLE 8.4: Semitones for superparticular ratios

	(X_1, X_2, X_3)	(X_1, X_4, X_2)	(X_5, X_4, X_2)
2/1	$X_1^5 X_2^4 X_3^3$	$X_1^2 X_4^3 X_2^7$	$X_5^2 X_4^5 X_2^5$
3/2	$X_1^3 X_2^2 X_3^2$	$X_1 X_4^2 X_2^4$	$X_5 X_4^3 X_2^3$
4/3	$X_1^2 X_2^2 X_3$	$X_1 X_4 X_2^3$	$X_5 X_4^2 X_2^2$
5/4	$X_1^2 X_2 X_3$	$X_1 X_4 X_2^2$	$X_5 X_4^2 X_2$
6/5	$X_1 X_2 X_3$	$X_4 X_2^2$	$X_4 X_2^2$
9/8	$X_1 X_3$	$X_4 X_2$	$X_4 X_2$
10/9	$X_1 X_2$	$X_1 X_2$	$X_5 X_4$
16/15	X_2	X_2	X_2
25/24	X_1	X_1	$X_5 X_4 X_2^{-1}$
81/80	$X_1^{-1} X_3$	$X_1^{-1} X_4$	$X_5^{-1} X_2$

and correspond to *octave, perfect fifth, perfect fourth, major third, minor third, major whole tone, minor whole tone, diatonic semitone, chromatic semitone,* and *comma of Didymus*, respectively.

The proof of the following theorem is easy.

Theorem 65 *See* Table 8.4, *where* $X_1 = 25/24$, $X_2 = 16/15$, $X_3 = 27/25$, $X_4 = 135/128$, $X_5 = 256/243$.

8.1.5   Classification of diatonic scales

We split the set of all 12-granule 3-quotient tone systems which contain C major scale into three classes. Denote them $\mathfrak{P}$, $\mathfrak{G}$, $\mathfrak{R}$, respectively. They can be characterized as follows: the class $\mathfrak{G}$ contains the Gypsy scale, the class $\mathfrak{R}$—the Redfield scale, and the class $\mathfrak{P}$—the Pythagorean heptatonic.

By a *diatonic scale* we mean usually a 7-tone scale within the octave in which the neighboring intervals are not smaller than a semitone and not greater than three semitones (= the *hiat*).

There are many different semitones in music. Each of them has its own good reason for existence (depending on the temperature of the scale). Some examples of semitones:

Pythagorean minor semitone (256/243),

Pythagorean major semitone (2 187/2 048),

Diatonic (16/15) and *Chromatic semitones* (25/24),

Praetorius minor semitone ($\sqrt[4]{78\,125}/16$),

Praetorius major semitone ($8/\sqrt[4]{3\,125}$,

Limma ascendant (135/128),

Alternate Renaissance semitone (27/25), and

Equal-tempered semitone ($\sqrt[12]{2}$), etc.

The appearance of other intervals between neighboring tones (the whole tones and/or hiats) in diatonic scales depends on the used semitones.

Various scales (diatonic or nondiatonic) use various numbers of different semitones. Equal temperaments use one semitone. Pythagorean System is constructed by two semitones, analogously Praetorius System (1/4-comma meantone). What about diatonic scales in general? It is known that the typical diatonic scale, the major scale, can be constructed, e.g., by the semitones: 16/15, 25/24, and 27/25. Thus, not greater that three semitones are needed for constructing of diatonic scales. But there exist more than one possibilities as we have seen in the previous sections (the triples (16/15, 25/24, 135/128) and (256/243, 16/15, 135/128) of semitones can also serve for constructing of the major diatonic scale).

From the dimensional point of view in the Euler musical space, Equal Temperament can be imaged in the line [the reper: the octave], Pythagorean System (Praetorius System) in the plane [the repers: the octave and perfect fifth (the octave and major third)], and the diatonic major scale [the repers: the octave, perfect fifth, major thirds] is a set of points in the 3-dimensional space. Note also that Just Intonation with the natural seventh, moreover, needs the fourth dimension, but it is not a diatonic tone system.

Although every musician understands what is a diatonic scale, we have found no mathematical definition in the literature. Certainly, the reason is the *big ambiguity* of this notion in music. We bring a definition of the notion of diatonic scale from the mathematical point of view.

We fix the structure of the 7-valued major scale (intervals between tones: 9/8, 10/9, 16/15, 9/8, 10/9, 9/8, 16/15) and extent it to 12-granule 3-quotient geometric nets so that the elements of the resulting tone system are of the form:

$$\Gamma_i = X_1^{\nu_{i,1}} X_2^{\nu_{i,2}} X_3^{\nu_{i,3}},$$

where $X_1, X_2, X_3 \in (2^{1/24}; 2^{3/24})$ (for instance; to have any reasonable boundaries) and i, $\nu_{i,1}$, $\nu_{i,2}$, $\nu_{i,3}$ are nonnegative integers such that

$$\nu_{i,1} + \nu_{i,2} + \nu_{i,3} = i$$

and

$$\cdots \le \nu_{1,j} \le \nu_{2,j} \cdots \le \nu_{n,j} \cdots,$$

where $j = 1, 2, 3$, and $i \in \mathbb{Z}$.

Further, we want to have the unison and octave in the tone system, i.e.,

$$\Gamma_0 = 1, \Gamma_{12} = 2,$$

and we suppose the *octave equivalency*, i.e.,

$$(\Gamma_{12i+0}, \Gamma_{12i+1}, \ldots, \Gamma_{12i+11}) = 2^i(\Gamma_0, \Gamma_1, \ldots, \Gamma_{11}), \quad i \in \mathbb{Z}.$$

The idea how to define the diatonic scales consists of (i) choosing a 7 valued variation from these generated geometric nets according to octave equivalency, and (ii) applying a homomorphism of geometric nets which do not destroy the diatonic structure.

(i) According to octave equivalency, consider a variation

$$\mathbf{D} = (\Gamma_{i_1}, \Gamma_{i_2}, \Gamma_{i_3}, \Gamma_{i_4}, \Gamma_{i_5}, \Gamma_{i_6}, \Gamma_{i_7}, \Gamma_{i_8})$$

from the set

$$\{\Gamma_i; \ i = 0, 1, \ldots, 11\}$$

such that

$$1 \le i_{n+1} - i_n \le 3, \quad n = 1, 2, 3, 4, 5, 6, 7.$$

(ii) Let $\langle S_i = X_1^{\nu_{i,1}} X_2^{\nu_{i,2}} X_3^{\nu_{i,3}} \rangle$, $\langle Q_i = Y_1^{\mu_{i,1}} Y_2^{\mu_{i,2}} Y_3^{\mu_{i,3}} \rangle$ be two geometric nets such that

$$S_0 = 1, \quad S_{12} = 2, \quad Q_0 = 1, \quad Q_{12} = 2,$$

and

$$X_1, X_2, X_3, Y_1, Y_2, Y_3 \in (2^{1/24}; 2^{3/24}).$$

A map

$$\theta : \langle S_i \rangle \to \langle Q_i \rangle$$

is a *homomorphism* of geometrical nets $\langle S_i \rangle$ and $\langle Q_i \rangle$ if for every $i \in \mathbb{Z}$, $S_i = X^{\nu_{i,1}} X_2^{\nu_{i,2}} X_3^{\nu_{i,3}} \Rightarrow Q_i = Y_1^{\nu_{i,1}} Y_2^{\nu_{i,2}} Y_3^{\nu_{i,3}}$, c.f. also Section 4.2.2

By the *diatonic scale* we understand a homomorphic image of variation **D**.

Table 8.1, Table 8.2, and Table 8.3 show the result of extention of the diatonic major scale $(C, D, E, F, G, A, B, C')$ to 12-granule 3-quotient scales. There are 96 12-granule 3-quotient scales such that they are restrictions of geometric nets. From these 12-granule scales we choose diatonic scales (not considering homomorphism).

Denote the classes of diatonic scales given by Table 8.2, Table 8.1, and Table 8.3 as $\mathfrak{R}, \mathfrak{G}, \mathfrak{P}$ corresponding triplets

$$(X_1, X_2, X_3), \quad (X_1, X_2, X_4), \quad (X_5, X_2, X_4),$$

respectively.

Theorem 66 *The class $\mathfrak{R}$ contains the Redfield diatonic scale.*

Proof. The Redfield diatonic scale is defined by the sequence of intervals between the neighbor tones:

$$(10/9, \ 9/8, \ 16/15, \ 9/8, \ 10/9, \ 9/8, \ 16/15).$$

We see, Table 8.2, that the sequence

$$(E_\flat, \ F, \ G, \ A_\flat, \ B_\flat, \ C', \ D', \ E'_\flat)$$

satisfies the requirement, where $D' = 2D$, $E_\flat = 2E'_\flat$. □

Theorem 67 *The class $\mathfrak{G}$ contains the Gypsy major and minor scales.*

Proof.

(a) The Gypsy major scale is defined by the sequence of intervals between the neighbor tones:

$$(16/15,\ 9/8 \cdot 25/24,\ 16/15,\ 9/8,\ 16/15,\ 9/8 \cdot 25/24,\ 16/15).$$

We see, Table 8.1, that the sequence

$$(C,\ D_\flat,\ E,\ F,\ G,\ A_\flat,\ B,\ C')$$

satisfies the requirement.

(b) The Gypsy minor scale is defined by the sequence of intervals between the neighbor tones:

$$(9/8,\ 16/15,\ 9/8 \cdot 25/24,\ 16/15,\ 16/15,\ 9/8 \cdot 25/24,\ 16/15).$$

We see, Table 8.1, that the sequence

$$(A,\ B,\ C',\ D'_\sharp,\ E',\ F',\ G'_\sharp,\ A')$$

satisfies the requirement, where $D'_\sharp = D_\sharp$, $E' = 2E$, $F' = 2F$. □

Theorem 68 *The class $\mathfrak{P}$ contains the Pythagorean heptatonic.*

Proof. The Pythagorean heptatonic is defined by the sequence of intervals between the neighbour tones: $(9/8, 9/8, 256/243, 9/8, 9/8, 9/8, 256/243)$. We see, Table 8.3, that the sequence

$$(D_\sharp,\ F,\ G,\ G_\sharp,\ B_\flat,\ C',\ D',\ D'_\sharp)$$

satisfies the requirement, where $D' = 2D$, $D'_\sharp = 2D_\sharp$. □

The following theorem can be verified directly.

Theorem 69

(1) *The class $\mathfrak{G}$ contains no Redfield scale and no Pythagorean heptatonic.*

(2) *The class $\mathfrak{R}$ contains no Gypsy scale and no Pythagorean heptatonic.*

(3) *The class $\mathfrak{P}$ contains no Redfield scale and no Gypsy scale.*

The other diatonic scales we obtain from classes $\mathfrak{G}$, $\mathfrak{P}$, $\mathfrak{R}$ via homomorphism (and specially, isomorphism). We do not describe them in general section and note now only some important special cases.

If $X_1 = X_2 = X_3 = X_4 = X_5 = \sqrt[12]{2}$, then Table 8.1, Table 8.2, and Table 8.3 define 12-tone Equal Temperament. Another interesting simplification of the general case via homomorphism we obtain in the following theorem which can be verified directly.

Theorem 70 *If*
    (a) $X_2 = X_3 = a$, $X_1 = b$ *(c.f. Table 8.1), or*
    (b) $X_1 = X_2 = a$, $X_4 = b$ *(c.f. Table 8.2), or*
    (c) $X_5 = X_2 = a$, $X_4 = b$ *(c.f. Table 8.3),*
and
    $a = 256/243$, $b = 2\,187/2\,048$ *(or $a = 8/\sqrt[4]{3\,125}$, $b = \sqrt[4]{78\,125}/16$),*
then Table 8.1, Table 8.2, *and* Table 8.3 *contain Pythagorean (or Praetorius) Tuning.*

Corollary 71 *Pythagorean System (reduced to a 12-valued one) and Praetorius System (1/4-comma meantone) are isomorphic. For another approach, c.f. [126]. The image of Praetorius System in the plane is mirror symmetrical with respect to the image of Pythagorean System, c.f. Figure 6.2. Pythagorean System and Praetorius System are negative and positive meantones, respectively.*

8.1.6 Application to partial monounary algebras

In this subsection we consider some partial monounary algebras on the system S_3 corresponding to matrix A_3 and representing corresponding transpositions in music. The tone systems corresponding to matrixes A_1 and A_2 are similar.

Let $\mathcal{A}$ be a nonempty set and $\mathcal{B} \subset \mathcal{A}$. Let $\Upsilon : \mathcal{B} \to \mathcal{A}$ be a mapping of $\mathcal{B}$ into $\mathcal{A}$. Then the couple $(\mathcal{A}, \Upsilon)$ is said to be a *partial monounary algebra*. More about this notion we can find, e.g., in [134].

Consider the partial functions Υ defined on some subsets of the set $S_3 = \{C,\ C_\sharp,\ D_\flat,\ D,\ D_\sharp,\ E_\flat,\ E,\ F,\ F_\sharp,\ G_\flat,\ G,\ G_\sharp,\ A_\flat,\ A,\ A_\sharp,\ B_\flat,\ B,\ C'\} \subset [1,2]$ as follows:

for some $u \in S_3$, $\Upsilon_\omega(u) = u \cdot \omega$ if $1 \le u \cdot \omega < 2$ or $\Upsilon_\omega(u) = u\omega/2$ otherwise, where

$\omega = \alpha_{1,11}, \beta_{1,11}, \gamma_{1,11};\ \alpha_{2,10},\ \beta_{2,10},\ \gamma_{2,10},\ \delta_{2,10},\ \epsilon_{2,10};\ \alpha_{3,9},\ \beta_{3,9},\ \gamma_{3,9},$ $\delta_{3,9},\ \epsilon_{3,9},\ \zeta_{3,9};\ \alpha_{4,8},\ \beta_{4,8},\ \gamma_{4,8},\ \delta_{4,8},\ \epsilon_{4,8};\ \alpha_{5,7},\ \beta_{5,7},\ \gamma_{5,7},\ \delta_{5,7};\ \alpha_{6,6},\ \beta_{6,6},$ $\gamma_{6,6},\ \delta_{6,6}$, respectively, where

$\alpha_{1,11} = X_1$, $\beta_{1,11} = X_2$, $\gamma_{1,11} = X_3$ (the minor seconds in music, semitones);

$\alpha_{2,10} = X_1 X_3$, $\beta_{2,10} = X_1 X_2$, $\gamma_{2,10} = X_1^2$, $\delta_{2,10} = X_2 X_3$, $\epsilon_{2,10} = X_1^2$ (the seconds, among them the major and minor whole tones);

$\alpha_{3,9} = X_1^1 X_2^1 X_3^1$, $\beta_{3,9} = X_1^2 X2^1 X_3^0$, $\gamma_{3,9} = X_1^1 X_2^2 X_3^0$, $\delta_{3,9} = X_1^2 X_2^1 X_3^0$, $\epsilon_{3,9} = X_1^0 X_2^2 X_3^1$, $\zeta_{3,9} = X_1^0 X_2^1 X_3^2$ (the minor thirds);

$\alpha_{4,8} = X_1^2 X_2^1 X_3^1$, $\beta_{4,8} = X_1^1 X_2^2 X_3^1$, $\gamma_{4,8} = X_1^2 X_2^0 X_3^2$, $\delta_{4,8} = X_1^2 X_2^2 X_3^0$, $\epsilon_{4,8} = X_1^3 X_2^1 X_3^0$ (the major thirds);

$\alpha_{5,7} = X_1^3 X_2^2 X_3^2$, $\beta_{5,7} = X_1^3 X_2^3 X_3^1$, $\gamma_{5,7} = X_1^2 X_2^3 X_3^2$, $\delta_{5,7} = X_1^4 X_2^2 X_3^1$ (the fourths);

$\alpha_{6,6} = X_1^2 X_2^2 X_3^2$, $\beta_{6,6} = X_1^3 X - 2^2 X_3^1$, $\gamma_{6,6} = X_1^2 X_2^3 X_3^1$, $\delta_{6,6} = X_1^3 X_2^1 X_3^2$ (tritones, the augmented fourths);

where $(X_1, X_2, X_3) = (25/24, 16/15, 27/25)$.

We define the following partial monounary algebras:

$(S_3, \Upsilon_{\alpha_{1,11}})$, $(S_3, \Upsilon_{\beta_{1,11}})$, $(S_3, \Upsilon_{\gamma_{1,11}})$;

$(S_3, \Upsilon_{\alpha_{2,10}})$, $(S_3, \Upsilon_{\beta_{2,10}})$, $(S_3, \Upsilon_{\gamma_{2,10}})$, $(S_3, \Upsilon_{\delta_{2,10}})$, $(S_3, \Upsilon_{\epsilon_{2,10}})$;

$(S_3, \Upsilon_{\alpha_{3,9}})$, $(S_3, \Upsilon_{\beta_{3,9}})$, $(S_3, \Upsilon_{\gamma_{3,9}})$, $(S_3, \Upsilon_{\delta_{3,9}})$, $(S_3, \Upsilon_{\epsilon_{3,9}})$, $(S_3, \Upsilon_{\zeta_{3,9}})$;

$(S_3, \Upsilon_{\alpha_{4,8}})$, $(S_3, \Upsilon_{\beta_{4,8}})$, $(S_3, \Upsilon_{\gamma_{4,8}})$, $(S_3, \Upsilon_{\delta_{4,8}})$, $(S_3, \Upsilon_{\epsilon_{4,8}})$;

$(S_3, \Upsilon_{\alpha_{5,7}})$, $(S_3, \Upsilon_{\beta_{5,7}})$, $(S_3, \Upsilon_{\gamma_{5,7}})$, $(S_3, \Upsilon_{\delta_{5,7}})$;

$(S_3, \Upsilon_{\alpha_{6,6}})$, $(S_3, \Upsilon_{\beta_{6,6}})$, $(S_3, \Upsilon_{\gamma_{6,6}})$, $(S_3, \Upsilon_{\delta_{6,6}})$,

c.f. Figures 8.1, 8.2, 8.3, 8.4, 8.5, and 8.6.

If we consider the inverse partial functions Υ_ω^{-1} (they exist; all arrows switch the direction in Figures 8.1, 8.2, 8.3, 8.4, 8.5, 8.6), then Figures 8.2, 8.3, 8.4, 8.5, 8.6 define also the partial monounary algebras

$(S_3, \Upsilon_{\alpha_{6,6}}^{-1})$, $(S_3, \Upsilon_{\beta_{6,6}}^{-1})$, $(S_3, \Upsilon_{\gamma_{6,6}}^{-1})$, $(S_3, \Upsilon_{\delta_{6,6}}^{-1})$;

$(S_3, \Upsilon_{\alpha_{5,7}}^{-1})$, $(S_3, \Upsilon_{\beta_{5,7}}^{-1})$, $(S_3, \omega_{\gamma_{5,7}}^{-1})$, $(S_3, \Upsilon_{\delta_{5,7}}^{-1})$;

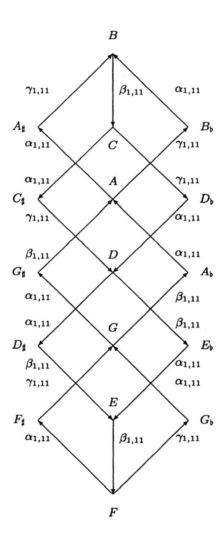

FIGURE 8.1: The minor seconds (the major sevenths)

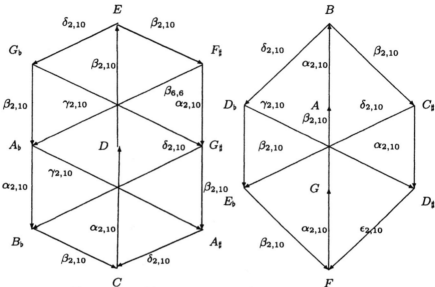

FIGURE 8.2: The major seconds (the minor sevenths)

$$(S_3, \Upsilon_{\alpha_{4,8}}^{-1}), (S_3, \Upsilon_{\beta_{4,8}}^{-1}), (S_3, \Upsilon_{\gamma_{4,8}}^{-1}), (S_3, \Upsilon_{\delta_{4,8}}^{-1}), (S_3, \Upsilon_{\epsilon_{4,8}}^{-1});$$
$$(S_3, \Upsilon_{\alpha_{3,9}}^{-1}), (S_3, \Upsilon_{\beta_{3,9}}^{-1}), (S_3, \Upsilon_{\gamma_{3,9}}^{-1}), (S_3, \Upsilon_{\delta_{3,9}}^{-1}), (S_3, \Upsilon_{\epsilon_{3,9}}^{-1}), (S_3, \Upsilon_{\zeta_{3,9}}^{-1});$$
$$(S_3, \Upsilon_{\alpha_{2,10}}^{-1}), (S_3, \Upsilon_{\beta_{2,10}}^{-1}), (S_3, \Upsilon_{\gamma_{2,10}}^{-1}), (S_3, \Upsilon_{\delta_{2,10}}^{-1}), (S_3, \Upsilon_{\epsilon_{2,10}}^{-1});$$
$$(S_3, \Upsilon_{\alpha_{1,11}}^{-1}), (S_3, \Upsilon_{\beta_{1,11}}^{-1}), (S_3, \omega_{\gamma_{1,11}}^{-1}).$$

The proof of the following theorem can be obtained easily and omitted.

Theorem 72 *The following partial monounary algebras are isomorphic:*

(i) $(S_3, \Upsilon_{\alpha_{5,7}}), (S_3, \Upsilon_{\alpha_{5,7}}^{-1}), (S_3, \Upsilon_{\alpha_{4,8}}),$ *and* $(S_3, \Upsilon_{\alpha_{4,8}}^{-1}),$

(ii) (S_3, Υ_{ω}) *and* $(S_3, \Upsilon_{\omega}^{-1}),$

where $\omega = \alpha_{1,11}, \ \beta_{1,11}, \ \gamma_{1,11}; \ \alpha_{2,10}, \ \beta_{2,10}, \ \gamma_{2,10}, \ \delta_{2,10}, \ \epsilon_{2,10}; \ \alpha_{3,9}, \ \beta_{3,9}, \ \gamma_{3,9}, \ \delta_{3,9}, \ \epsilon_{3,9}, \ \zeta_{3,9}; \ \alpha_{4,8}, \ \beta_{4,8}, \ \gamma_{4,8}, \ \delta_{4,8}, \ \epsilon_{4,8}; \ \alpha_{5,7}, \ \beta_{5,7}, \ \gamma_{5,7}, \ \delta_{5,7}; \ \alpha_{6,6}, \ \beta_{6,6}, \ \gamma_{6,6}, \ \delta_{6,6}.$

Example 66 *INDIA(MAGRAMA)-7, INDIA(A)-7, INDIA(B)-7, INDIA(C)-7, INDIA(D)-7, INDIA(E)-7.*

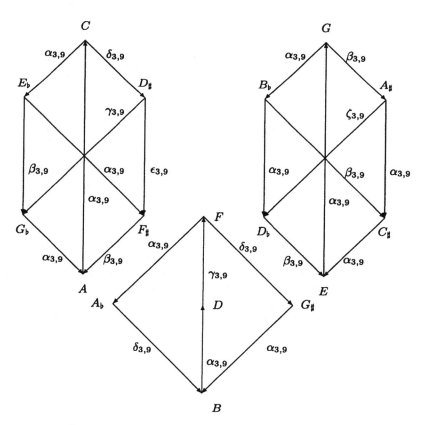

FIGURE 8.3: The minor thirds (the major sixths)

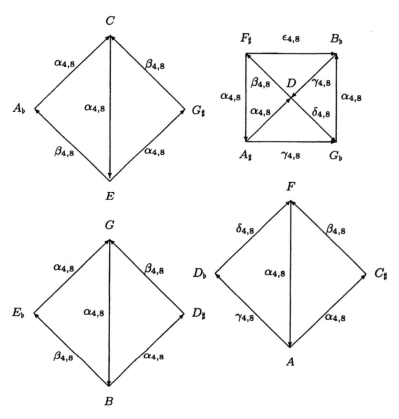

FIGURE 8.4: The major thirds (the minor sixths)

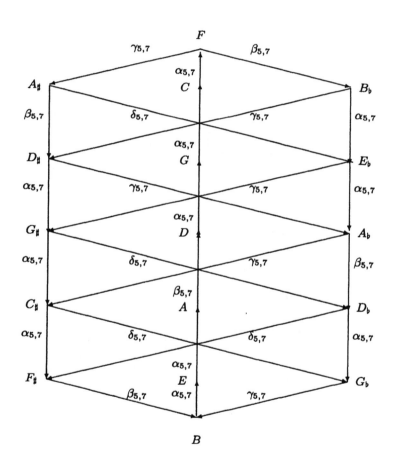

FIGURE 8.5: The fourths (the fifths)

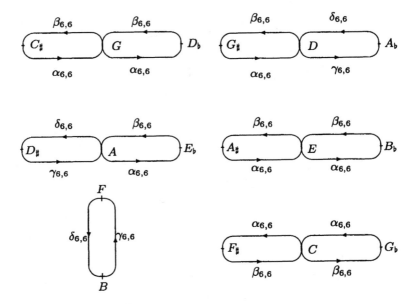

FIGURE 8.6: Tritones

8.2   10 granulation

8.2.1   Gamelan

There are two predominant scales in traditional Indonesian music, Slendro and Pelog. Each has a unique character identified by the relationships of intervals between tones. Though both of *these scales vary considerably in actual measurement from one gamelan to another*, their basic form is consistent throughout Indonesia and much of Southeast Asia. Both Slendro and Pelog can be derived from the master scale $E_{10} = \{0, 120, 240, 360, 480, 600, 720, 840, 960, 1\,080, 1\,200\}$ cent). This tone scale is never played as such and only serves as a basis for scale theory. So, we see the 10 granulation within octave which is principally different from 12 granulation of European music.

Slendro   In common practice, Slendro is a five-tone scale consisting of, approximately, whole step and minor third intervals. The basic order of intervals from low to high is $(W_1, M_2, W_3, W_4, M_5)$, where W_1, W_3, W_4 are approximate whole steps and M_2, M_5 approximate minor thirds. A few different variations of the ten tones of E_{10} may be derived which sound like Slendro, but each will have a slightly different character which may be best suited for one "modality" (term used loosely) or another. When building a Gamelan, one of these derivations is chosen as best suited for its purpose and the instrument maker must commit to that interpretation of the Slendro scale. Every tuner also has their own preferences within those derived scales, adding to the wide variety of interpretations of both Slendro and Pelog from Gamelan to Gamelan. String, vocal, and wind instruments have the ability to transcend this limitation and explore other interpretations of the scale by virtue of their unfixed pitch.

Pelog   Pelog is a seven-tone scale, although only five of the seven pitches are typically used at one time. Pelog may also originate from the master scale E_{10} mentioned above. The basic seven intervals of Pelog are $(H_1, W_2, M_3, H_4, H_5, W_6, W_7)$, where H_1, H_4, H_5 are approximate half steps, W_2, W_6, W_7 are approximate whole steps, and

M_3 is an approximate minor third. There are wide variations of Pelog. Some may be closer to the interval structure

$$(W_1, H_2, M_3, H_4, H_5, W_6, W_7).$$

In real practice, Pelog is best described as a combination of mostly approximate half steps and major thirds. The presence of a half step, or lack of it, is an easy indicator for distinguishing by ear between Pelog and Slendro. If we number the pitches 1 through 7, the actual five tone Pelog scales (modes) used in Gamelan are predominantly $(1, 2, 3, 5, 6)$, $(1, 2, 4, 5, 6$, and $(2, 3, 5, 6, 7)$. Calling an approximate major third T and a near fourth F, this gives relationships of intervals, i.e.

$$(H_1, W_2, T_3, H_4, T_5), (H_1, F_2, H_3, H_4, T_5), (W_1, T_2, H_3, W_4, W_5),$$

respectively. These scales sometimes intermesh and overlap, particularly in Java. Balinese Pelog is also theoretically seven tones, but most instruments only have the pitches $(1, 2, 3, 5, 6)$ available. There are only a few types of Balinese instruments capable of producing seven tones, mostly classical ensembles, such as Semar Pegulingan and Gong Suling, not in great use anymore. Playing these intervals on a piano or other instrument, if no gamelan is available, can help to gain a sense of the character of the different scales and how the pitch substitutions occur.

There is no conclusive proof to say whether Slendro or Pelog was introduced first or if they may have arisen concurrently. The commonly held belief in Indonesia is that Slendro is the predecessor. Slendro is the primary scale for all of the oldest rites of importance including shadow play in both Bali and Java. The previously mentioned ten-tone master scale is sometimes referred to as Slendro but usually not Pelog. Both share a common origin, so it doesn't really matter too much which was first to be accepted. By population statistics it would appear that Pelog is actually the elder scale as it is in some ways more popular and widespread. If we assume that usage density and dispersion equals age, this may be true. Debates will likely continue on this matter as long as there is an interest in Indonesian music.

The Origins of Indonesian Scales Indonesian scales and systems have many roots, some from China, others from India, and others from the islands themselves. Chinese influences are best seen in the structure of the scales, while the Indian influence is more obvious in how the scales are utilized. One strong link to China is the shape of intervals of the Slendro scale which is very close to common scales of the mainland. Another China link, although not so apparent, is the Huang Chong. The Huang Chong is an instrument of pitch measurement. It consists of a stopped bamboo tube of specific length and girth which is blown across the end to produce a set standard pitch and harmonics from which other music is to be derived. The pitch of the Huang Chong, 366 Hz, is common within a few cents, to a pitch in most Slendro Gamelan, usually pitch 3 in Java and pitch 2 in Bali. Most Gamelan makers today may not realize that they use it at all and the Huang Chong is not a hard and fast rule. In fact, many Gamelan have no pitches anywhere near 366 Hz. It is explored only as a matter of historical reference. To best observe India's influence on the use of scales in Indonesia, it helps to have a cursory understanding of the "modal" systems called Patet in Java and Saih in Bali. The Indian Raga system is a set of rules of motion, often ascending the scale in one form and descending with a different set of pitch options. Indonesia has only a descending rule for each "mode." This descending pitch motion is what defines the "mode." It contributes greatly to the musical effect of "falling" to key structural points or gongs. The word mode appears in quotes because it is a misleading term if used in the Western sense. Indonesian "modes" are actually more like treatments of scale, rather than a set of pitches starting on a certain tone. Patet and Saih are sets of pitches *ending* on a certain tone and getting there in a particular way.

Example 67

UDAN-MAS(GAMELAN)-12; Gamelan Udan Mas, approximately,
s - Slendro, p - Pelog: s6, p6, p7, s1, p1, s2, p2, p3, s3, p4, s5, p5;
1, 10/9, 7/6, 32/25, 47/35, 32/23, 3/2, 20/13, 16/9, 16/9, 23/12, 2

GAMELAN(NO.1)-12; Other Music gamelan;
15/14, 9/8, 7/6, 5/4, 4/3, 7/5, 3/2, 14/9, 5/3, 7/4, 15/8, 2

FORTUNA(GAMELAN)-12; From [17], *Slendro on black, Pelog on white keys;*
101.100, 233.000, 332.500, 410.500, 512.600, 572.600, 649.900, 800.600, 958.600, 1 056.800, 1 087.700, 1 200

GAMELAN(NO.4)-12; Pelog white, Slendro black;
137, 228, 446, 575, 484, 687, 728, 820, 960, 1 098, 1 200

KAYOLONIA-19; 19-tone 5-limit Kayolonian scale;
128/125, 16/15, 9/8, 75/64, 6/5, 5/4, 32/25, 4/3, 512/375, 64/45, 3/2, 25/16, 8/5, 5/3, 128/75, 16/9, 15/8, 125/64, 2

KAYOLONIA(F)-9; Kayolonian scale F;
16/15, 75/64, 5/4, 4/3, 3/2, 8/5, 128/75, 15/8, 2

KAYOLONIA(P)-9; Kayolonian scale P;
16/15, 75/64, 5/4, 4/3, 3/2, 8/5, 225/128, 15/8, 2

KAYOLONIA(S)-9; Kayolonian scale S;
1 125/1 024, 75/64, 5/4, 5 625/4 096, 3/2, 8/5, 225/128, 15/8, 2

KAYOLONIA(T)-9; Kayolonian scale T;
16/15, 256/225, 4 096/3 375, 4/3, 8 192/5 625, 8/5, 128/75, 2 048/1 125, 2

KAYOLONIA(Z)-9; Kayolonian scale Z;
16/15, 256/225, 5/4, 4/3, 3/2, 8/5, 128/75, 2 048/1 125, 2

PACIFIC-5; Observed south pacific pentatonic xylophone scale;
202, 370, 685, 903, 1 200

JAVA(PE)-7; Observed Javanese Pelog scale;
137, 446, 575, 687, 820, 1098, 1 200

KENGETAN(PE)-11; Gamelan selunding from Kengetan, South Bali (Pelog), a' = 141 Hz;
124.347, 63/47, 68/47, 805.224, 2/1, 101/47, 126/47, 136/47, 449/141, 4/1, 202/47

PENTATRIAD-11; 4:5:6 Pentatriadic scale;
10/9, 9/8, 5/4, 4/3, 45/32, 3/2, 5/3, 27/16, 16/9, 15/8, 2

SIAMESE-12; Siamese Tuning, after [17];
49.800, 172, 215, 344, 515, 564.800, 685.800, 735.800, 857.800, 914.800, 1 028.800, 1 200.000

SINGAPORE-7; An observed xylophone scale from Singapore;
187, 356, 526, 672, 856, 985, 1 200

JAVA(SLE)-5; Observed Javanese Slendro scale;
228, 484, 728, 960, 1 200

SLE/PEL(16-TET/SLE/PE)-12; 16-TET Slendro and Pelog;
150, 150, 225, 300, 450, 675, 675, 750, 825, 900, 1 200

SLE/PE(23-TET)-12; 23-TET Slendro and Pelog;
208.696, 208.696, 156.522, 469.565, 313.043, 730.435, 730.435, 678.261, 939.130, 834.783, 1 200.000

PLIATAN(SLE)-5; Gender wayang from Pliatan, South Bali (Slendro), a′ = 305.5 Hz;
235.419, 453.560, 704.786, 927.453, 1 200.000

THAILAND-7; Observed scale from Thailand;
129, 277, 508, 726, 771, 1 029, 1 200

TRANH(NO.3)-6; Sa Mac Dan Tranh scale;
17/14, 4/3, 3/2, 38/21, 51/28, 2

8.2.2 Superparticular pentatonics

Deal with pentatonics $1 = \omega_1 \leq \omega_2 \leq \omega_3 \leq \omega_4 \leq \omega_5 \leq \omega_6 = 2$, $\omega_i \in \mathbb{Q}$, whose one-step intervals ω_{i+1}/ω_i, $i = 1, 2, 3, 4, 5$, are superparticular, i.e., of the form $(n + 1)/n$ for some $n \in \mathbb{N}$.

Excluding unison and assuming octave equivalence, every pentatonic contains 20 intervals, not all necessarily different: 5 one-step intervals, 5 two-step intervals, 5 three-step intervals, and 5 four-step intervals.

The aim of this subsection is to show how the idea about superparticular ratios works in the case of Slendro (and Pelog consequently), c.f. Table 8.5. Evaluations of I. E. Hofman and D. Canright, c.f. [88], are very apt for this purpose.

Hofman's algorithm   The I. E. Hofman's, c.f. [88], recursive algorithm for dividing a given interval r, $1 < r < 2$ into a certain number n of superparticular steps is easy:

(1) try the biggest superparticular step (say s) that will fit in r;

(2) then you need to find how to divide r/s into $n - 1$ steps;

(3) then try the next biggest step, etc.;

(4) until the first step is small enough that n of them are smaller than r, and you are done.

So, this algorithm shows that

Corollary 73 *There is only a finite number of superparticular pentatonics; after all, there are an infinite number of superparticular intervals. The limiting factor was that exactly n superparticular steps must make up the given interval (say 2, or 4/3), and while there is no shortage of tiny superparticular intervals, there are only a few big ones.*

Bring now some examples.

Example 68 Applying this procedure to finding all superparticular divisions of 4/3 into 3 steps does indeed give the 26 different ways tabulated by Hofmann.

Example 69 Say we want to divide a 5/4 into two superparticular steps. The largest superparticular interval that will fit into the 5/4 is 6/5, leaving 25/24, which is also superparticular, so one division is 5/4=(6/5)(25/24). The next largest superparticular that will fit into 5/4 is 7/6, leaving 15/14; this makes a second superparticular division. Trying the next largest, 8/7, leaves 35/32, but that's not superparticular. The next works, giving the third division 9/8 and 10/9. The next largest after 9/8 is 10/9, but that will just give the last case (in a different order) so we are done. Altogether we get three divisions of 5/4 into two superparticular steps, and each could be taken in two different orders.

Example 70 Let us divide 2 into three superparticular steps, or in other terms, let us find superparticular triads. The largest superparticular that could fit is 3/2, leaving 4/3 to be split into two steps. Proceeding as above gives three ways to do that, into (5/4)(16/15), or (6/5)(10/9), or (7/6)(8/7). The next largest step that could fit into 2 is 4/3, leaving 3/2 to be divided into two steps, which can be done only two ways: (4/3)(9/8), or (5/4)(6/5). Then the next largest step that fits is 5/4, but that just gives two of the previous cases (in different orders), hence we are done. So altogether there are five different ways to divide a 2/1 into three superparticular steps, disregarding the order of the steps. For each such division, we could order the three steps in $3! = 3 \cdot 2 \cdot 1 = 6$ different ways (except the 4/3, 4/3, 9/8 division has two steps the same, so only $3!/2! = 3$ different orders); if we count all possible orderings we get $4 \cdot 6 + 1 \cdot 3 = 27$ different triads. Then again, we may consider different inversions (cyclic permutations) of a triad to be the same, and a triad has three inversions, for a total of $27/3 = 9$ different triads.

Reduction to intervals between 65 536/65 535 and 3/2 Having the Hofman's algorithm, we find all superparticular pentatonics (divisions of 2 into 5 steps). We were rather taken aback to find that there are 876 of them! The first one the algorithm came up with is $(3/2)(5/4)(17/16)(257/256)(65\,536/65\,535)$, which is not musically useful as a pentatonic. For one thing, the smallest interval, $65\,536/65\,535$, is only 0.026 cents, which is smaller than human pitch discrimination! For another, the largest interval, 3/2, is much bigger than one would expect for a "step" of a pentatonic; the whole scale is poorly proportioned. In fact, of the 876 superparticular pentatonics, 593 of them have a largest step of 3/2, and another 229 of them have a largest step of 4/3, which is still pretty large to be considered a step in a pentatonic. (Another 46 have 5/4 as largest, 7 have 6/5, and only 1 has 7/6 as largest. There are 308 different smallest steps possible.) So for Table 8.5, we only include those we consider musically useful as pentatonics, whose largest steps are smaller than 4/3, and whose smallest steps are at least as large as a comma of Didymus, 81/80 (21.5 cents), though none of those remaining actually have a step quite that small. Of course, this eliminates some interesting possibilities, such as the scale 1/1, 9/8, 5/4, 11/8, 3/2, and superparticular pentatonics similar to LaMonte Young's "dream chords," such as 4/3, 25/24, 26/25, 27/26, 4/3.

Those restrictions reduce the number to 40 pentatonics, an almost digestible quantity, until one thinks about all the different orderings. The divisions with all five steps different (pentatonics S_i in Table 8.5, $i = 4, 5, 6, 9, 10, 12, 16, 18, 23, 27, 38$) have $5! =$

120 different orderings, those with a pair the same have $5!/2! = 60$ orderings, those with two pairs (pentatonics S_i in Table 8.5, $i = 1$, 3, 13, 34) have $5!/(2! \cdot 2!) = 30$, and the one with three of a kind (S_{20}) gets $5!/3! = 20$; in this sense, Table 8.5 represents 2 960 pentatonic scales. If one counts the five different cyclic permutations (modes) of a scale as one then we get 24 scales for divisions with all steps different, 12 with a pair, 6 with two pairs, and 4 with a triplet, for a total of 592 different scales in Table 8.5.

Any classification? There are a number of ways to classify, evaluate, and measure scales, by which to make comparisons of all these possibilities. Here we will discuss only a few; for many more see the book [7]. Perhaps the most basic comparison is in terms of step size. For example, a pentatonic whose smallest step is a whole tone will have a different character than another containing one or more semitones; the small steps give a scale a sort of inherent tension. Hence in Table 8.5 the pentatonics are arranged in order of smallest step size, from largest to smallest, or "mellowest" to "tensest." Another property based on interval size is called "propriety"; a scale is "proper" if all the one-step intervals are smaller than all the two-step intervals, which in turn are smaller than all the three-step intervals, etc. Propriety depends on the ordering of the steps (independent of cyclic permutations), for example, pentatonic S_3 in the given order is improper, since one of the two-step intervals, $(6/5)(6/5) = 36/25$, is larger than one of the three-step intervals, $(10/9)(10/9)(9/8) = 25/18$, though other orderings of the same steps (that separate the two 6/5's) give proper scales. One could also judge a scale by the total number of superparticular intervals it contains; this depends on step order but not on cyclic permutation. Of course, not all of the 20 intervals of a pentatonic can be superparticular, because the octave complement of a superparticular interval is not superparticular (except 3/2 and 4/3).

A basic harmonic property of a set of intervals is the set of prime factors involved; the highest prime factor is often called the prime limit of a scale. The underlying premise is that any just interval can be analyzed in terms of prime harmonies, the (octave-reduced) intervals of the prime numbers in the harmonic series, and that each prime

TABLE 8.5: Superparticular pentatonics

Pentatonic						Prime factors
S_1	9/8	8/7	8/7	7/6	7/6	[3, 7]
S_2	10/9	9/8	8/7	7/6	6/5	[3, 5, 7]
S_3	10/9	10/9	9/8	6/5	6/5	[3, 5]
S_4	12/11	11/10	8/7	7/6	5/4	[3, 5, 7, 11]
S_5	12/11	11/10	10/9	6/5	5/4	[3, 5, 11]
S_6	14/13	13/12	8/7	6/5	5/4	[3, 5, 7, 13]
S_7	15/14	8/7	7/6	7/6	6/5	[3, 5, 7]
S_8	15/14	10/9	7/6	6/5	6/5	[3, 5, 7]
S_9	16/15	9/8	8/7	7/6	5/4	[3, 5, 7]
S_{10}	16/15	10/9	9/8	6/5	5/4	[3, 5]
S_{11}	16/15	12/11	11/10	5/4	5/4	[3, 5, 11]
S_{12}	16/15	15/14	7/6	6/5	5/4	[3, 5, 7]
S_{13}	16/15	16/15	9/8	5/4	5/4	[3, 5]
S_{14}	20/19	19/18	6/5	6/5	5/4	[3, 5, 19]
S_{15}	21/20	8/7	8/7	7/6	5/4	[3, 5, 7]
S_{16}	21/20	10/9	8/7	6/5	5/4	[3, 5, 7]
S_{17}	21/20	16/15	8/7	5/4	5/4	[3, 5, 7]
S_{18}	22/21	12/11	7/6	6/5	5/4	[3, 5, 7, 11]
S_{19}	25/24	8/7	7/6	6/5	6/5	[3, 5, 7]
S_{20}	25/24	10/9	6/5	6/5	6/5	[3, 5]
S_{21}	25/24	16/15	6/5	6/5	5/4	[3, 5]
S_{22}	26/25	14/13	8/7	5/4	5/4	[5, 7, 13]
S_{23}	28/27	9/8	8/7	6/5	5/4	[3, 5, 7]
S_{24}	28/27	15/14	6/5	6/5	5/4	[3, 5, 7]
S_{25}	32/31	31/30	6/5	5/4	5/4	[3, 5, 31]
S_{26}	36/35	8/7	7/6	7/6	5/4	[3, 5, 7]
S_{27}	36/35	10/9	7/6	6/5	5/4	[3, 5, 7]
S_{28}	36/35	16/15	7/6	5/4	5/4	[3, 5, 7]
S_{29}	36/35	28/27	6/5	5/4	5/4	[3, 5, 7]
S_{30}	40/39	13/12	6/5	6/5	5/4	[3, 5, 13]
S_{31}	40/39	26/25	6/5	5/4	5/4	[3, 5, 13]
S_{32}	46/45	24/23	6/5	5/4	5/4	[3, 5, 23]
S_{33}	49/48	8/7	8/7	6/5	5/4	[3, 5, 7]
S_{34}	50/49	7/6	7/6	6/5	6/5	[3, 5, 7]
S_{35}	55/54	12/11	6/5	6/5	5/4	[3, 5, 11]
S_{36}	56/55	11/10	8/7	5/4	5/4	[5, 7, 11]
S_{37}	56/55	22/21	6/5	5/4	5/4	[3, 5, 7, 11]
S_{38}	64/63	9/8	7/6	6/5	5/4	[3, 5, 7]
S_{39}	64/63	21/20	6/5	5/4	5/4	[3, 5, 7]
S_{40}	76/75	20/19	6/5	5/4	5/4	[3, 5, 19]

harmony has a unique character that it contributes to any composite interval of which it forms a part. For example, pentatonic S_1, which involves primes 3 and 7, combines the power of perfect fifths (3/2) with the bluesiness of natural sevenths (7/4), but does not have any of the sweetness of major thirds (5/4). The prime factors (excluding 2) of each pentatonic are given in brackets in Table 8.5. Traditional harmony is based on primes 3 and 5, and possibly 7 for dominant-seventh chords; harmonies using 11, 13, and beyond will definitely sound unconventional.

A more detailed form of harmonic analysis involves plotting the notes of the scale on a harmonic lattice, with one dimension for each different prime harmony (in other words, each prime is a harmonic dimension). Here the order of the steps is important, although cyclic permutations give the same structure. For example, when the steps of S_3 are taken in the order 9/8, 10/9, 6/5, 10/9, 6/5, the resulting scale is 1/1, 9/8, 5/4, 3/2, 5/3 (a "major" pentatonic), which is diagrammed as

$$
\begin{array}{ccccc}
5/3 & \longrightarrow & 5/4 & & \\
 & & \uparrow & & \\
1/1 & \longrightarrow & 3/2 & \longrightarrow & 9/8
\end{array}
$$

where the horizontal direction shows the 3/2 prime harmonies and the vertical shows 5/4. Another common order for S_3 is 6/5, 10/9, 9/8, 6/5, 10/9 giving the scale 1/1, 6/5, 4/3, 3/2, 9/5 ("minor" version) with the structure

$$
\begin{array}{ccccc}
4/3 & \longrightarrow & 1/1 & \longrightarrow & 3/2 \\
 & & & & \uparrow \\
 & & 6/5 & \longrightarrow & 9/5
\end{array}
$$

which in a sense is just the inverse of the other scale. A third order is 6/5, 10/9, 9/8, 10/9, 6/5 giving 1/1, 6/5, 4/3, 3/2, 5/3 ("Dorian"?) or

$$5/3$$
$$\uparrow$$
$$4/3 \longrightarrow 1/1 \longrightarrow 3/2$$
$$\uparrow$$
$$6/5$$

The other three (cyclically distinct) orderings give different harmonic structures that are less harmonically compact, that is, not as well connected by prime harmonies.

Now we would like to discuss a few of the specific pentatonics in Table 8.5, mainly near the top of the list.

Specific pentatonics Pentatonic S_1 is the closest to five equal steps, and gives just versions of the Indonesian *Slendro* scale; some of the resulting scales were discussed by Jacques Dudon in "Seven-limit Slendro Mutations." One such ordering would be 7/6, 8/7, 9/8, 7/6, 8/7 (a mode of Dudon's *N*-scale), giving the scale 1/1, 7/6, 4/3, 3/2, 7/4, one of the favorites for improvisation. And, like S_3 discussed above, the bottom pair of steps and/or the top pair can be reversed while still retaining a strong harmonic structure. This pentatonic is the only one on the list based on primes 3 and 7 without using 5.

Pentatonic S_2 (in reverse order) gives the scale 1/1, 6/5, 7/5, 8/5, 9/5; this is five consecutive steps of the harmonic series, 5 to 10, so the frequency differences from one tone to the next are all equal. Another (cyclic) way to look at it is as a dominant-seventh ninth chord (with the ninth brought down into the octave): 1/1, 9/8, 5/4, 3/2, 7/4, which makes clear its strongly tonal nature. Of course, the opposite order gives the subharmonic version. Again, if we look at it as 1/1, 7/6, 4/3, 3/2, 5/3, we can reverse the bottom and/or top pairs of steps or even swap the two pairs and still keep fairly strong harmonic structure with the 4/3, 1/1, 3/2 backbone. Other orderings (and since all five steps are different, there are a lot of them) are less tonal.

Pentatonic S_3 was discussed previously; this is what might be called the "standard" pentatonic, being based on traditional harmony, with no semitones. Many traditional melodies are based on the "major" and "minor" versions.

The fourth one, S_4, can be arranged to give 1/1, 5/4, 11/8, 3/2, 7/4, whose tones are all prime harmonies (the 1/1 represents the prime 2). Here, not only are all the steps different, but all 20 intervals of the scale are different, and this is true regardless of the ordering of the steps (the other pentatonics with this property are S_i, $i = $ 6, 18, 23, and 38). This is the first using the prime 11, definitely an adventurous harmony, and the smallest step, 12/11 (150.6 cents), is halfway between a whole step and a half step. S_6 is similar to S_4 in many ways, having the prime scale 1/1, 5/4, 3/2, 13/8, 7/4, all intervals different, involving adventurous 13, and with a smallest step of 13/12 (138.6 cents).

With S_5, it seems difficult to make a scale with a nice harmonic structure; perhaps the best is 1/1, 5/4, 3/2, 5/3, 11/6.

The smallest step of S_7, 15/14 (119.4 cents), is getting into the semitone range; one nice scale of S_7 is 1/1, 7/6, 5/4, 3/2, 7/4.

The next S_8 is somewhat similar.

Pentatonic S_9 includes the scale 1/1, 5/4, 11/8, 3/2, 15/8, which has a nice character, similar to some Indonesian *Pelog* scales; the 15/11 interval (537.0 cents) is enough larger than a perfect fourth that it does not sound dissonant, just unusual.

The first pentatonic with a semitone that uses only traditional harmony (prime factors 3 and 5) is S_{10}, one scale of which is 1/1, 5/4, 4/3, 3/2, 5/3;

S_{13} has two semitones, and includes the scale 1/1, 9/8, 6/5, 3/2, 8/5.

Traditional-harmony scales with small semitones of 25/24 (70.7 cents) are S_{20}, the only one with three steps the same, and S_{21}, which also has a larger semitone. (T. Riley makes good use of the 25/24 steps of his 5-limit piano scale on the album "The Harp of New Albion.")

On the non-traditional side, there are two pentatonics that cannot have any perfect fifths, being free of the prime factor 3: S_{22} and S_{36}, though the latter, with a step of 56/55 (31.2 cents) may be too tense to use. The highest-limit harmony occurs in S_{25}, which involves the prime 31.

Example 71 HIRADOSHI(NO.2)-5.

8.2.3 Pacific Ocean region, the fuzzy tone systems

The Chinese fifth was substantially less then 3/2. The small fifths we can commonly observe also going from the Asia continent to the direction of the Pacific ocean region. Moreover, there are two further phenomena which are important: the detuned octave and fuzzy tone systems. We quote [40]:

"In Java musicians and gamelan makers, when queried about the small fifth, explained it as coming from nature but gave no specific examples. At dawn one morning in a small village in Java, a bird was singing with a call of an octave and a small fifth. This I recorded. The same song was frequently heard later and again recorded for confirmation of interval. It is a fascinating bit of fancy that the fifth in the bird call when measured on the stroboscope varied only two or three cents from the one of 678 cents."

In [40], pp. 33, the gamelan scale is described:

"Three basic tones and two or four secondary tones are the background of the gamelan tonal system. The main tone (called *dong* in Bali) is supported by two tones, one a fifth above (called *dang*) and the other a fifth below (called *dung*). The secondary tones are a fifth above (*deng*) and a fifth below (*ding*) the supporting tones. By bringing the five tones within an octave, the following scale results: dong, deng, dung, dang, ding."

But it should be underlined that the Oriental fifths are variable in size and do not correspond to a Western fifth. In the Balinese-Javanese five-tone scale, a large interval, approximately a minor third, which will vary in size from one gamelan to the next, occurs between the second and third and the fourth and fifth degrees. Using a fifth of 678 cents, we can generate scales, c.f. *SLE/PE(23--TET)*-12, *TETRAGAM(SLE/PE/N0.2)*-12.

The fuzziness of tuning is a specific phenomena of this Earth region. The fuzziness is a result of "continuous mistuning" of the whole orchestra, a gamelan. Gamelan tone systems and detuned octaves are typical in Central Javanese gamelan in particular. We bring graphics showing the fuzziness of some well-known gamelan.

The data that forms the basis of these graphics comes from [64]. Surjodiningrat, Sudarjana, and Susanto measured the pitches of the central octave saron of 76 respected gamelan from Central Java. They also measured all the instruments in a few select gamelan, those shown below. Hood measured all the instruments of Kyahi Mendung, a Solonese gamelan. Alves

(http://www2.hmc.edu/~ alves/index.html)

graphed their data on scales of cents, relative to the average pitch of tone 1 of the middle octave.

Here are the Slendro and Pelog gamelan tone systems Kyahi Kanyutmesem of the Mangkunegaran Palace in Surakarta, c.f. Figure 8.7 and Figure 8.8. The different octaves are shown in parallel columns. The range of variation of a particular pitch between instruments is shown by a vertical line spanning that variation. The average pitch between all the instruments measured is shown by a horizontal line. With this arrangement, the octave detunings become clear.

Remarks and ideas for exploring

16 For further extended file of pentatonics, c.f. Appendix A.2.

17 For further extended file of heptatonics, c.f. Appendix A.3.

18 $(S_3, \Upsilon_{\alpha_{1,11}})$, $(S_3, \Upsilon_{\beta_{1,11}})$, $(S_3, \Upsilon_{\gamma_{1,11}})$ are partial monounary algebras of the minor seconds (the chromatic $(25/24)$, diatonic $(16/15)$ semitones and the complement of the chromatic semitone to the major whole tone $(27/25 = 9/8 : 25/24)$, respectively),

$(S_3, \Upsilon_{\alpha_{2,10}})$, $(S_3, \Upsilon_{\beta_{2,10}})$ are partial monounary algebras of the major seconds (the major $(9/8)$ and minor $(10/9)$ whole tones, respectively),

$(S_3, \Upsilon_{\alpha_{3,9}})$ is a partial monounary algebra of the minor thirds $(6/5)$,

$(S_3, \Upsilon_{\alpha_{4,8}})$ is a partial monounary algebra of the major thirds $(5/4)$,

$(S_3, \Upsilon_{\alpha_{5,7}}^{-1})$ is a partial monounary algebra of the perfect fourths $(4/3)$,

$(S_3, \Upsilon_{\alpha_{6,6}})$ is a partial monounary algebra of the tritones—the augmented fourths $(25/18)$,

$(S_3, \Upsilon_{\alpha_{6,6}}^{-1})$ is a partial monounary algebra of the tritones—the diminished fifths $(36/25)$,

$(S_3, \Upsilon_{\alpha_{5,7}})$ is a partial monounary algebra of the perfect fifths $(3/2)$,

$(S_3, \Upsilon_{\alpha_{4,8}}^{-1})$ is a partial monounary algebra of the minor sixths $(8/5)$,

$(S_3, \Upsilon_{\alpha_{3,9}}^{-1})$ is a partial monounary algebra of the major sixths $(5/3)$,

$(S_3, \Upsilon_{\alpha_{2,10}}^{-1})$, $(S_3, \Upsilon_{\beta_{2,10}}^{-1})$ are some partial monounary algebras of the minor sevenths,

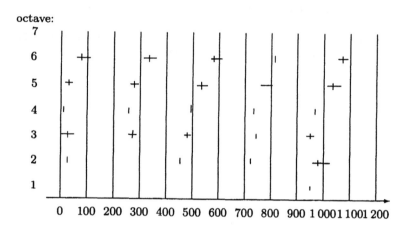

FIGURE 8.7: Slendro gamelan Kyahi Kanyutmesem

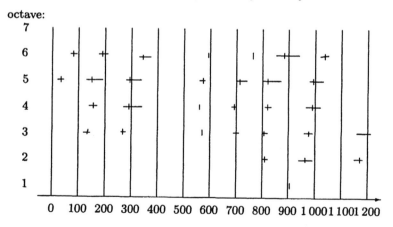

FIGURE 8.8: Pelog gamelan Kyahi Kanyutmesem

$(S_3, \Upsilon_{\alpha_{1,11}^{-1}}, (S_3, \Upsilon_{\beta_{1,11}}^{-1}), (S_3, \Upsilon_{\gamma_{1,11}}^{-1})$ are partial monounary algebras of the major sevenths.

19 C.f. Table 8.5. Comparable tables for superparticular hexatonic (6-tone) contain 380 scales and superparticular heptatonic (7-tone) 2908 scales.

20 There are more 5-tuples, c.f. Table 8.5, of superparticular intervals which can serve for constructing Pelog scales. Find them.

Chapter 9

Ptolemy System

9.1 Tetrachords

In general, tetrachords are not granulated tone systems. However, considering the set of all superparticular tetrachords, we will see the interesting connection between 10- and 12- granulations on the basis of octave symmetry. So, tetrachords are also the bridge between systems which are non-granulated and granulated ones.

The importance of considering of tetrachords consists also of the fact that they are the simplest used structured tone systems. Discrete tetrachords and pentatonics (they are supersets of tetrachords) are still so simple so we are able to consider these tone systems classes in their complexity.

We show that the set of all tetrachords is naturally structured as a lattice. There are 156 superparticular tetrachords, where nine of them can be nontrivially expressed in the form of the geometric net. We show how these 9 tetrachords construct tone systems of 12 or 10 granules.

Under the *tetrachord* we understand the following tone system: $S = \{\omega_i;\ i = 0, 1, 2, 3,\ \text{and}\ 1 = \omega_0 < \omega_1 < \omega_2 < \omega_3 = 4/3\}$.

Tetrachords, the archaic tone systems, also have a philosophical context of their construction. The *Pythagoreans* built tone systems based on the numbers (1), 2 and 3, because they asserted that only these numbers are perfect. Tone systems of the *Aristoxenos School* contain also musical intervals based on the prime factors 5, 7, etc. There is also a large group of tetrachords coming from the Islamic world (*Al-Farabi, Mohajira*). For a present music theory survey of tetrachords and tetrachordal scales, c.f. [7].

9.1.1   The set of all tetrachords

Consider the subset of all geometric nets given by 4 quotients and the Diophantine equation system:

$$\begin{aligned} \nu_{1,k} &\le \nu_{2,k} \le \nu_{3,k} \le \nu_{4,k}, \\ \nu_{k,1} + \nu_{k,2} + \nu_{k,3} + \nu_{k,4} &= k, \\ \nu_{j,k} \in \mathbb{N}, \quad j,k &= 1,2,3,4. \end{aligned} \tag{9.1}$$

Denote by

$$\mathbb{A} = \begin{bmatrix} \nu_{1,1} & \nu_{1,2} & \nu_{1,3} & \nu_{1,4} \\ \nu_{2,1} & \nu_{2,2} & \nu_{2,3} & \nu_{2,4} \\ \nu_{3,1} & \nu_{3,2} & \nu_{3,3} & \nu_{3,4} \\ \nu_{4,1} & \nu_{4,2} & \nu_{4,3} & \nu_{4,4} \end{bmatrix}$$

the matrix of integers satisfying (9.1). According to the symmetry, the all $4! \cdot 15 = 360$ possibilities of $\mathbb{A}$ satisfying (9.1) are reduced to the following 15 matrixes:

$$\mathbb{A}_1 = \begin{bmatrix} 1 & 0 & 0 & 0 \\ 2 & 0 & 0 & 0 \\ 3 & 0 & 0 & 0 \\ 4 & 0 & 0 & 0 \end{bmatrix}, \quad \mathbb{A}_2 = \begin{bmatrix} 1 & 0 & 0 & 0 \\ 2 & 0 & 0 & 0 \\ 3 & 0 & 0 & 0 \\ 3 & 1 & 0 & 0 \end{bmatrix}, \quad \mathbb{A}_3 = \begin{bmatrix} 1 & 0 & 0 & 0 \\ 2 & 0 & 0 & 0 \\ 2 & 1 & 0 & 0 \\ 3 & 1 & 0 & 0 \end{bmatrix},$$

$$\mathbb{A}_4 = \begin{bmatrix} 1 & 0 & 0 & 0 \\ 2 & 0 & 0 & 0 \\ 2 & 1 & 0 & 0 \\ 2 & 2 & 0 & 0 \end{bmatrix}, \quad \mathbb{A}_5 = \begin{bmatrix} 1 & 0 & 0 & 0 \\ 2 & 0 & 0 & 0 \\ 2 & 1 & 0 & 0 \\ 2 & 1 & 1 & 0 \end{bmatrix}, \quad \mathbb{A}_6 = \begin{bmatrix} 1 & 0 & 0 & 0 \\ 1 & 1 & 0 & 0 \\ 2 & 1 & 0 & 0 \\ 3 & 1 & 0 & 0 \end{bmatrix},$$

$$\mathbb{A}_7 = \begin{bmatrix} 1 & 0 & 0 & 0 \\ 1 & 1 & 0 & 0 \\ 2 & 1 & 0 & 0 \\ 2 & 2 & 0 & 0 \end{bmatrix}, \quad \mathbb{A}_8 = \begin{bmatrix} 1 & 0 & 0 & 0 \\ 1 & 1 & 0 & 0 \\ 2 & 1 & 0 & 0 \\ 2 & 1 & 1 & 0 \end{bmatrix}, \quad \mathbb{A}_9 = \begin{bmatrix} 1 & 0 & 0 & 0 \\ 1 & 1 & 0 & 0 \\ 1 & 2 & 0 & 0 \\ 2 & 2 & 0 & 0 \end{bmatrix},$$

$$\mathbb{A}_{10} = \begin{bmatrix} 1 & 0 & 0 & 0 \\ 1 & 1 & 0 & 0 \\ 1 & 2 & 0 & 0 \\ 1 & 3 & 0 & 0 \end{bmatrix}, \quad \mathbb{A}_{11} = \begin{bmatrix} 1 & 0 & 0 & 0 \\ 1 & 1 & 0 & 0 \\ 1 & 2 & 0 & 0 \\ 1 & 2 & 1 & 0 \end{bmatrix}, \quad \mathbb{A}_{12} = \begin{bmatrix} 1 & 0 & 0 & 0 \\ 1 & 1 & 0 & 0 \\ 1 & 1 & 1 & 0 \\ 2 & 1 & 1 & 0 \end{bmatrix},$$

$$A_{13} = \begin{bmatrix} 1 & 0 & 0 & 0 \\ 1 & 1 & 0 & 0 \\ 1 & 1 & 1 & 0 \\ 1 & 2 & 1 & 0 \end{bmatrix}, \quad A_{14} = \begin{bmatrix} 1 & 0 & 0 & 0 \\ 1 & 1 & 0 & 0 \\ 1 & 1 & 1 & 0 \\ 1 & 1 & 2 & 0 \end{bmatrix}, \quad A_{15} = \begin{bmatrix} 1 & 0 & 0 & 0 \\ 1 & 1 & 0 & 0 \\ 1 & 1 & 1 & 0 \\ 1 & 1 & 1 & 1 \end{bmatrix}.$$

With the matrix A we associate the following equation system:

$$\begin{aligned} \omega_1 &= X_1^{\nu_{1,1}} X_2^{\nu_{1,2}} X_3^{\nu_{1,3}} X_4^{\nu_{1,4}} \\ \omega_2 &= X_1^{\nu_{2,1}} X_2^{\nu_{2,2}} X_3^{\nu_{2,3}} X_4^{\nu_{2,4}} \\ \omega_3 &= X_1^{\nu_{3,1}} X_2^{\nu_{3,2}} X_3^{\nu_{3,3}} X_4^{\nu_{3,4}} = 4/3 \\ \omega_4 &= X_1^{\nu_{4,1}} X_2^{\nu_{4,2}} X_3^{\nu_{4,3}} X_4^{\nu_{4,4}} = 3/2 \\ & \quad 1 < \omega_1 < \omega_2 < \omega_3. \end{aligned} \qquad (9.2)$$

The proofs of the following theorems are easy.

Theorem 74 *According to the symmetry, the Diophantine equation system (9.1) has the unique solution A, such that $\det A \neq 0$, namely A_{15}.*

Theorem 75 *Let $A = A_1, A_2, \ldots, A_{15}$, $1 < \omega_1 < \omega_2 < 4/3$. Then the equation system (9.2) has the solutions collected in Table 9.1 (the intervals for the parameters t and v are collected in the third column of Table 9.2).*

Theorem 76 *For every tetrachord S_i, $i = 2, 3, \ldots, 15$, c.f. Table 9.2 there exists $k \in \{1, 2, 3, 4\}$, such that $X_k = 9/8$ (the Pythagorean whole tone).*

Theorem 77 *Let $S_1, S_2, \ldots, S_{15}$ be tetrachords corresponding to A_1, $A_2, \ldots, A_{15}$, respectively.*
Then the expressions of $S_1, S_2, \ldots, S_{15}$ are collected in Table 9.2.

Theorem 78 *For arbitrary ω_1, ω_2, $1 < \omega_1 < \omega_2 < 4/3$, and A_{15}, the unique solution of the equation system (9.2) is the following:*

$$(X_1, X_2, X_3, X_4) = (\omega_1, \omega_2/\omega_1, 4/(3\omega_2), 9/8).$$

Thus, we obtained the expression (possible, not unique) of each tetrachord by a geometric net. Since A_1, A_2, $\ldots$, A_{15} are (according to the symmetry) all the possibilities of powers for geometric net, solving the system (9.2) for A_1, A_2, $\ldots$, A_{15}, respectively, we obtain *classification of all tetrachords*, i.e., Table 9.2.

TABLE 9.1: Constructing intervals

	X_1	X_2	X_3	X_4
A_1	–	–	–	–
A_2	$\sqrt[3]{4/3}$	$9/8$	–	–
A_3	$9/8$	$508/243$	–	–
A_4	$\sqrt{32/27}$	$9/8$	–	–
A_9	$9/8$	$\sqrt{32/27}$	–	–
A_{10}	$256/243$	$9/8$	–	–
A_6	$9/8$	$508/243$	–	–
A_7	$\sqrt{32/27}$	$9/8$	–	–
A_5	t	$4/(3t^2)$	$9/8$	–
A_8	t	$4/(3t^2)$	$9/8$	–
A_{11}	t	$\sqrt{4/(3t)}$	$9/8$	–
A_{12}	$9/8$	$8t/9$	$4t/3$	–
A_{13}	t	$9/8$	$32/(27t)$	–
A_{14}	t	$32/(27t)$	$9/8$	–
A_{15}	t	v/t	$4/(3v)$	$9/8$

TABLE 9.2: The set of all tetrachords

	ω_1	ω_2	
S_1	–	–	
S_2	$\sqrt[3]{4/3}$	$\sqrt[3]{16/9}$	
S_3	$9/8$	$81/64$	
S_4	$\sqrt{32/27}$	$32/27$	
S_6	$9/8$	$32/27$	
S_7	$\sqrt{32/27}$	$\sqrt{3/2}$	
S_9	$9/8$	$\sqrt{3/2}$	
S_{10}	$256/243$	$32/27$	
S_5	t	t^2	$1 < t < \sqrt{4/3}$
S_8	t	$4/(3t)$	$1 < t < 4/3$
S_{11}	t	$\sqrt{4t/3}$	$1 < t < 4/3$
S_{12}	$9/8$	t	$9/8 < t < 4/3$
S_{13}	t	$9t/8$	$9/8 < t < 4/3$
S_{14}	t	$32/27$	$1 < t < 32/27$
S_{15}	t	v	$1 < t < v < 4/3$

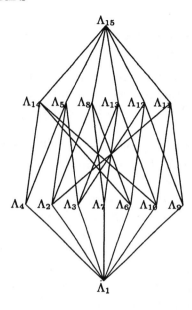

FIGURE 9.1: Lattice of *F*-tetrachords

9.1.2 Tetrachord lattice

Definition 29 The vector function $\Theta(\cdot) = (1, \omega_1(\cdot), \omega_2(\cdot), 4/3)$ is said to be the *F-tetrachord* (fuzzy tetrachord) if the quadruple

$$\Lambda(M) = (1, \omega_1(M), \omega_2(M), 4/3)$$

is a tetrachord for every $M \in D_\Lambda$, where D_Λ is the support of Λ ($\emptyset$, 0-, 1-, or 2- dimensional, c.f. Table 9.2).

Let $\mathbf{S} = \{\Lambda_1, \Lambda_2, \ldots \Lambda_{15}\}$, where $\Lambda_i = \{S_i(M);\ M \in D_{\Lambda_i}\}$, D_{Λ_i} is the support of Λ_i, Λ_i corresponds to S_i in Table 9.2, $i = 1, 2, \ldots,$ 15. For $\Phi, \Psi \in \mathbf{S}$, we define $\Phi \leq \Psi$ if and only if $\Phi \subset \Psi$. In Table 9.2, we see that each couple of elements of $\mathbf{S}$ has both the supremum $\wedge$ and infimum $\vee$.

Theorem 79 *The set* $\mathbf{S}$ *equipped with the operations* $\wedge, \vee$ *is a non-modular atomic lattice with atoms* Λ_2, Λ_3, Λ_4, Λ_6, Λ_7, Λ_9, Λ_{10}.

Proof. We see that $\mathbf{S}$ is a lattice, c.f. Figure 9.1. Since it contains a pentagon, it is a non-modular (hence non-distributive) lattice. The assertion about atoms is easy. $\qquad\square$

The *maximal* F-tetrachord Λ_{15} is given by $\mathbb{A}_{15}$, c.f. Table 9.2. The *minimal* F-tetrachord Λ_1 is given by $\mathbb{A}_1$ and it is the empty set of tetrachords. Indeed, $\mathbb{A}_1$ yields a contradiction: $X_1 = 4/3, X_1^4 = 3/2$.

The F-tetrachords Λ_1 and Λ_{15} we will call to be *trivial*.

The support of Λ_1 is $\emptyset$ set and $S \in \Lambda_1$ have no geometric net constructing intervals.

The *nontrivial* 13 F-tetrachords $\Lambda_2, \Lambda_3, \ldots, \Lambda_{14}$ are of two types. The first one contains seven F-tetrachords ($\Lambda_2, \Lambda_3, \Lambda_4, \Lambda_6, \Lambda_7, \Lambda_9, \Lambda_{10}$) with one individual tetrachord with 2 geometric net constructing intervals, i.e. X_1, X_2. Each of these F-tetrachords has a 0-dimensional support.

The second type contains 6 F-tetrachords ($\Lambda_5, \Lambda_8, \Lambda_{11}, \Lambda_{12}, \Lambda_{13}, \Lambda_{14}$) with the 1-dimensional support. Individual tetrachords have 3 geometric net constructing intervals, i.e. X_1, X_2, X_3.

The support of Λ_{15} is a 2-dimensional set. Individual tetrachords $T \in S_{15}$ have 4 geometric net constructing intervals, i.e. X_1, X_2, X_3, X_4. $\qquad\square$

9.1.3 Superparticular tetrachords

We say that a tetrachord S is *superparticular* if $\omega_1, \omega_2/\omega_1, 4/(3\omega_2)$ are of the form $(n+1)/n$ for some $n = 1, 2, \ldots$, c.f. [130]. Tetrachords S_4, S_5, S_6, S_8, and S_{10} in Subsection 9.1.1 are superparticular.

Clearly Λ_{15} contains all superpatricular tetrachords, Λ_1 none, c.f. Table 9.2. But what about nontrivial F-tetrachords $\Lambda_2, \ldots, \Lambda_{14}$?

Definition 30 A superparticular tetrachord which can be expressed as a 2- or 3- geometric net we will call the S-*tetrachord*. The set of all S-tetrachords denote by W.

If we define the *tetrachord sequence* $\langle \omega_i \rangle$ recursively as follows:

$$(\omega_{4k}, \omega_{4k+1}, \omega_{4k+2}, \omega_{4k+3}) = (3/2)^k S,$$

(where S is a tetrachord and k an integer number), we obtain a tone system with the period 3/2 (i.e., the perfect fifth).

TABLE 9.3: Numbers constructing superparticular tetrachords

5/4	17/16	256/255	6/5	15/14	28/27
5/4	18/17	136/135	6/5	16/15	25/24
5/4	19/18	96/95	6/5	19/18	20/19
5/4	20/19	76/75	7/6	9/8	64/63
5/4	21/20	64/63	7/6	10/9	36/35
5/4	22/21	56/55	7/6	12/11	22/21
5/4	24/23	46/45	7/6	15/14	16/15
5/4	26/25	40/39	8/7	8/7	49/48
5/4	28/27	36/35	8/7	9/8	28/27
5/4	31/30	32/31	8/7	10/9	21/20
6/5	11/10	100/99	8/7	13/12	14/13
6/5	12/11	55/54	9/8	10/9	16/15
6/5	13/12	40/39	10/9	11/10	12/11

Definition 31 We say that two tetrachords $S = (\omega_0, \omega_1, \omega_2, \omega_3)$ and $L = (\lambda_0, \lambda_1, \lambda_2, \lambda_3)$ are *equivalent*, we write $L \approx T$, if for the corresponding tetrachord sequences $\langle \omega_i \rangle$ and $\langle \lambda_i \rangle$, there exist integer numbers p, q and $k \in (0, \infty)$ such that

$$(\omega_p, \omega_{p+1}, \omega_{p+2}, \omega_{p+3}) = k \cdot (\lambda_q, \lambda_{q+1}, \lambda_{q+2}, \lambda_{q+3}).$$

It is easy to see that Definition 31 introduces a relation of *equivalency* on the set of all tetrachords (tetrachord sequences).

Theorem 80 *According to the equivalency of tetrachords, all the S-tetrachords are the following nine:*

$$(1, 9/8, 5/4, 4/3), \quad (1, 9/8, 6/5, 4/3) \quad (1, 16/15, 6/5, 4/3)$$
$$(1, 9/8, 7/6, 4/3), \quad (1, 9/8, 21/16, 4/3), \quad (1, 9/8, 8/7, 4/3),$$
$$(1, 7/6, 21/16, 4/3), \quad (1, 8/7, 9/7, 4/3), \quad (1, 9/8, 9/7, 4/3).$$

Proof. In Table 9.3, there are collected all 26 superparticular ratios, the combinations of the numbers $[\omega_1, \ \omega_2/\omega_1, \ 4/(3\omega_2)]$. To obtain this table, we used the general algorithm for dividing a given interval into a certain number of superparticular steps, [88], where Table 9.3 is attributed to I. E. Hofmann. One triple $[\omega_1, \ \omega_2/\omega_1, \ 4/(3\omega_2)]$ yields $3 \times 2 \times 1 = 6$ permutations. So, the set $\mathfrak{S}_*$ consists

TABLE 9.4: Superparticular tetrachords

S_{12}	S_{14}	S_{13}	prime factors
$(9/8, 21/16)$	$(7/6, 32/27)$	$(7/6, 21/16)$	$2, 3, 7$
$(9/8, 8/7)$	$(64/63, 32/27)$	$(64/63, 8/7)$	$2, 3, 7$
$(9/8, 9/7)$	$(8/7, 32/27)$	$(8/7, 9/7)$	$2, 3, 7$
$(9/8, 7/6)$	$(28/27, 32/27)$	$(28/27, 7/6)$	$2, 3, 7$
$(9/8, 5/4)$	$(10/9, 32/27)$	$(10/9, 5/4)$	$2, 3, 5$
$(9/8, 6/5)$	$(16/15, 32/27)$	$(16/15, 6/5)$	$2, 3, 5$

of $6 \times 26 = 156$ superparticular tetrachords. We have to find the intersection

$$W = \mathfrak{S}_* \cap \bigcup_{i=2,\ldots,14} \bigcup_{M \in D_{\Lambda_i}} S_i(M),$$

where $D_{\Lambda.}$ is the support of $\Lambda. \in \mathbf{S}$.

(a) *Superparticular tetrachords.*

The interval 9/8 appears in the F-tetrachord S_{12} and also in the combinations [7/6, 9/8, 64/63], [8/7, 9/8, 28/27], [9/8, 10/9, 16/15], c.f. Table 9.3. So we obtain the first column of Table 9.4.

Analogously, for S_{13}, S_{14}, we obtain the second and third column of S-tetrachords in Table 9.4.

It can be verified that the following pairs of tetrachords are equivalent:

$$\begin{aligned}
(1, 9/8, 7/6, 4/3) &\approx (1, 28/27, 32/27, 4/3) \\
(1, 9/8, 21/16, 4/3) &\approx (1, 7/6, 32/27, 4/3) \\
(1, 9/8, 8/7, 4/3) &\approx (1, 64/63, 32/27, 4/3) \\
(1, 9/8, 9/7, 4/3) &\approx (1, 8/7, 32/27, 4/3) \\
(1, 9/8, 5/4, 4/3) &\approx (1, 10/9, 32/27, 4/3) \\
(1, 9/8, 6/5, 4/3) &\approx (1, 16/15, 32/27, 4/3) \\
(1, 7/6, 21/16, 4/3) &\approx (1, 64/63, 8/7, 4/3) \\
(1, 8/7, 9/7, 4/3) &\approx (1, 28/27, 7/6, 4/3) \\
(1, 16/15, 6/5, 4/3) &\approx (1, 10/9, 5/4, 4/3).
\end{aligned}$$

(b) *Non superparticular tetrachords.*

$4/(3\omega_2)$ is not superparticular for S_3.

ω_1 is not superparticular for S_2, S_4, S_7, S_{10}.

ω_2/ω_1 is not superparticular for S_6, S_9.

If $t = (n+1)/n$, then $t^2 = (n^2+2n+1)/n^2$ is not superparticular, so S_5 is not superparticular.

If $t = (n + 1)/n$, then $\omega_2/\omega_1 = \sqrt{4n/(3n + 3)}$. Suppose $n = k^2$, then $2k = 1 + \sqrt{3k^2 + 3}$ which has not an integer solution. Hence S_{11} is not superparticular.

ω_2/ω_1 is not superparticular for S_8. Indeed, $\omega_2/\omega_1 = 4/(3t^2)$. Hence $4 - 3t^2 = 1$ implies $t = -1$ or $+1$, a contradiction.

We proved also the following statements:

Theorem 81 *No S-tetrachord is Pythagorean (based only on prime numbers 2 and 3).*

Theorem 82 *S-tetrachords are based either on the triplet [2, 3, 5] or [2, 3, 7].*

Theorem 83 *Other primes or triplets of primes than [2,3,5] and [2,3,7] or n-tuples of primes cannot construct S-tetrachords.*

9.1.4 Symmetry: Slendro versus 12 granulation

Definition 32 We say that a tone system

$$S = \{\omega_1, \omega_2, \ldots, \omega_n; \ \omega_1 = 1 < \omega_2 < \cdots < \omega_{n-1} < \omega_n = 2\}$$

has the center of symmetry $\sqrt{2}$ if for every $\omega' \in S$, there exists $\omega'' \in S$ such that $\omega'\omega'' = 2$.

We granule all superparticular tetrachord sequences based on the number triple $[2, 3, 5]$ into a new tone system as follows:

(1) take two tetrachord periods ($k = 0, 1$);

(2) group the neighboring values with their ratio less than or equal to $81/80$ (Comma of Didymus) into granules (qualitative musical degrees);

(3) the L-length between the tag points of two different clusters is equal or greater than the diatonic semitone ($16/15$);

(4) consider the octave equivalence for the new tone system.

This way, we obtain the following 11 granules:

$$G_1 = \quad \{16/15\};$$
$$G_2 = \quad \{10/9, 9/8\};$$
$$G_3 = \quad \{32/27, 6/5\};$$
$$G_4 = \quad \{5/4\};$$
$$G_5 = \quad \{4/3\};$$
$$G_6 = \quad \{-\};$$
$$G_7 = \quad \{3/2\};$$
$$G_8 = \quad \{8/5\};$$
$$G_9 = \quad \{5/3, 27/16\};$$
$$G_{10} = \quad \{16/9, 9/5\};$$
$$G_{11} = \quad \{15/8\};$$
$$G_{12} = \quad \{2 \equiv 1\},$$

i.e. the diatonic semitone; major and Pythagorean tones, respectively; Pythagorean and pure minor thirds, respectively; pure major third; perfect fourth;—; perfect fifth; just minor sixth; pure and Pythagorean minor sevenths, respectively; Pythagorean and pure minor sevenths, respectively; pure major seventh; unison and octave, respectively.

We denote the resulting tone system (the many valued 12-granulated system) as follows:

$$D_{2,3,5} = \{1, 16/15, (10/9, 9/8), (32/27, 6/5), 5/4, 4/3,$$

$$3/2, 8/5, (5/3, 27/16), (16/9, \ 9/5), \ 2\},$$

where (...) denotes the clustering.

Consider also the 12-granulated many valued tone system:

$$D_{2,3,5}^+ = \{1, 16/15, (10/9, 9/8), (32/27, 6/5), 5/4, 4/3, (45/32, 64/45),$$

$$3/2, 8/5, (5/3, 27/16), (16/9, \ 9/5), 15/8, \ 2\}.$$

It is easy to see that the tone system $D_{2,3,5}$ has not the center of symmetry $\sqrt{2}$.

We say that the system D^+ is the *minimal extension of the system D with respect to a property* (*) if (1) $D \subseteq D^+$; (2) D^+ fulfills the property (*); (3) every proper subset D', $D \subseteq D' \subset D^+$ does not fulfill the property (*).

Theorem 84 *The tone system* $D_{2,3,5}^+$ *is the minimal extension of* $D_{2,3,5}$ *such that it has the center of symmetry* $\sqrt{2}$.

Analogously, if we granule all superparticular tetrachord sequences based on the number triple [2,3, 7] into a new tone system as follows:

(1) take two tetrachord periods $(k = 0, 1)$;

(2) group values having their ratio less or equal than 49/48 into granules (qualitative musical degrees);

(3) the L-length between the tag points of two different clusters is greater than the chromatic semitone (25/24);

(4) consider the octave equivalence for the new tone system.

This way, we obtain the following 5 granules:

$$\begin{aligned}
G_1 &= \{9/8, 8/7, 7/6, 32/27\}; \\
G_2 &= \{9/7, 21/16, 4/3\}; \\
G_3 &= \{3/2, 32/21, 14/9\}; \\
G_4 &= \{27/16, 12/7, 7/4, 16/9\}; \\
G_5 &= \{27/14, 63/32, 2 \approx 1, 64/63, 28/27\}.
\end{aligned}$$

Or, in cents:

$$\begin{aligned}
G_1 &= \{203, 231, 266, 294\}; \\
G_2 &= \{435, 470, 498\}; \\
G_3 &= \{702, 729, 765\}; \\
G_4 &= \{905, 933, 968, 996\}; \\
G_5 &= \{1\,137, 1\,173, 1\,200 \approx 0, 27, 63\}.
\end{aligned}$$

Compare with the E_{10} gamelan "master scale": $\sqrt[10]{2^i} = \underline{0}$, 120, $\underline{240}$, 360, $\underline{480}$, 600, $\underline{720}$, 840, $\underline{960}$, 1080, $\underline{1\,200}$ cents, $i = 0, 1, 2, \ldots, 10$. The resulting 5 granules (the many valued pentatonic) is then as follows:

$$D_{2,3,7} = \{(1, 64/63, 28/27), (9/8, 8/7, 7/6, 32/27), (9/7, 21/16, 4/3),$$

$$(3/2, 32/21, 14/9), (27/16, 12/7, 7/4, 16/9), (27/14, 63/32, 2)\},$$

where $(\ldots)$ denotes the granules.

Theorem 85
The tone system $D_{2,3,7}$ has the center of symmetry $\sqrt{2}$.

Note that tone systems with 12 granules are characteristic for the European cultural zone, while Slendro are typical tone systems in Asia, Africa, Polynesia, Micronesia, Malaysia, etc.

Example 72 *INDIA-SAGRAMA-7.*

Example 73 There are all tetrachords listed in Subsection 9.1.5

Example 74 There is a list of other tetrachords (two periods).
AVERAGE-7; Average Bac System;
10/9, 20/17, 4/3, 3/2, 5/3, 30/17, 2
BARBOUR(CHRO/NO.1)-7; Barbour's No. 1 Chromatic;
55/54, 10/9, 4/3, 3/2, 55/36, 5/3, 2
BARBOUR(CHRO/NO.2)-7; Barbour's No. 2 Chromatic;
40/39, 10/9, 4/3, 3/2, 20/13, 5/3, 2
BARBOUR(CHRO/NO.3)-7; Barbour's No. 3 Chromatic;
64/63, 8/7, 4/3, 3/2, 32/21, 12/7, 2
BARBOUR(CHRO/NO.3/P)-7; Permuted Barbour's No. 3 Chromatic;
9/8, 8/7, 4/3, 3/2, 27/16, 12/7, 2
BARBOUR(CHRO/NO.3/P2)-7; Permuted Barbour's No. 3 Chromatic;
7/6, 32/27, 4/3, 3/2, 7/4, 16/9, 2
BARBOUR(CHRO/NO.4)-7; Barbour's No. 4 Chromatic;
81/80, 9/8, 4/3, 3/2, 243/160, 27/16, 2
BARBOUR(CHRO/NO.4/P)-7; Permuted Barbour's No. 4 Chromatic;
10/9, 9/8, 4/3, 3/2, 5/3, 27/16, 2
BARBOUR(CHRO/NO.4/P2)-7; Permuted Barbour's No. 4 Chromatic;
32/27, 6/5, 4/3, 3/2, 16/9, 9/5, 2
BOETH(CHRO)-7; Boethius's Chromatic. The CI is 19/16;
256/243, 64/57, 4/3, 3/2, 128/81, 32/19, 2

CHRO/INT-7; Intense Chromatic genus 4 + 8 + 18 parts;
200/3, 200, 500, 700, 2 300/3, 900, 1 200

CHRO/NEW-7; Chromatic genus 4.5 + 9 + 16.5;
75, 225, 500, 700, 775, 925, 1 200

CHRO/NEW(NO.2)-7; Chromatic genus 14/3 + 28/3 + 16 parts;
77.778, 700/3, 500, 700, 777.778, 2 800/3, 1 200

CHRO/SOFT(NO.1)-7; 100/81 Chromatic. This genus is a good approximation to the soft chromatic;
27/26, 27/25, 4/3, 3/2, 81/52, 81/50, 2

CHRO/SOFT(NO.2)-7; 1:2 Soft Chromatic;
44.444, 400/3, 500, 700, 744.444, 2 500/3, 1 200

CHRO/SOFT(NO.3)-7; Soft chromatic genus is from Schlesinger's modified Mixolydian Harmonia;
28/27, 14/13, 4/3, 3/2, 14/9, 21/13, 2

DIAT/CHRO-7; On the border between the chromatic and diatonic genera;
15/14, 15/13, 4/3, 3/2, 45/28, 45/26, 2

DUDON(A)-7; Dudon Tetrachord A, From Jacques Dudon, an Islamic or "Mohajira" type;
59/54, 11/9, 4/3, 3/2, 59/36, 11/6, 2

DUDON(B)-7; Dudon Tetrachord B, From Jacques Dudon, another Islamic or Mohajira tetrachord;
13/12, 59/48, 4/3, 3/2, 13/8, 59/32, 2

ENH(NO.2)-7; 1:2 Enharmonic. New genus 2 + 4 + 24 parts;
100/3, 100, 500, 700, 2 200/3, 800, 1 200

ENH(NO.14)-7; 14/11 Enharmonic;
44/43, 22/21, 4/3, 3/2, 66/43, 11/7, 2

EULER(ENH)-7; Euler's Old Enharmonic, From Tentamen Novae Theoriae Musicae;
128/125, 256/243, 4/3, 3/2, 192/125, 128/81, 2

HEM/CHRO13-7;
13'al Hemiolic Chromatic or neutral-third genus has a CI of 16/13;
26/25, 13/12, 4/3, 3/2, 39/25, 13/8, 2

HEM/CHRO2-7; 1:2 Hemiolic Chromatic genus 3 + 6 + 21 parts;
50, 150, 500, 700, 750, 850, 1 200

HIPKINS-7; Hipkins' Chromatic;
256/243, 8/7, 4/3, 3/2, 128/81, 12/7, 2

HOFMANN(NO.1)-7; Hofmann's Enharmonic No. 1, Dorian mode;
256/255, 16/15, 4/3, 3/2, 128/85, 8/5, 2

HOFMANN(NO.2)-7; Hofmann's Enharmonic No. 2, Dorian mode;
136/135, 16/15, 4/3, 3/2, 68/45, 8/5, 2

HOFMANN(CHRO)-7; Hofmann's Chromatic;
100/99, 10/9, 4/3, 3/2, 50/33, 5/3, 2

HUSMANN-7; Tetrachord division according to Husmann;
256/243, 9/8, 32/27, 19 683/16 384, 81/64, 4/3

HYPER/ENH)-7; 13/10 Hyper Enharmonic. This genus is at the limit of usable tone systems;
80/79, 40/39, 4/3, 3/2, 120/79, 20/13, 2

HYPER/ENH2-7; Hyper Enharmonic genus from Schlesinger's enharmonic Phrygian Harmonia;
48/47, 24/23, 4/3, 3/2, 72/47, 36/23, 2

LIU(MAJ)-7; Linus Liu's Major Scale, c.f. Linus Liu: "Intonation Theory", 1978;
10/9, 100/81, 4/3, 3/2, 5/3, 50/27, 2

MID/ENH(NO.1)-7; Mid-Model Enharmonic, permutation of Archytas's with the 5/4 lying medially;
36/35, 9/7, 4/3, 3/2, 54/35 27/14, 2

NEUTR/DIAT-7; Neutral Diatonic, 9 + 9 + 12 parts;
150, 300, 500, 700, 850, 1 000, 1 200

NEUTR/DIAT(NO.2)-7; Neutral Diatonic, 9 + 12 + 9 parts;
150, 350, 500, 700, 850, 1 050, 1 200

NEW/DIAT/SOFT-7; New Soft Diatonic genus with equally divided Pyknon, Dorian Mode, 1:1 pyknon;
250, 375, 500, 700, 950, 1 075, 1 200

NEW/ENH)-7; New Enharmonic;
81/80, 16/15, 4/3, 3/2, 243/160, 8/5, 2

NEW/ENH(NO.2)-7; P2 New Enharmonic;
5/4, 81/64, 4/3, 3/2, 15/8, 243/128, 2

PARACHRO-7; Parachromatic, new genus 5 + 5 + 20 parts;
250/3, 500/3, 500, 700, 2 350/3, 2 600/3, 1 200

MARWA(PERM/ENH)-7; Permuted Enharmonic, After Wilson's Marwa Permutations;
28/27, 16/15, 4/3, 3/2, 14/9, 16/9, 2

PERRETT/TARTINI/PACHYMERES(ENH)-7; Enharmonic;
21/20, 16/15, 4/3, 3/2, 63/40, 8/5, 2

PERRETT(CHRO)-7; Perrett's Chromatic;
21/20, 9/8, 4/3, 3/2, 63/40, 27/16, 2

REDFIELD-7; Redfield New Diatonic;
10/9, 5/4, 4/3, 3/2, 5/3, 15/8, 2

SALINAS(ENH)-7; Salinas's Enharmonic;
25/24, 16/15, 4/3, 3/2, 25/16, 8/5, 2

HARRISON(SOFT/DIAT)-7; From L. Harrison, a soft diatonic;
21/20, 6/5, 4/3, 3/2, 63/40, 9/5, 2

SOFT/DIAT(NO.2)-7; New Soft Diatonic genus with equally divided Pyknon, Dorian Mode, 1:1 pyknon;
125, 375, 500, 700, 825, 1075, 1 200

VERTEX/CHRO(NO.1)-7; Vertex tetrachord, 66.7 + 266.7 + 166.7;
200/3, 1 000/3, 500, 700, 2 300/3, 2 800/3, 1 200

VERTEX/CHRO(NO.2)-7; Vertex tetrachord, 83.3 + 283.3 + 133.3;
250/3, 1 100/3, 500, 700, 2 350/3, 3 200/3, 1 200

VERTEX/CHRO(NO.3)-7; Vertex tetrachord, 87.5 + 287.5 + 125;
87.500, 375, 500, 700, 787.500, 1 075, 1 200

VERTEX/CHRO(NO.4)-7; Vertex tetrachord, 88.9 + 288.9 + 122.2;
88.900, 377.800, 500, 700, 788.900, 1 077.800, 1 200

VERTEX/CHRO(NO.5)-7; Vertex tetrachord, 133.3 + 266.7 + 100;
400, 400, 500, 700, 2 500/3, 1 100, 1 200

VERTEX/DIAT(NO.1)-7; Vertex tetrachord, 233.3 + 133.3 + 133.3;
700/3, 1 100/3, 500, 700, 2 800/3, 3 200/3, 1 200

VERTEX/DIAT(NO.10)-7; Vertex tetrachord, 212.5 + 162.5 + 125;
425, 375, 500, 700, 1 825/2, 1 075, 1 200

VERTEX/DIAT(NO.11)-7; Vertex tetrachord, 212.5 + 62.5 + 225;
425, 275, 500, 700, 1 825, 975, 1 200

VERTEX/DIAT(NO.12)-7; Vertex tetrachord, 200 + 125 + 175;
200, 325, 500, 700, 900, 1 025, 1 200

VERTEX/DIAT(NO.2)-7; Vertex tetrachord, 233.3 + 166.7 + 100;
233.333, 400.000, 500, 700, 2 800/3, 1 100.000, 1 200

VERTEX/DIAT(NO.3)-7; Vertex tetrachord, 75 + 225 + 200;
75, 300, 500, 700, 775, 1 000, 1 200

VERTEX/DIAT(NO.4)-7; Vertex tetrachord, 225 + 175 + 100;
225, 400, 500, 700, 925, 1 100, 1 200

VERTEX/DIAT(NO.5)-7; Vertex tetrachord, 87.5 + 237.5 + 175;
87.500, 325.000, 500, 700, 787.500, 1 025.000, 1 200

VERTEX/DIAT(NO.7)-7; Vertex tetrachord, 200 + 75 + 225;
200, 275, 500, 700, 900, 975, 1 200

VERTEX/DIAT(NO.8)-7; Vertex tetrachord, 100 + 175 + 225;
100, 275, 500, 700, 800, 975, 1 200

VERTEX/DIAT(NO.9)-7; Vertex tetrachord, 212.5 + 137.5 + 150;
425, 350, 500, 700, 1825, 1 050, 1 200

VERTEX/S/DIAT(NO.1)-7; Vertex tetrachord, 87.5 + 187.5 + 225;
175/2, 275, 500, 700, 1575/2, 975.000, 1 200

VERTEX/S/DIAT(NO.2)-7; Vertex tetrachord, 75 + 175 + 250;
75, 250, 500, 700, 775, 950, 1 200

VERTEX/S/DIAT(NO.3)-7; Vertex tetrachord, 25 + 225 + 250;
25, 250, 500, 700, 725, 950, 1 200

VERTEX/S/DIAT(NO.4)-7; Vertex tetrachord, 66.7 + 183.3 + 250;
200/3, 250, 500, 700, 2 300/3, 950, 1 200

VERTEX/S/DIAT(NO.5)-7; Vertex tetrachord, 233.33 + 16.67 + 250;
233.333, 250, 500, 700, 2 800/3, 950, 1 200

WIL(ENH)-7; Wilson's Enharmonic & 3rd new Enharmonic on Hofmann's list of superp. 4chords;
96/95, 16/15, 4/3, 3/2, 144/95, 8/5, 2

WIL(ENH/NO.2)-7; Wilson's 81/64 Enharmonic, a strong division of the 256/243 pyknon;
64/63, 256/243, 4/3, 3/2, 32/21, 128/81, 2

TABLE 9.5: Modes mapped to our present system

Aeolian	A	B	C	D	E	F	G
Mixolydian	B	C	D	E	F	G	A
Lydian	C	D	E	F	G	A	B
Phrygian	D	E	F	G	A	B	C
Dorian	E	F	G	A	B	C	D
Hypolydian	F	G	A	B	C	D	E
Hypophrygian	G	A	B	C	D	E	F

TABLE 9.6: Modes mapped into the key of C

Aeolian	C	D	E_b	F	G	A_b	B_b
Mixolydian	C	D_b	E_b	F	G_b	A_b	B_b
Lydian	C	D	E	F	G	A	B
Phrygian	C	D	E_b	F	G	A	B_b
Dorian	C	D_b	E_b	F	G	A_b	B_b
Hypolydian	C	D	E	$F_\sharp$	G	A	B
Hypophrygian	C	D	E	F	G_b	A	B_b

WIL(HYP/ENH)-7; Wilson's Hyperenharmonic, this genus has a CI of 9/7;

56/55, 28/27, 4/3, 3/2, 84/55, 14/9, 2

9.1.5   Ancient Greece

We owe the basis for the present European musical systems (mainstream) to the ancient Greeks.

It seems that the music culture of the ancient Greeks had also the following roots which interacted: culture of Middle Asia, India, and the home sources. Specially, the 7 degree tone systems, *modes*, used until the Middle Ages are nothing else as maquams, c.f. Tab. 9.5, Tab. 9.6, and *maqams* table. Similarly, Pythagorean System we can search in Indian Srutis. Tetrachords with the special fifth periode are rather original Greek.

For the concrete frequency values we can observe, e.g., the Byzantine Liturgical modes (two tetrachord periods, 500 cents = 30 parts):

Example 75

SAVAS(BAR/DIAT)-7; Savas's Byzantine Liturgical mode, 8 + 12 + 10 parts;
400/3, 1 000/3, 500, 700, 2 500/3, 3 400/3, 1 200

SAVAS(BAR/ENH)-7; Savas's Byzantine Liturgical mode, 8 + 16 + 6 parts;
400/3, 400.000, 500, 700, 2 500/3, 1 100, 1 200

SAVAS(CHRO)-7 ;
Savas's Chromatic, Byzantine Liturgical mode, 8 + 14 + 8 parts;
400/3, 1 100/3, 500, 700, 2 500/3, 3 200/3, 1 200

SAVAS(DIAT)-7; Savas's Diatonic, Byzantine Liturgical mode, 10 + 8 + 12 parts;
500/3, 300, 500, 700, 2 600/3, 1 000, 1 200

SAVAS(PALACE)-7; Savas's Byzantine Liturgical mode, 6 + 20 + 4 parts;
100, 1 300/3, 500, 700, 800, 3 400/3, 1 200

TIBY(NO.1)-7; Tiby's 1st Byzantine Liturgical genus, 12 + 13 + 3 parts;
211.765, 441.176, 494.118, 705.882, 917.647, 1 147.059, 1 200.000

TIBY(NO.2)-7; Tiby's second Byzantine Liturgical genus, 12 + 5 + 11 parts;
211.765, 300.000, 494.118, 705.882, 917.647, 1 005.882, 1 200.000

TIBY(NO.3)-7; Tiby's third Byzantine Liturgical genus, 12 + 9 + 7 parts;
211.765, 370.588, 494.118, 705.882, 917.647, 1 076.471, 1 200.000

TIBY(NO.4)-7; Tiby's fourth Byzantine Liturgical genus, 9 + 12 + 7 parts;
158.824, 370.588, 494.118, 705.882, 864.706, 1 076.471, 1 200.000

XENAKIS(CHRO)-7; Xenakis's Byzantine Liturgical mode, 5 + 19 + 6 parts;
250/3, 400, 500, 700, 2 350/3, 1 100, 1 200

XENAKIS(DIAT)-7; Xenakis's Byzantine Liturgical mode, 12 + 11 + 7 parts;
200, 2 500/3, 500, 700, 900, 3 250/3, 1 200

*XENAKIS(S/CHRO)-7; Xenakis's Byzantine Liturgical mode, 7 +
16 + 7 parts;*
350/3, 2 500/3, 500, 700, 2 450/3, 3 250/3, 1 200

The mathematician Pythagoras (probably 582–492 B.C.) realized
that simple intervals could be formed by dividing a stretched string
such that the divisions were in whole-number ratios; divide it in half
and you get an octave; divide by a ratio of 3:2 and you get a fifth and
so on. The Pythagorean Systems, c.f. Table 6.3, are derived from
perfect fifths alone (no Pramana). The intonation of the Renaissance
period in Europe used eight ascending and three descending fifths
(= 3 ascending fourths).

Obviously, it was known rather in Babylon, maybe in Summer,
but Pythagoras additionally asked a very important question: "Why
is consonance determined by the ratio of small whole numbers?,"
meaning that as the ratio numbers get larger, the interval gets more
dissonant. Now at that time, the Greeks were into the natural sciences; they were probably amongst the first people to study natural phenomena and to seek an explanation of the way the universe worked through empirical observations and deductions, but
they were also noted for their aesthetic appreciation of the arts.

A common instrument from those times was the *tetrachord*, a
kind of four-stringed harp. These essentially open-tuned instruments
spanned a fourth on the outer pair of strings with the inner pair
tuned to intervals ranging from quarter tones to two tones.

Pythagoras certainly spent much time experimenting with different ways of tuning tetrachord instruments in an attempt to formalize
the procedure. One of his major contributions to music was extend-
ing the range of the scale by using two tetrachords tuned a fifth apart.
This then gave a range of an octave and, consequently, a much wider
variation of scales became possible which were to become *Modes*.
Observe that there are 12 principally different musical degrees when
integrating all modes into one octave, c.f. Tables 9.5 and 9.6.

Since these scales now spanned an octave, some method was
needed to tune them in a consistent fashion. For the exact description of the algorithm of Pythagorean System, c.f. Chapter 6.

It is perhaps unfortunate for us that Pythagoras did not proceed to divide the string into fifths, sevenths and so on otherwise music might have taken a very different course. The problem stems from Pythagoras' use of the numbers 1, 2, and 3. He asserted that only these three numbers are perfect.

Example 76

PYTH-12; 12-tone Pythagorean scale;
2 187/2 048, 9/8, 32/27, 81/64, 4/3, 729/512, 3/2, 6 561/4 096, 27/16, 16/9, 243/128, 2

PYTH-27; 27-tone Pythagorean scale;
531 441/524 288, 256/243, 2 187/2 048, 65 536/59 049, 9/8, 4 782 969/4 194 304, 32/27, 19 683/16 384, 8 192/6 561, 81/64, 4/3, 177 147/131 072, 1 024/729, 729/512, 262 144/177 147, 3/2, 1 594 323/1 048 576, 128/81, 6 561/4 096, 32 768/19 683, 27/16, 14348907/8 388 608, 16/9, 59 049/32 768, 4 096/2 187, 243/128, 2

PYTH-31; 31-tone Pythagorean scale;
531 441/524 288, 256/243, 2 187/2 048, 1 162 261 467/1 073 741 824, 9/8, 4 782 969/4 194 304, 32/27, 19 683/16 384, 341.055, 8 192/6 561, 81/64, 43 046 721/33 554 432, 4/3, 177 147/131 072, 1 024/729, 729/512, 387 420 489/268 435 456, 3/2, 1 594 323/1 048 576, 128/81, 6 561/4 096, 839.100, 27/16, 14 348 907/8 388 608, 16/9, 59 049/32 768, 1043.010, 4 096/2 187, 243/128, 129 140 163/67 108 864, 2

PYTH(CHRO)-8; Dorian mode of the so-called Pythagorean chromatic, recorded by Gaudentius;
256/243, 9/8, 4/3, 3/2, 128/81, 27/16, 16/9, 2

Aristoxenus, a student of Aristotle, considered that aesthetic appreciation was more important than the mathematics. As the basis for his scale, he took two tetrachords with the top note of the upper tetrachord tuned a perfect fifth above the top note of the lower one.

There are examples of tetrachords (two periods) from the Arabic world:

Example 77

AL-FARABI-7, AL-FARABI(CHRO)-7,
AL-FARABI(DIAT/NO.1)-7, AL-FARABI(DIAT/NO.2)-7,
AL-FARABI(DOR/NO.1)-7, AL-FARABI(DOR/NO.2)-7,
AL-FARABI(TET)-12, AL-HWARIZMI-6, AL-KINDI-6.

TABLE 9.7: Pythagoras, Aristoxenus, Zarlino

	Pythagoras	Aristoxenus	Zarlino
C	0	0	0
D	9/8	9/8	9/8
E	81/64	5/4	5/4
F	4/3	4/3	4/3
G	3/2	3/2	3/2
A	27/16	27/16	5/3
B	243/128	15/8	15/8
C'	2	2	2

Since each tetrachord spans a fourth, the disjunctive tone (the interval between the top note of the lower tetrachord and the bottom note of the upper one) is the difference between a perfect fifth (702 cents) and a perfect fourth (498 cents) or 204 cents. If our notes of the lower tetrachord are C, D, E, F, D is a Pythagorean tone (204 cents) from C, F the perfect fourth and E the interval we now know as the Major third (386 cents). If we add the same group of four notes starting from G we arrive the following intervals, c.f. Table 9.7, the third column (Aristoxenus).

This has a better Major 3rd and Major 7th and is closer to the pure tone temperament in that some of the the interval ratios in this scale are multiples of the prime numbers 2, 3 and 5; the only problem now is that the tones are of different sizes and are known historically as Major and Minor whole Tones (9/8 and 10/9, respectively).

Thus, it seems that the Hindu Srutis was overtaken by ancient Greeks as two split systems—as the Pythagorean System and as the Just Intonation.

The Just (perfect) scale, c.f. Table 9.7 (Zarlino), is a beat-canceling integral-ratio scale commonly used until the eighteenth century in Western music. It is still used by some performers of ancient music for the sake of authenticity and in certain contexts by other performers. The actual pitches of the Just scale depend on which note is taken as the tonic of the scale. The ratios, however, remain constant regardless of the tonic.

The astronomer, mathematician and geographer Ptolemy (approx. 90–160 A.D.) and another theoretician Archytas had proposed the use of 5:4 and 8:7 ratios as consonant intervals thus advocating the use of all ratios of integer numbers, c.f. *LAMBDOMA(PRIM)-56*, *LAMBDOMA*(5 × 12)-*42*. This system (i.e. the set of all nonnegative rational numbers) is named the *Ptolemy's System*, resp. the (general) Just Intonation.

However, the further development went by "specific (not Hindu Srutis!) integration" the Pythagorean and Aristoxenus directions. The very ramified Ptolemy's tone system stopped its development for about 2000 years and starts to live again after prevailing of the 12 Tone Equal Temperament—in the 20th century it was reborn used in connection with the development of computer music, c.f. Subsection 9.2. This is a very interesting moment.

Example 78 Tetrachords: c.f. Example A.1.

Example 79

ARCHY(MULT)-12; Multiple Archytas;
28/27, 16/15, 5/4, 9/7, 4/3, 112/81, 3/2, 14/9, 8/5, 15/8, 27/14, 2

ARCHY/PTO(/NO.1)-12; Archytas/Ptolemy Hybrid No. 1;
28/27, 16/15, 10/9, 32/27, 4/3, 112/81, 3/2, 14/9, 8/5, 5/3, 16/9, 2

ARCHY/PTO(/NO.2)-12; Archytas/Ptolemy Hybrid No. 2;
28/27, 16/15, 9/8, 6/5, 4/3, 112/81, 3/2, 14/9, 8/5, 27/16, 9/5, 2

ARCHY(SEPT)-12; Archytas Septimal;
28/27, 16/15, 9/8, 32/27, 4/3, 112/81, 3/2, 14/9, 8/5, 27/16, 16/9, 2

ARCHY(DOR)-8;
Dorian mode of Archytas' Chromatic, added 16/9;
28/27, 9/8, 4/3, 3/2, 14/9, 16/9, 27/16, 2

ARCHY(ENH/NO.2)-8; Archytas' Enharmonic with added 16/9;
28/27, 16/15, 4/3, 3/2, 14/9, 16/9, 8/5, 2

LAMBDOMA(5 × 12)-*42; 5 × 12 Lambdoma;*
1/12, 1/11, 1/10, 1/9, 1/8, 1/7, 1/6, 2/1, 1/5, 2/9, 1/4, 3/11, 2/7, 3/10, 1/3, 4/11, 3/8, 2/5, 5/12, 3/7, 4/9, 5/11, 1/2, 5/9, 4/7, 3/5, 5/8, 2/3, 5/7, 3/4, 4/5, 5/6, 1/1, 5/4, 4/3, 3/2, 5/3, 2/1, 5/2, 3/1, 4/1, 5/1

LAMBDOMA(PRIM)-56; Prime Lambdoma;
1/31, 1/29, 1/23, 1/19, 1/17, 2/31, 2/29, 1/13, 2/23, 1/11, 3/31, 3/29, 2/19, 2/17, 3/23, 1/7, 2/13, 3/19, 5/31, 5/29, 3/17, 2/11, 1/5, 5/23, 7/31, 3/13, 7/29,

5/19, 3/11, 2/7, 5/17, 7/23, 1/3, 7/19, 5/13, 2/5, 7/17, 3/7, 5/11, 1/2, 7/13, 3/5, 7/11, 2/3, 5/7, 1/1, 7/5, 3/2, 5/3, 2/1, 7/3, 5/2, 3/1, 7/2, 5/1, 7/1

TERPANDER-6; One modern guess at the scale of the ancient Greek poet Terpander,

11/10, 11/9, 11/8, 11/7, 11/6, 2

9.2 Ptolemy System in the 20th century

The Ptolemy System, as the most complex one, led in Ancient Greece, in fact, to granulation of tone systems in general. We demonstrate this claim in Section 9.3.4. In the 20th century, we observe the reverse tendency: Ptolemy System is bounded with the so called microtonal music which is non-granulated and (seemingly?) free in creating mathematical structures of tone systems in the frame of rational numbers.

Just intonation enthusiasts trace the origin of their movement to H. Partch, H. Helmholtz and the British and German just intonation theorist/instrument-builders of the late 19th and early 20th century: Sh. Tanaka, C. Eitz, Th. Cahill, W. Perrett, K. Schlesinger, and M. Meyer.

Just intonation composers typically construct their own acoustic instruments and/or write the software for their own digital computer-based instruments. New York microtonalists include harmonic series instrument-builder G. Branca and E. Sharp, who takes an industrial-music approach to microtonality with tone systems derived from Fibonacci series. B. Hero has built her own generalized *MIDI* keyboard and works with the Lambdoma, a set of just intonation with roots in Greek and Babylonian number theory. S. Wayne composes microtonal music based on the tunings of Native Americans. Northern California microtonal activity centers primarily around just intonation. In the San Francisco Bay area, many xenharmonic composers work with real-time algorithmic synthesizer composition.

One of the defining figures of Northern California microtonality is L. Harrison, who has composed much music for gamelan alone, as well as gamelan mixed with European instruments. Harrison is credited with sparking the American gamelan movement, and he appears to have been the first composer to tune a Javanese-style orchestra of metallophones to just intonation. W. Colvig, a close associate of L. Harrison, was responsible for inventing and popularizing the "tubulong," a set of xenharmonically tuned tubes typically cut from aluminum electrical conduit. He has built many microtonal zithers and monochords as well as a wide variety of gamelan-like instruments. Colvig remains one of the most talented and imaginative instrument builders in the United States, and his work has exerted a definitive influence throughout the entire instrument-building community in America.

Instrument builder and just intonation composer C. Forster has spent many years perfecting a set of extended JI instruments using a 13-limit extension of H. Partch's tone system.

Elsewhere in Los Angeles, S. Wilkinson performs on both acoustic and electronic instruments in just intonation.

M. Hobbs, also in the Los Angeles area, works in just intonation as well as various equal temperaments using the sophisticated Kyma parallel-processing real-time digital synthesizer designed by C. Scaletti and K. Hebel at the University of Illinois. Hobbs has recently written software to "morph" between various Wilson CPS scales in real time, as well as to change the value of the "octave" in an equal-tempered composition in real time.

Swimming against the regional tide, Chico California composer M. McCoskey uses IBM PC sound cards to obtain extremely pitch-accurate just intonation, and composes imaginative and intricate contrapuntal music in 5-limit just intonation.

The San Diego-based theorist J. Chalmers has been called "one of the gurus" of microtonal music, and his 1993 book *Divisions of the Tetrachord*, c.f. [7], stands for musicians as a definitive examination of Greek scale construction and just intonation theory.

9.2.1   Ptolemy System

Ptolemy System (generalized Just Intonation) is an arbitrary finite subset of rational numbers m/n with the octave equivalency (and preferably with small irreducible m, n). The musically reasonable subsets of numbers arrange the nowadays musicians into diagrams

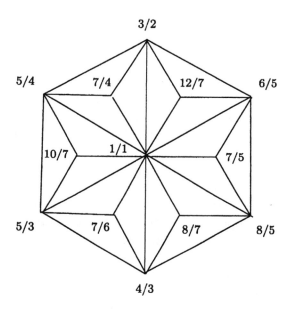

FIGURE 9.2: Projection along the octave axe to the plane

(which they call poetically "lattice diagrams," "diamonds," etc.). The parallel diagram edges of the same length connecting points in the plane (representing rational numbers) denote the same musical intervals. For instance, in the Figure 9.2, there is a projection of the tone system SQUARE-13 in Euler space along the octave axe to the plane; the pure fifths are parallel and have the equal length with the edge $3/2 \equiv (6/5) : (8/5) = (5/4) : (5/3)$, analogously for the major thirds $(5/4)$, minor thirds $(6/5)$, natural sevenths $(7/4)$, etc.

More precisely, these diagrams are nothing else than various parallel projections of the Euler n-dimensional space into a plane. Specially, in the case of octave equivalency, Euler points may be projected along the octave axis.

Also, we can consider discrete geometric nets as linear transformations as projections of Euler space. Projecting linearly independent vectors into the plane, independence of vectors will be lost but the relations among vectors is saved using parallel vectors expressing these relations.

9.2.2   Set of all superparticular ratios

In this section we describe the set of all superparticular ratios.

The following superparticular ratios 2/1, 3/2, 4/3, 5/4, 6/5, 9/8, 10/9, 16/15, 25/24, 81/80 are assigned without hesitation to the musical intervals that are the basis of traditional Western music. They has been known since the time of Zarlino and Descartes, [8]. With inversions, they account for all the common intervals except the tritone.

The unstable character of the tritone sets in apart, as discussed, for example, in [28]. It can be expressed as a ratio by compounding suitable superparticular ratios. Whether it is assigned the ratio 64/45 or 45/32, depending on the musical context, or indeed some other ratio, it is not superparticular, which is in keeping with its unique role in music.

Silver [172] implies that the above ratios, limited to contain prime factors of 2, 3, and 5, are a finite sequence. It has been long known that the sequence actually terminates with 81/80: this was provided in 1897 by C. Størmer [174]. Størmer also proved a more general theorem as follows.

Theorem 86 ([175]) *Let A, B, M_1, M_2, ..., M_m, N_1, N_2, ..., N_n be given positive integers. Then the equations*

$$AM_1^{x_1} M_2^{x_2} \ldots M_m^{x_n}, -BN_1^{y_1} N_2^{y_2} \ldots, N_n^{y_n} = \pm 1, \pm 2$$

admit only a finite number of solutions, all of which can be computed from the smallest positive solutions u_k of Pell's equation

$$t_1^2 - D_1 u_1^2 = 1, \quad \ldots, \quad t_r^2 - D_r u_r^2 = 1$$

for certain D_k's that can be written down in terms of A, B, M_i, and N_j.

D. H. Lehmer [146] has recently given a new proof of Størmer's theorem for prime M_j's and $A = B = 1$ (excluding ± 2 on the right side) and published complete tables for the primes 2, 3, 5, ..., 41.

It is of interest to give a following derivation of Størmer's theorem, because all possible superparticular ratios derived from the first three primes were long ago identified by musical theory.

The following theorem plays the role of the fundamental theorem concerning superparticular ratios.

Theorem 87 *All the pairs of integers* $(x, x + 1)$ *for which* x *and* $x + 1$ *are divisible only by 2, 3, or 5 are* $(1, 2)$, $(2, 3)$, $(3, 4)$, $(4, 5)$, $(5, 6)$, $(8, 9)$, $(9, 10)$, $(15, 16)$, $(24, 25)$, $(80, 81)$.

Proof. We establish the result by checking all possible cases. We first note that if

$$2^a 3^b 5^c - 2^{a'} 3^{b'} 5^{c'} = \pm 1, \tag{9.3}$$

(all exponents are nonnegative integers), then $aa' = bb' = cc' = 0$, since the left side has absolute value at least $2^{|a-a'|}$ if $a \neq a'$. A moment's thought shows that the only possible solutions of the equation (9.3) are the following, where a, b, c denote positive integers:

$$1 = 2^1 - 1, \tag{9.4}$$

$$2^a = 3^b \pm 1, \tag{9.5}$$

$$3^a = 5^b \pm 1, \tag{9.6}$$

$$2^a 5^b = 3^c \pm 1, \tag{9.7}$$

$$2^a = 5^b \pm 1, \tag{9.8}$$

$$2^a 3^b = 5^c \pm 1, \tag{9.9}$$

$$3^a 5^b = 2^c \pm 1, \tag{9.10}$$

For the equation (9.5), we know the solutions $(a, b) = (1, 1)$, $(1, 0)$, $(2, 1)$, $(3, 2)$. We shall show that there are no others. Assuming that there are other solutions, we may suppose that $a > 3$ and that a is the least value that yields a solution of the equation in (9.5). Plainly we have $b > 2$, and so $2^a \equiv \pm 1 (\text{mod } 9)$. Since $2^a \equiv +1 (\text{mod } 9)$ if and only if $a \equiv 0 (\text{mod } 6)$ and $2^a \equiv -1 (\text{mod } 9)$ if and only if $a \equiv 0 (\text{mod } 3)$, then $a = 3a'$. Thus we have

$$2^{3a'} \pm 1 = (2^{a'} \pm 1)x = 3^b$$

for some positive integer x, and so unique factorizing shows that $2^{a'} \pm 1 = 3^{b'}$. The minimum condition on a and the restriction $a > 3$

show that $a' = 2$ or 3, i.e., $a = 6$ or 9. Since $2^6 \pm 1 = 63, 65$ and $2^9 \pm 1 = 511, 513$, we see that (9.5) admits no solution besides those listed above.

For the two equations (9.6), we know one solution, namely $(a, b) = (2, 1)$. If there are others, suppose that we have the least exponent $b > 1$. Thus we have $5^b \equiv \pm 1 \pmod 8$, and since $5^{2k+1} \equiv 5 \pmod 8, 5^{2k} \equiv 1 \pmod 8$, we see that $2^a = 5^b + 1$ has no solution. For $2^a = 5^b - 1$, we get $2^a = 5^{2b'} - 1, 2^a = (5^{b'} + 1)(5^{b'} - 1)$. Now we argue as in the discussion of equation (9.5).

The equation (9.7) trivially has no solution since one side is even and the other is odd.

In the case of equation (9.8) consider the equation $2^a 3^b = 5^c - 1$. We know the solution $(a, b, c) = (3, 1, 2)$. Plainly we must have $c > 1$, and so

$$2^a 3^b = 2^2 \sum_{j=0}^{c-1} 5^j,$$

which implies that $a \geq 2$. Since $\sum_{j=0}^{c-1} 5^j \equiv 0 \pmod 3$, c has to be even, $c = 2c'$, and we have

$$2^a 3^b = (5^{c'} - 1)(5^{c'} + 1).$$

The number $5^{c'} + 1$ is congruent to 2 mod 4. Unique factorizing and the last equality yields

$$5^{c'} + 1 = 2 \cdot 3^{b'}, 5^{c'} - 1 = 2^{a-1} \cdot 3^{b-b'}$$

for some integer b' such that $1 \leq b' \leq b$. Subtracting, we find

$$1 = 3^b - 2^{a-2} 3^{b-b'}.$$

Plainly we must have $a > 2$, and also either $b' = 0$ or $b = b'$. Since $b' \geq 1$, we have

$$5^{c'} - 1 = 2^{a-1},$$

which by the above solution of (9.6) implies that $a - 1 = 2, c' = 1$. Thus $2^3 3^1 = 5^2 - 1$ is the only solution of (9.8-).

Next consider the equation (9.8+), for which we know the solution $2^1 3^1 = 5^1 + 1$. Assuming that there is a solution with $c > 1$, we may suppose that we have the solution with the least value of $c > 1$. Since the right side is congruent to 2 mod 4, we must have $a = 1$. Since $2 \cdot 3^b \equiv 1 \bmod 5$ if and only if $b \equiv 1 \pmod 4$, we have $b = 4b' + 1$ with $b' \geq 0$. Since $5^c \equiv 1 \pmod 3$ if and only if c is odd, we have $c = 2c' + 1$, with $c' \geq 0$, and so our equation is

$$23^{4b'+1} = 5^{2c'+1} + 1 = 6 \sum_{j=0}^{2c'} (-1)^j 5^j,$$

i.e.,

$$3^{4b'} = \sum_{j=0}^{2c'} (-1)^j 5^j.$$

If $b' = 0$, we have $c' = 0$ and we are at our known solution $a = b = c = 1$. If $b' > 0$, we argue as follows. Since $-5 \equiv 1 \pmod 3$, we have

$$\sum_{j=0}^{2c'} (-1)^j 5^j \equiv 2c' + 1 \pmod 3,$$

and so $2c' + 1 \equiv 1 \pmod 3$. That is, c has the form $3(2d + 1)$, and our original equation has the form

$$2 \cdot 3^{4b'+1} = 5^{3(2d+1)} + 1 = (5^{2(2d+1)} + 1)(5^{2(2d+1)} - 5^{2d+1} + 1).$$

Applying the unique factorizing, we see that there is a b'' such that

$$2 \cdot 3^{4b''+1} = 5^{2d+1} + 1.$$

Since $c = 3(2d + 1)$ is the least value of $c > 1$ yielding a solution of (9.8+), we see that $2d + 1 = 1, c = 3$. Since $5^3 + 1 = 2 \cdot 3^2 \cdot 7$, we have proved that (9.8+) has only one solution, $2^1 \cdot 3^1 = 5^1 + 1$.

For the equation (9.9) we have only the solutions $(a, b, c) = (4, 1, 4)$ and $(1, 1, 2)$.

The equation (9.10−) has only the solution $(1, 1, 4)$ and the equation (9.10+) has no solutions at all. The proofs are like those gone through above and are omitted. The theorem is proved.   □

Hindemith [28] uses two ratios, 7/5 and 10/7, in a tentative analysis of the dominant seventh chord, and he ascribes these ratios to the tritone. Although these ratios are not superparticular, the interval that characterizes their difference is superparticular, 50/49.

Theorem 88 (Størmer [174]) *The list solutions of Størmer equation for the primes* {2, 3, 5, 7,} *and* ±1 *are:*
(6, 7), (7, 8), (14, 15), (20, 21), (27, 28), (35, 36), (48, 49), (49, 50), (63, 64), (125, 126), (224, 225), (2 400, 2 401), (4 374/4 375).
These are the only adjacent pairs for the primes 2, 3, 5, 7.

Lehmer [146] has a complete table for primes 2, 3, 5, 7, ..., 41.

Moreover, there is a generalization of part of Størmer's theorem, which follows readily from a theorem of A. Barker [73].

Theorem 89 *Given any finite set* $\mathcal{P} \subset \mathbb{P}$ *of primes and any fixed positive integer* a, *there are only a finite number of pairs* (x, y) *of positive integers such that* $|x - y| \le a$ *and* x *and* y *admit as prime factors only numbers from* $\mathcal{P}$.

In the following example, there are tone systems based specially on superparticular ratios.

Example 80

SUPER-17;
22/21, 23/21, 8/7, 25/21, 26/21, 9/7, 4/3, 25/18, 13/9, 3/2, 11/7, 23/14, 12/7, 25/14, 13/7, 27/14, 2

SUPER-19;
25/24, 13/12, 9/8, 7/6, 29/24, 5/4, 13/10, 27/20, 7/5, 29/20, 3/2, 39/25, 81/50, 42/25, 87/50, 9/5, 28/15, 29/15, 2

SUPER1-19;
26/25, 27/25, 28/25, 29/25, 6/5, 56/45, 58/45, 4/3, 25/18, 13/9, 3/2, 39/25, 81/50, 42/25, 87/50, 9/5, 28/15, 29/15, 2

SUPER-22;
29/28, 15/14, 31/28, 8/7, 33/28, 17/14, 5/4, 31/24, 4/3, 11/8, 17/12, 35/24, 3/2, 87/56, 45/28, 93/56, 12/7, 62/35, 64/35, 66/35, 68/35, 2

SUPER1-22;
26/25, 27/25, 28/25, 29/25, 6/5, 87/70, 9/7, 93/70, 48/35, 99/70, 51/35, 3/2, 31/20, 8/5, 33/20, 17/10, 7/4, 9/5, 37/20, 19/10, 39/20, 2

SUPER-24;
31/30, 16/15, 11/10, 17/15, 7/6, 6/5, 37/30, 19/15, 13/10 4/3, 11/8, 17/12, 35/24, 3/2, 31/20, 8/5, 33/20, 17/10, 7/4, 9/5, 37/20, 19/10, 39/20, 2

9.3 Commas

To take the simplest example, start at the lowest C on the piano and go upward by fifths to G, D, A, and so forth. When you come back to C after twelve fifths, you will have covered the seven complete octaves on the keyboard. But this is only possible because we have agreed to use equally-tempered intervals on the piano, where a fifth is 700 cents, or $\sqrt[12]{2^7}$. A pure 3/2 fifth turns out to be $1\,200 \cdot \log_2(3/2) \approx 701.96$ cent. That is, we make fifths on the piano about 2 cents flatter (narrower) than pure, so that twelve of them equal seven octaves. The total discrepancy if we used pure fifths would be $\mathcal{K} = (3/2)^{12}/2^7 = 3^{12}/2^{19} \approx 23.46$ cent. This interval is called the *Pythagorean comma*, and the amount by which equal- tempered fifths are flattened is one-twelfth of it.

Another "historic" comma is the *Didymus (or syntonic) comma* D. It is most quickly defined in the key of C as the difference between the major 2nd between F and G and the major 2nd between E_b and F. In frequency terms, it is $\mathcal{D} = ((3/2)/(4/3)) : ((4/3)/(6/5))$, or $(9/8) : (10/9)$. Or, it is the ratio (Pythagorean major 3rd):(major 3rd) = (Pythagorean minor 3rd):(minor 3rd) = 81/80.

There are two different understandings of the comma notion when generalizing the Pythagorean and Didymus commas to larger classes of commas. We explain in short these two comma concepts.

9.3.1 Aestetics of ratios of small natural numbers

Let us consider the sequence S of the subgroups of the multiplicative group Q^+ which are generated by the intervals between a fundamental and its overtones:

$$S = \{\langle 2^a 3^b\rangle, \langle 2^a 3^b 5^c\rangle, \ldots, \langle 2^a 3^b 5^c \ldots p^d\rangle, \ldots\},$$

where $a, b, c, d \in \mathbb{Z}$, $p \in \mathbb{N}$. All these groups are dense subsets of $\mathbb{R}^+$ (Kronecker's theorem, [34], [35]).

Definition 33 We say that the irreducibile fraction

$$K = \frac{P}{Q} \in \langle 2^a 3^b \cdots p^c\rangle$$

is a *comma* in $\langle 2^a 3^b \cdots p^c \rangle$ (or, *the best approximation of* 1), c.f. [137], if $K \neq 1$ and for every $\frac{p}{q} \in \langle 2^a 3^b \cdots p^c \rangle$, the following implication holds:

$$\left| \log\left(\frac{p}{q}\right) \right| < \left| \log\left(\frac{P}{Q}\right) \right| \Rightarrow \max(p, q) > \max(P, Q). \qquad (9.11)$$

The following assertion about superparticular ratios is a criterion for how to find commas. The proof is obvious.

Lemma 90 *Let* $\langle 2^a 3^b \cdots p^c \rangle \in \mathbf{S}$. *If* $\frac{n+1}{n} \in \langle 2^a 3^b \cdots p^c \rangle$ *for some natural* n, *then* $\frac{n+1}{n}$ *is a comma in* $\langle 2^a 3^b \cdots p^c \rangle$.

Theorem 91 *The comma sequences of the groups* $\langle 2^a 3^b \rangle$, $\langle 2^a 3^b 5^c \rangle$, $\langle 2^a 3^b 5^c 7^d \rangle$ *are:*

$$\cdots, \frac{2^{84}}{3^{53}}, \frac{3^{41}}{2^{65}}, \frac{2^{19}}{3^{12}}, \frac{3^5}{2^8}, \frac{2^3}{3^2}, \frac{3}{2^2}, \frac{2}{3}, \frac{1}{2},$$

$$\frac{2}{1}, \frac{3}{2}, \frac{2^2}{3}, \frac{3^2}{2^3}, \frac{2^8}{3^5}, \frac{3^{12}}{2^{19}}, \frac{2^{65}}{3^{41}}, \frac{3^{53}}{2^{84}}, \cdots,$$

and

$$\cdots, \frac{3^2 \cdot 5^{15}}{2^{38}}, \frac{2^{15}}{3^8 \cdot 5}, \frac{2^6 \cdot 3^5}{5^6}, \frac{3^4 \cdot 5^2}{2^{11}}, \frac{2^4 \cdot 5}{3^4}, \frac{2^3 \cdot 3}{5^2}, \frac{3 \cdot 5}{2^4}, \frac{3^2}{2 \cdot 5}, \frac{2^3}{3^2},$$

$$\frac{2^2}{5}, \frac{3}{2^2}, \frac{2}{3}, \frac{1}{2}, \frac{2}{1}, \frac{3}{2}, \frac{2^2}{3}, \frac{5}{2^2}, \frac{3^2}{2^3},$$

$$\frac{2 \cdot 5}{3^2}, \frac{2^4}{3 \cdot 5}, \frac{5^2}{2^3 \cdot 3}, \frac{3^4}{2^4 \cdot 5}, \frac{2^{11}}{3^4 \cdot 5^2}, \frac{5^6}{2^6 \cdot 3^5}, \frac{3^8 \cdot 5}{2^{15}}, \frac{2^{38}}{3^2 \cdot 5^{15}}, \cdots,$$

and

$$\cdots, \frac{2^5 \cdot 3 \cdot 5^2}{7^4}, \frac{3^2 \cdot 5^2}{2^5 \cdot 7}, \frac{3^4}{2^4 \cdot 5}, \frac{2^6}{3^2 \cdot 7}, \frac{7^2}{2^4 \cdot 3}, \frac{2^2 \cdot 3^2}{5 \cdot 7}, \frac{3 \cdot 7}{2^2 \cdot 5}, \frac{2 \cdot 5}{3^2},$$

$$\frac{7}{2 \cdot 3}, \frac{2 \cdot 3}{5}, \frac{2^2}{3}, \frac{2}{1}, \frac{2}{1}, \frac{2^2}{3}, \frac{2 \cdot 3}{5}, \frac{7}{2 \cdot 3}, \frac{2 \cdot 5}{3^2},$$

$$\frac{3 \cdot 7}{2^2 \cdot 5}, \frac{2^2 \cdot 3^2}{5 \cdot 7}, \frac{7^2}{2^4 \cdot 3}, \frac{2^6}{3^2 \cdot 7}, \frac{3^4}{2^4 \cdot 5}, \frac{3^2 \cdot 5^2}{2^5 \cdot 7}, \frac{2^5 \cdot 3 \cdot 5^2}{7^4}, \cdots,$$

respectively.

The comma concept presented in this subsection is apt when comparing two Ptolemy tunings. Pythagorean tuning serves then like a master tuning, c.f. also [129]).

9.3.2 Mean tone aesthetics

Various meantone systems attempt to average out the differences in the major seconds 9/8 and 10/9. For a better illumination, c.f. Figure 6.13.

In a *quarter-comma meantone*, all but one of the fifths are flattened from the pure 3/2 ratio of 701.96 cents by one-fourth of the syntonic comma $\mathcal{D}$, or 5.38 cents. So fifths are tuned to 696.58 cents, except for the *wolf* interval of 737.64 cents. Here the wolf is shown inverted as the fourth of 462.36 cents, between $G_\sharp$ and $E_\flat$. All other fourths are 503.42 cents. Quarter-comma meantone gives pure major thirds in most cases (except for four wide ones) and evens out the major seconds (except for two wide ones from $C_\sharp$ to $D_\sharp$ and $G_\sharp$ to $A_\sharp$) as half the pure major third. This is accomplished at the expense of minor seconds of two widely different sizes, a very sour wolf fifth, and even worse mistunings for the wide major thirds and three narrow minor thirds.

In a *fifth-comma meantone*, the fifths are flattened from the pure 3/2 ratio of 701.96 cents by one-fifth of the syntonic comma, or 4.30 cents. So fifths are tuned to 697.65 cents, except for the wolf of the size 725.81 cents. Here the wolf is shown inverted as the fourth of 474.19 cents, between $G_\sharp$ and $E_\flat$. All other fourths are 502.35 cents. Once again, we have major seconds that are half the size of the major thirds in most cases. The difference is that we have allowed the major thirds to expand from pure to one-fifth comma wider than pure. This reduces most of the other deviations from pure intervals; only the "nice" minor thirds are a little farther from pure than in quarter-comma.

In a *sixth-comma meantone*, the fifths are flattened from the pure 3/2 ratio of 701.96 cents by one-sixth of the syntonic comma, or 3.58 cents. So fifths are tuned to 698.37 cents, except for the wolf which must be 717.92 cents. Here the wolf is shown inverted as the fourth of 482.08 cents between $G_\sharp$ and $E_\flat$. All other fourths are 501.63 cents.

By this point, we think the trend should be clear. The wolves are almost tamed; the major thirds are not too wide, but they are getting farther off, and the minor thirds are going flat. But what about the comma notion? The intervals in meantone systems are expressed as irrational numbers, so the comma definition (9.11) does not work. Since the Pythagorean tone system is one of the meantone systems, generalize the definition of the Pythagorean comma:

Definition 34 Let $\tau_{3/2} \approx 1$ be the fifth temperature. Put

$$K = \left(\frac{3}{2} \cdot \tau_{3/2} \right)^{12} : 2^7.$$

We say that the real number K is a comma (or the best approximation of 1) of the meantone system with the generalized fifth $\Lambda_{3/2,\tau_{3/2}} = \tau_{3/2} \cdot 3/2$.

Specially, for $\mathcal{D}/n$-meantone systems, we have the following comma sequence $K_n, n = 4, 5, 6, \ldots$, where

$$K_n = \tau_{3/2}^{12/n} \cdot \frac{3^{12}}{2^{19}}, \quad \tau_{3/2} = \frac{1}{\mathcal{D}} = 80/81.$$

We see that for every $\tau_{3/2} > 0$,

$$\lim_{n \to \infty} K_n = \lim_{n \to \infty} \tau_{3/2}^{12/n} \frac{3^{12}}{2^{19}} = \frac{3^{12}}{2^{19}}.$$

So, for every $\tau_{3/2} > 0$, these sequences do tend to Pythagorean comma (and not to 1 as for the comma sequences according to Definition 33). The only $\tau_{3/2}$ for which $c = 1$, is $\tau_{3/2} = 2 \sqrt[12]{2}/3$, the temperature of the *12-tone Equal Temperament* E_{12}. From this viewpoint, the 12-tone Equal Temperament is a special case of meantone systems.

9.3.3   Approximations of temperaments

Definition 35 Let $\varepsilon > 0$. Let ρ be a metric on the family of all subsets of the class $\mathcal{T}$ of tones, $\Omega : \mathcal{T} \to \mathbb{R}$ be the pitch function. We say that the tone system $(\mathbb{T}_1, \Omega)$ (ε, ρ)-approximates a tone system $(\mathbb{T}_2, \Omega)$ (and, symmetrically, the tone system $(\mathbb{T}_2, \Omega)$ (ε, ρ)-approximates the tone system $(\mathbb{T}_1, \Omega)$) if $\rho(\Omega(\mathbb{T}_1), \Omega(\mathbb{T}_2)) < \varepsilon$.

There arise various problems about the tone system approximations. For instance, to find the most equal superparticular N-tone scales, where $N \in \mathbb{N}$, continuous fraction approximations, etc.

Example 81

MC-LAREN(CONT-FRAC/NO.1)-14; Continued fraction scale 1, c.f. McLaren in Xenharmonikon 15, pp.33-38;
17.586, 35.324, 144.501, 170.140, 262.822, 393.347, 466.181, 591.807, 692.773, 770.284, 818.652, 932.366, 1 080.764, 1 115.066

MC-LAREN(CONT-FRAC/NO.2)-15; Continued fraction scale 2, c.f. McLaren in Xenharmonikon 15, pp.33-38;
46.003, 135.968, 165.005, 257.376, 272.981, 400.028, 422.067, 518.453, 646.633, 704.876, 845.880, 871.569, 1 024.875, 1 064.813, 1 187.312

SUPER-6;
Approximation of 6-TET by a superparticular 6-tone scale;
9/8, 5/4, 11/8, 3/2, 7/4, 2

SUPER-8;
Approximation of 8-TET by a superparticular 8-tone scale;
11/10, 6/5, 13/10, 7/5, 3/2, 5/3, 11/6, 2

SUPER-9;
Approximation of 9-TET by a superparticular 9-tone scale;
11/10, 6/5, 13/10, 7/5, 3/2, 13/8, 7/4, 15/8, 2

SUPER-10;
Approximation of 10-TET by a superparticular 10-tone scale;
13/12, 7/6, 5/4, 4/3, 17/12, 3/2, 13/8, 7/4, 15/8, 2

SUPER-11;
Approximation of 11-TET by a superparticular 11-tone scale;
13/12, 7/6, 5/4, 4/3, 17/12, 3/2, 8/5, 17/10, 9/5, 19/10, 2

SUPER(NO.1)-12;
Approximation of 12-TET by a superparticular 12-tone scale;
15/14, 8/7, 17/14, 9/7, 19/14, 10/7, 3/2, 8/5, 17/10, 9/5, 19/10, 2

SUPER(NO.2)-12; Two but approximation of 12-TET by a super-particular 12-tone scale;
15/14, 8/7, 17/14, 9/7, 19/14, 10/7, 3/2, 45/28, 12/7, 38/21, 40/21, 2

SUPER(NO.3)-12;
Approximation of 12-TET by a superparticular 12-tone scale;
16/15, 17/15, 6/5, 19/15, 4/3, 64/45, 68/45, 8/5, 76/45, 16/9, 17/9, 2
SUPER-13;
Approximation of 13-TET by a superparticular 13-tone scale;
17/16, 9/8, 19/16, 5/4, 21/16, 11/8, 23/16, 3/2, 8/5, 17/10, 9/5, 19/10, 2
SUPER-14;
Approximation of 14-TET by a superparticular 14-tone scale;
17/16, 9/8, 19/16, 5/4, 21/16, 11/8, 23/16, 3/2, 19/12, 5/3, 7/4, 11/6, 23/12, 2
SUPER-15;
Approximation of 15-TET by a superparticular 15-tone scale;
19/18, 10/9, 7/6, 11/9, 23/18, 4/3, 25/18, 35/24, 55/36, 115/72, 5/3, 7/4, 11/6, 23/12, 2
CET88(APPR)-22; non-octave 88-equal temperament, scale approximated;
256/243, 10/9, 7/6, 49/40, 9/7, 256/189, 10/7, 3/2, 128/81, 5/3, 7/4, 448/243, 27/14, 128/63, 15/7, 9/4, 64/27, 5/2, 21/8, 224/81, 35/12, 49/16

Approximations of *JI* are widely considered. Approximation of *JI* by other *JI* scale are also used (e.g., a Petzval system of the 2nd type).

Example 82 *WIL(MOS/34)-11; Wilson 11 of 34-TET, G=9, Chain of minor & major thirds with Kleismatic fusion;*
70.588, 247.059, 317.647, 494.118, 564.706, 635.294, 811.765, 882.353, 952.941, 1 129.412, 1 200.000

MOS(17)-12; MOS 12 of 17, generator 7;
70.588, 141.176, 282.353, 352.941, 494.118, 564.706, 635.294, 776.471, 847.059, 988.235, 1 058.824, 1 200.000

MOS(22)-12; MOS 12 of 22, contains nearly just, recognizable diatonic, and pentatonic scales;
163.636, 218.182, 381.818, 436.364, 490.909, 654.545, 709.091, 872.727, 927.273, 1 090.909, 1 145.455, 1 200

MOS(22)-13; MOS 13 of 22, contains 5 and 9 tone MOS as well. G= 5 or 17;
109.091, 218.182, 327.273, 381.818, 490.909, 600, 654.545, 763.636, 872.727, 927.273, 1 036.364, 1 145.455, 1 200

MOS(22)-15; MOS 15 in 22, contains 7 and 8 tone MOS as well. G = 3 or 19;

109.091, 163.636, 272.727, 327.273, 436.364, 490.909, 600, 654.545, 763.636, 818.182, 927.273, 981.818, 1 090.909, 1 145.455, 1 200

9.3.4 Pythagorean approximation of Just Intonation

In this section, we show a Pythagorean approximation (i.e., using only pure fifth 3/2) of the Just Intonation which generalizes Equal Temperament. The intervals causing a dilemma are the second and the minor seventh and the tritone because they are unambiguous in Just Intonation (the relative frequencies 10/9, 9/8, 8/7 and 7/4, 16/9, 18/10 and 45/32, 64/45, respectively). If we do not consider the second and seventh with the relative frequencies 8/7 and 7/4, respectively, all the music intervals in this approximation either coincide with the Just Intonation interval values (the octave, fifth, fourth, second (9/8) and the minor seventh (16/9)) or are exactly one comma distant from the corresponding Just Intonation intervals. This comma is $32\,805/32\,768 \approx 1.001\,129\,15$, which is less than the ratio of frequencies of the perfect and the equal tempered fifths ($\approx 1.001\,129\,89$).

The proof of the following lemma is easy.

Lemma 92 *Let $q = \mathrm{const} \in (0, \infty)$.*
Then $\rho_q(u, v) = \left| \log_2 \sqrt[q]{u/v} \right|$, $u, v \in (0, \infty)$, is a metric on $\mathbb{L}$.

Denote

$$
\begin{aligned}
E_{12} &= \{1, \sqrt[12]{2}, \sqrt[12]{2^2}, \ldots, \sqrt[12]{2^{12}} = 2\} \\
&= \{C^{(E_{12})}, C_\sharp^{(E_{12})}, \ldots, B^{(E_{12})}, C'^{(E_{12})}\}.
\end{aligned}
$$

Lemma 93 *Let $X, Y, a \in (1;\ \infty)$, $m, n \neq 0$, $m, n \in \mathbb{R}$. Then the following equalities are equivalent:*
(1) $a = X^m Y^n$,   (2) $1 = m/\log_X a + n/\log_Y a$.

Proof. $a = X^m Y^n \iff \log a = m \log X + n \log Y \iff 1 = m\frac{\log X}{\log a} + n\frac{\log Y}{\log a} \iff 1 = \frac{m}{\log_X a} + \frac{n}{\log_Y a}.$ □

TABLE 9.8: Just Intonation

$C^{(JI)}$	$2^0 3^0 5^0$	1/1	1 000
$C_\sharp^{(JI)}, D_\flat^{(JI)}$	$2^4 3^{-1} 5^{-1}$	16/15	1.066 667
	$2^1 3^{-2} 5^1$	10/9	1.111 111
$D^{(JI)}$	$2^{-3} 3^2$	9/8	1.125 000
	$2^3 7^{-1}$	8/7	1.142 857
$D_\sharp^{(JI)}, E_\flat^{(JI)}$	$2^1 3^1 5^{-1}$	6/5	1.200 000
$E^{(JI)}$	$2^{-2} 5^1$	5/4	1.250 000
$F^{(JI)}$	$2^2 3^1$	4/3	1.333 333
$F_\sharp^{(JI)}$	$2^{-5} 3^2 5^1$	45/32	1.406 250
$G_\flat^{(JI)}$	$2^6 3^{-2} 5^{-1}$	64/45	1.422 222
$G^{(JI)}$	$2^{-1} 3^1$	3/2	1.500 000
$G_\sharp^{(JI)}, A_\flat^{(JI)}$	$2^3 5^{-1}$	8/5	1.600 000
$A^{(JI)}$	$3^{-1} 5^1$	5/3	1.666 667
	$2^{-2} 7^1$	7/4	1.750 000
$A_\sharp^{(JI)}, B_\flat^{(JI)}$	$2^4 3^{-2}$	16/9	1.777 778
	$3^2 5^{-1}$	9/5	1.800 000
$B^{(JI)}$	$2^{-3} 3^1 5^1$	15/8	1.875 000
$C'^{(JI)}$	2^1	2/1	2 000

Recall that the unique two rational intervals for 12-granule 2-quotient $(2,3)$-scales are $X = 2^8/3^5$ and $Y = 3^7/2^{11}$, c.f. Theorem 20. Fix these values X, Y (the minor and major Pythagorean semitones).

The map ϑ, c.f. Figure 6.1, is an isomorphism. The image of Pythagorean System Π_{17}, Π_{31} under the isomorphism ϑ is in Figure 4.3. Define the following Pythagorean approximation of Just Intonation:

Definition 36 The *Pythagorean approximation $\mathcal{A}$ of Just Intonation* is defined by Table 9.9.

The following theorem is easy to verify, c.f. also Figure 9.3.

Theorem 94 *Let* $a_j = 2$, *15/8, 9/5, 7/4, 16/9, 5/3, 8/5, 3/2, 64/45, 45/32, 4/3, 5/4, 6/5, 8/7, 9/8, 10/9, 16/15, where* $j = 12$, *11, 10, 10, 10, 9, 8, 7, 6, 6, 5, 4, 3, 2, 2, 2, 1, respectively. For*

TABLE 9.9: Pythagorean approximation of Just Intonation

$C^{(A)}$	X^0Y^0	2^03^0	1/1	1 000
$C_\sharp^{(A)}, D_\flat^{(A)}$	X^0Y^1	$2^{-11}3^7$	2 187/2 048	1.067 871
	X^2Y^0	$2^{16}3^{-10}$	65 536/59 049	1.109 858
$D^{(A)}$	X^1Y^1	$2^{-3}3^2$	9/8	1.125 000
	X^0Y^2	$2^{-22}3^{14}$	4 782 969/4 194 304	1.140 349
$D_\sharp^{(A)}, E_\flat^{(A)}$	X^1Y^2	$2^{-14}3^9$	19 683/16 384	1.201 355
$E^{(A)}$	X^3Y^1	$2^{13}3^{-8}$	8 192/6 561	1.248 590
$F^{(A)}$	X^3Y^2	2^23^{-1}	4/3	1.333 333
$F_\sharp^{(A)}$	X^4Y^2	$2^{10}3^{-6}$	1 024/729	1,404 664
$G_\flat^{(A)}$	X^3Y^3	$2^{-9}3^6$	729/512	1.423 828
$G^{(A)}$	X^4Y^3	$2^{-1}3^1$	3/2	1.500 000
$G_\sharp^{(A)}, A_\flat^{(A)}$	X^4Y^4	$2^{-12}3^8$	6 561/4 096	1.601 807
$A^{(A)}$	X^6Y^3	$2^{15}3^{-9}$	32 768/19 683	1.664 787
	X^7Y^3	$2^{23}3^{-14}$	8 388 608/4 782 969	1.753 850
$A_\sharp^{(A)}, B_\flat^{(A)}$	X^6Y^4	2^43^{-2}	16/9	1.777 778
	X^5Y^5	$2^{-15}3^{10}$	59 049/32 768	1.802 032
$B^{(A)}$	X^7Y^4	$2^{12}3^{-7}$	4 096/2 187	1.872 885
$C^{(A)}$	X^7Y^5	2^13^0	2	2 000

$X = 2^8/3^5, Y = 3^7/2^{11}$, *the system of linear equations*

$$\frac{m_j}{\log_X a_j} + \frac{n_j}{\log_Y a_j} = 1,$$
$$m_j + n_j = j$$

has the following integer solutions $(m_{12}, n_{12}) = (7, 5)$, $(m_{10}, n_{10}) = (6, 4)$, $(m_7, n_7) = (4, 3)$, $(m_5, n_5) = (3, 2)$, $(m_2, n_2) = (1, 1)$, *for* $a_j = 2, 16/9, 3/2, 4/3, 9/8$, *where* $j = 12, 10, 7, 5, 2$, *respectively.*

Theorem 95 *The Approximation A, JI, and 12-tone Equal Temperament are symmetric with respect to the center point* $(7/2, 5/2)$ *in the map* ϑ_*.

Proof. For Approximation $\mathcal{A}$, c.f. Figure 9.3. $\qquad\qquad\square$

Prove the assertion for JI. Observe that

$$m_j = \frac{j \log Y - \log a_j}{\log Y - \log X}, \qquad n_j = \frac{\log a_j - j \log X}{\log Y - \log X}$$

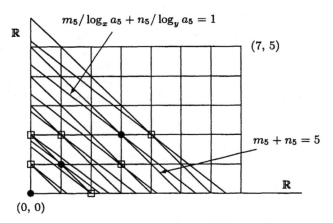

FIGURE 9.3: Projection of *JI* in the plane

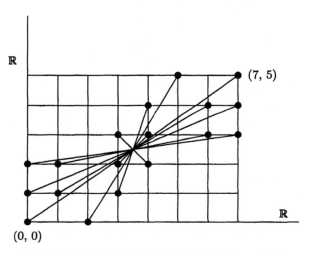

FIGURE 9.4: The *JI* Approximation: Symmetry

is the solution of the system in Theorem 94. Further, $a_{12-j} = 2/a_j, j = 0, 1, \ldots, 12$, c.f. Table 9.9. Since $2 = X^7 Y^5$, we have:

$$
\begin{aligned}
m_{12-j} &= \frac{(12-j)\log Y - \log a_{12-j}}{\log Y - \log X} \\
&= \frac{12 \log Y - j \log Y - \log(2/a_j)}{\log Y - \log X} \\
&= \frac{-j \log Y + \log a_j}{\log Y - \log X} + \frac{\log Y^{12} - \log X^7 Y^5}{\log Y - \log X} \\
&= -m_j + \frac{\log(Y^{12}/X^7 Y^5)}{\log(y/x)} = -m_j + 7.
\end{aligned}
$$

Analogously, $n_{12-j} = -n_j + 5$.

Equal Temperament is also symmetric in the map ϑ_* with respect to the center point $(7/2, 5/2)$, c.f. Figure 9.5.

So, the point $(7/2, 5/2)$ is a common center of symmetry for Equal Temperament, *JI*, and the *JI* Approximation $\mathcal{A}$. Note that $X^{7/2} Y^{5/2} = (\sqrt[12]{2})^6 = \sqrt{2}$ (the tritone in Equal Temperament). $\square$

The proof of the following theorem is implied by Table 9.10, the construction of the *JI* Approximation $\mathcal{A}$.

Theorem 96 *The JI Approximation $\mathcal{A}$ is a 12-granule 2-quotient $(2, 3)$-system.*

The fact that the *JI* Approximation $\mathcal{A}$ is a many-valued generalization of Equal Temperament E_{12}, is shown in the following theorem which states that the image of Equal Temperament is a projection of the *JI* Approximation $\mathcal{A}$ to a line in the map ϑ_*.

Theorem 97 *Let p, q be two lines in the complex plane, such that $(0, 0), (7, 5) \in p$; $(12, 0), (7, 5) \in q$, respectively. Let $\pi : \mathbb{C} \to p$ be the projection of the plane $\mathbb{C}$ into the line p along the line q. Let S be an arbitrary 12-granule 2-quotient $(2, 3)$-system. Then $E_{12} = \vartheta_*^{-1}(\pi(\eta(\vartheta(S))))$.*

Proof. C.f. Figure 9.5.

Without loss of generality, we prove the assertion for $S = \mathcal{A}$.

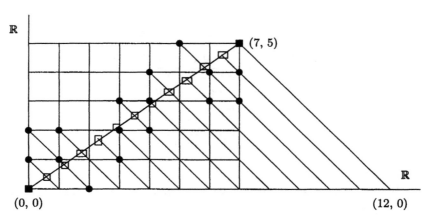

FIGURE 9.5: ϑ_*-image of *JI* Approximation $\mathcal{A}$ and E_{12}

Denote by $w_0 = (0,0), w_1 = (\frac{7}{12}, \frac{5}{12}), w_2 = 2 \cdot (\frac{7}{12}, \frac{5}{12}), \ldots, w_{12} = 12 \cdot (\frac{7}{12}, \frac{5}{12})$.

We have:

$$
\begin{aligned}
w_0 &= \pi(\eta(\vartheta(C^{(\mathcal{A})}))), & \vartheta_*^{-1}(w_0) &= X^0 Y^0 & &= & 1, \\
w_1 &= \pi(\eta(\vartheta(C_\sharp^{(\mathcal{A})}))), & \vartheta_*^{-1}(w_1) &= X^{\frac{7}{12}} Y^{\frac{5}{12}} & &= & \sqrt[12]{2}, \\
w_2 &= \pi(\eta(\vartheta(D^{(\mathcal{A})}))), & \vartheta_*^{-1}(w_2) &= X^{2\cdot\frac{7}{12}} Y^{2\cdot\frac{5}{12}} & &= & \sqrt[12]{2^2},
\end{aligned}
$$

$$\cdots$$

$$w_{12} = \pi(\eta(\vartheta(C'^{(\mathcal{A})}))), \quad \vartheta_*^{-1}(w_{12}) = X^{12\cdot\frac{7}{12}} Y^{12\cdot\frac{5}{12}} = \sqrt[12]{2^{12}}.$$

So, $E_{12} = \vartheta_*^{-1}(\pi(\eta(\vartheta(S))))$. $\qquad\qquad\qquad\qquad\qquad\qquad$ □

9.3.5   Comma 32 805/32 768 (Schizma)

Now we show that the *JI* Approximation $\mathcal{A}$ is closer to *JI* than Equal Temperament. As a unit q for imagination of measure values, we take a logarithm of the temperature $\tau_{3/2} = G^{(E_{12})}/G^{(JI)}$, i.e.

$$q = \log_2(G^{(E_{12})}/G^{(JI)}) = \log_2 3 - (19/12) \approx 0.001\,629\,167.$$

For the metric ρ_q, c.f. Lemma 1 and also [10], [11]. Clearly,

$$\rho_q(G^{(E_{12})}/G^{(JI)}) = \rho_q(\tau_{3/2}) = 1.$$

TABLE 9.10: *JI* Approximation $\mathcal{A}$ versus E_{12}

	$\rho_q(.^{(JI)},.^{(\mathcal{A})})$	$\rho_q(.^{(JI)},.^{(E_{12})})$
C	0.000 000	0.000 000
$C_\sharp = D_\flat$	0.999 343	6.000 641
D 10/9	0.999 343	9.000 634
D 9/8	0	1.999 996
D 8/7	1.945 816	15.945 784
$D_\sharp, E_\flat$	0.999 343	7.000 637
E	0.999 343	7.000 637
F	0.000 000	1.000 000
$F_\sharp$	0.999 343	5.000 641

Note that the equal tempered fifth is the best approximated (to *JI*) music interval in Equal Temperament (not mentioning trivially the unison and octave).

Theorem 98 *For every granule C, $C_\sharp$, ..., B, C', the $\mathcal{A}$ approximation value is closer to JI value than the Equal Temperament value in the metric ρ_q with $q = \log_2 3 - (19/12)$.*

Proof. C.f. Table 9.10 □

The following lemma shows that the *JI* Approximation $\mathcal{A}$ is characterized by a comma

$$\kappa = 5 \cdot 3^8 \cdot 2^{-15} = 32\,805/32\,768 \approx 1.001\,129 \approx 1.95 \text{ cents},$$

known under the name *Schizma*. We immediately obtain:

Theorem 99 *Let $\mathcal{K} = 531\,441/524\,280, \mathcal{D} = 81/80$ be the Pythagorean and Didymus commas, respectively. Then*

$$\kappa = K/\mathcal{D}.$$

Theorem 100 *Let $\kappa = 32805/32\,768$. Let $q = \log_2 3 - (19/12)$.*

Then

$$\log_2 \sqrt[q]{\kappa} = \rho_q(C_\sharp^{(A)}, C_\sharp^{(JI)}) = \rho_q(D_\sharp^{(A)}, D_\sharp^{(JI)})$$
$$= \rho_q(E^{(A)}, E^{(JI)}) = \rho_q(F_\sharp^{(A)}, F_\sharp^{(JI)})$$
$$= \rho_q(D^{(A)}(10/9), D^{(JI)}(10/9))$$
$$= \rho_q(B^{(A)}(9/5), B^{(JI)}(9/5))$$
$$= \rho_q(G_\flat^{(A)}, G_\flat^{(JI)}) = \rho_q(A_\flat^{(A)}, A_\flat^{(JI)})$$
$$= \rho_q(A^{(A)}, A^{(JI)}) = \rho_q(B^{(A)}, B^{(JI)})$$
$$< 0.999\,345 < 1 = \rho_q(G^{(E_{12})}, G^{(JI)}).$$

We collect the results of this subsection into the following theorem.

Theorem 101 *Let* $X = 2^8/3^5$, $Y = 3^7/2^{11}$. *Let* $\kappa = 32\,805/32\,768$. *Then the JI Approximation*

$$\mathcal{A} = \left\{ (X^4Y^3)^k (X^7Y^5)^{-h}; \ (k, h) = \right.$$

$$[-14, -9],$$
$$(-10, -6), \ (-9, -6), \ (-8, -5), \ (-7, -5), \ (-6, -4),$$
$$\{-2, -2\}, \ \{-1, -1\}, \ \{0, -1\}, \ \{0, 0\}, \ \{1, 0\}, \ \{2, 1\},$$
$$(6, 3), \ (7, 4), \ (8, 4), \ (9, 5), \ (10, 5),$$
$$[14, 8]\},$$

where

$\{\cdot, \cdot\}$ *denotes approximated intervals based on numbers 2 and 3*

$$(B_\flat^{(A)}(16/9), \ F^{(A)}, \ C'^{(A)}; \ C^{(A)}, \ G^{(A)}, \ D^{(A)}(9/8),$$

respectively) and the JI Approximation $\mathcal{A}$ *coincides with the JI value;*

$(\cdot, \cdot)$ *denotes approximated intervals based on numbers 2, 3, and 5*

$$(D^{(A)}(10/9), \ A^{(A)}, \ E^{(A)}, \ B^{(A)}, \ F_\sharp^{(A)};$$

$$G_\flat^{(A)}, \ C_\sharp^{(A)}, \ A_\flat^{(A)}, \ E_\flat^{(A)}, \ B_\flat^{(A)}(9/5),$$

respectively) and the intervals are κ-*approximated;*

[·, ·] *denotes the approximated intervals based on numbers 2 and 7*

$$(B_b^{(\mathcal{A})}(7/4);\ D^{(\mathcal{A})}(8/7),$$

respectively).

Proof. The vector (k, h) is a solution of the following system of equations:

$$4k + (-7)h = \alpha, \quad 3k + (-5)h = \beta,$$

where α, β are exponents of X, Y, respectively, c.f. Table 9.9 (e.g., $(3, 2)$ for F). Observe that

$$\begin{vmatrix} 4 & -7 \\ 3 & -5 \end{vmatrix} = 1,$$

so $k, h \in \mathbb{Z}$ and $(k, h) = (-5\alpha + 7\beta, -3\alpha + 4\beta)$. Rearrange now the solutions (k, h) as above. $\qquad\square$

Remarks and ideas for exploring

21 For further extended file of tetrachords are collected in Appendix A.1.

22 The classical Euler space, c.f. [45], is defined only for $\{2, 3, 5, 7, \ldots, p_n\} = \{2, 3, 5, 7\}$. It is obvious that our generalization brings no qualitative complications or new problems from the mathematical viewpoint.

The set $\{2, 3, 5, 7, \ldots, p_n\}$ represent the *consonance spectrum of tones*.

23 For the extended file of just intoned tone systems with 6, 8, 9, 10, and 11 tones per octave, c.f. Appendix A.4. For the extended file of just intoned tone systems with more than 12 tones per octave, c.f. Appendix A.6.

24 Note that Definition 33 is strongly based on the consonance theory expressed shortly with the words "ratios of small numbers." Following these aesthetics, we can choose some other appropriate functions $f(r, s)$ instead of $\max(r, s)$, e.g., $f(r, s) = r + s$, $f(r, s) = r^2 + s^2$, etc., in Definition 33.

25 Enlarging the scales as usually by the *octave equivalency* (i.e., multiplying the scale values by 2^i, $i \in \mathbb{Z}$) to the whole $\mathbf{L}$, we see that the tritones in Equal Temperament and also octaves (i.e., $2^i\sqrt{2}$, 2^i, respectively) are all common centers of symmetry of the approximation $\mathcal{A}$ of *JI*, *JI*, and E_{12}.

26 A tone system based on the (-14)th, (-8)th, 0th, 1st, and 9th fifths is called the *Petzval tone system of the Second class*, c.f. Section 7. We see that the *JI* Approximation $\mathcal{A}$ is the many-valued system of this type.

Bibliography

Books

[1] Adelung, W.: Einführung in der Orgelbau [Introduction to the organ building], (in German). VEB Breitkopf & Härtel, Leipzig 1972.

[2] Barbour, J. M.: Tuning and Temperament: A Historical survey. Michigan State College Press, East Lancing 1951. Reprint: Da Capo, New York, 1973.

[3] Benade, A.: Fundamentals of Musical Acoustics. Oxford University Press, 1976. Reprint: Dover, New York, 1990.

[4] Berliner, P. F.: The Soul of Mbira. University of California Press, 1978, 1981.

[5] Bunch, B.: Reality's mirror: exploring the mathematics of symmetry. Wiley Science Editions. John Wiley & Sons, Inc., New York, 1989, 286 pp. MR 91b:00004.

[6] Bonsi, D.—Gonzalez, D.—Stanzial, D. (eds.): Musical sounds from past milenia. Proceedings of the Int. Symposiom on Musical Acoustics, (September 10–14, 2001, Perugia), The Musical and Architectural Acoustics Laboratory FSSG-CNR c/o Fondazione Giorno Cini, Venice, 2001, vol. I (pp. 1–348) and II. (pp. 349–640).

[7] Chalmers, J.: Divisions of the Tetrachord. Frog Peak Music, Hanover, 1993.

[8] Colles H. C. (ed.): Grove's Dictionary of music and musicians, 3th edition, Vol. V, Macmillian, New York 1936.

[9] Darvas, Gy.—Nagy, D.—Pardavi Horvath, M. (eds.): Symmetry: natural and artificial. 1, 2, 3. International Society for the Interdisciplinary Study of Symmetry, Budapest, 1995, pp. 1–188, 186–375, 377–563.

[10] Duy, P: Musics of Vietnam. Southern Illinois University Press 1975.

[11] d'Erlanger, R.: La musique arabe, vol. I–VI, Librairie Orientaliste Paul Geuthner, Paris, 1930, 1935, 1938, 1939, 1949, 1953.

[12] Erményi, L.: Petzvals Theorie der Tonsysteme. B. G. Teubner, Leipzig, 1904.

[13] Feferman, S.: The number systems. Addison–Wesley Publ. Co., Mass. –Palo Alto–London, 1963.

[14] Feichtinger, H.—Dörfler, M. (eds.): Diderot forum on mathematics and music, Österreichische Computer Gesellschaft, Vienna 1999.

[15] Fletcher, N. H.—Rossing, T. D.: The Physics of musical instruments. Springer-Verlag, 1991.

[16] Forte, A.: The Structure of Atonal music. Yale University Press, New Haven 1973.

[17] Fortuna, C.: Microtone Guide. Self-published, 1989.

[18] Garbuzov, N. A.: The zonal nature of the human aural perception (in Russian), Izdatelstvo Akademii Nauk SSSR, Moscow–Leningrad 1948.

[19] Garbuzov, N. A.: The zonal nature of tempo and rythm (in Russian), Izdatelstvo Akademii Nauk SSSR, Moscow 1950.

[20] Gewirtz, A.—Quintas, L. V. (eds.): Second International Conference on Combinatorial Mathematics. (New York, April 4–7, 1978). Annals of the New York Academy of Sciences, 319. New York Academy of Sciences, New York 1979, 602 pp. MR 80j:05003.

[21] Goldberg, S.: Genetic Algorithms in Search, Optimization, and Machine Learning, Addison-Wesley, New York, NY, 1989.

[22] Gotze, H.—Wille, R. (eds.): Music and mathematics (in German). Springer Verlag, Berlin–New York, 1985, 97 pp. MR 86f:00032.

[23] Hába, A.: Harmonic foundation of the 1/4-tone system (in Czech). Hodební Matice Umělecké besedy, Prague 1922. pp. 40 + 12 musical applications.

[24] Haluška, J. (ed.): Ambiguity and Music. Polygrafia, Bratislava 2000.

[25] Haluška, J. (ed.): *Harmonic Analysis and Tone Systems*. Tatra Mt. Math. Publs. **23**(2002), a special issue. Math. Inst. of the Slovak Acad. of Sci., Bratislava, 171 pp.

[26] Helmholtz, H.: Die Lehre von den Tonemphindungen als phychologische Grundlage für die Theorie der Musik. Braunschweig, 1863. [translation into English: On the Sensations of Tones, translator: J. Ellis, Dover, New York, 1954.]

[27] Herlinger, J. W.: The Lucidarium of Marchetto of Padua (1317–18). University of Chicago Press, Chicago 1985.

[28] Hindemith P.: The Crafts of Musical Composition, Book I. Associated Music Publishers, New York 1945.

[29] Hood, M: Slendro and Pelog Redefined, Selected Reports I/1 (1966):28.

[30] Jorgenson, O. H.: Tuning: Containing the Perfection of 18th-century Temperament; the Lost Art of 19th-century Temperament; and the Science of Equal Temperament. East Lancing, Michigan State University Press, 1991.

[31] Kamleiter, P.: Carl Stumps Theorie der Tonverschmelzung als Erklärung des Konsonanzphänomens. Dissertation, Universität Würzburg, 1993.

[32] Kelley, J. L.: General topology. New York, D. Van Nostrand, 1955.

[33] Klir, G. J.—Wierman, M. J.: Uncertainty-based information, elements of generalized information theory. Creighton University, Omaha 1997.

[34] Kronecker, L.: Werke III, Teubner, Leipzig–Berlin, 1899.

[35] Kuipers, L.—Niederreiter, H.: Uniform distribution of sequences, Wiley, London, 1974.

[36] Kulli, V. R. (ed.): Advances in graph theory. Vishwa International Publications, Gulbarga, 1991, 315 pp.

[37] Kurkela, Kari: Note and tone. (A semantic analysis of conventional music notation). Dissertation, University of Helsinki, Helsinki, 1986. Acta Musicologica Fennica, 15. The Musicological Society of Finland, Helsinki, 1986, 161 pp. MR 88m:00028.

[38] Leman, M.: Music and schema theory. Springer Series in Information Sciences, 31. Springer Verlag, Berlin, 1995, 234 pp.

[39] Levarie, S.—Levy, E.: Tone. A study in Musical Acoustics. Kent State University Press, 1968.

[40] Lentz, D. A.: The Gamelan Music of Java and Bali. University of Nebraska Press LCCCN 65-10545, 1965.

[41] Liénard, J. S.: Speech analysis and reconstruction using short-time, elementary maveforms, ICASSP, 1987.

[42] Liu, C. L.: Introduction to Combinatorial Mathemathics. McGraw–Hill, New York 1973.

[43] Madan, R. N. (ed.): Chua's circuit: a paradigm for chaos. World Scientific Series on Nonlinear Science. Series B: Special Theme Issues and Proceedings, 1. World Scientific Publishing Co., Inc., River Edge, 1993, 1043 pp.

[44] Mazzola, G.: Groups and categories in music (Sketch of a mathematical music theory) (in German). R & E Research and Exposition in Mathematics, 10. Heldermann Verlag, Berlin, 1985, 205 pp. MR 88f:00025.

[45] Mazzola, G.: Geometry of Tones: Elements of the mathematical theory of music (in German), Birkhäuser, Basel–Boston–Berlin, 1990.

[46] Meyer, Y.: Wavelets. Society for Industrial and Applied Mathematics (SIAM), Philadelphia, 1993, 133 pp.

[47] Mostowski, A.: The present state about consideration of the fundaments of mathematics (in Russian). Issledovanija matematicheskikh nauk 61(1964), tom 9, no. 3.

[48] Murray S. K. (ed.): Mathematical modelling. Classroom notes in applied mathematics. Society for Industrial and Applied Mathematics (SIAM), Philadelphia, 1987, 338 pp. MR 89f:00047.

[49] Neumaier, W.: What is a tone system? (in German). Verlag Peter D. Lang, Frankfurt am Main, 1986, 261 pp. MR 89d:01007.

[50] Neuwirth, E.: Musical Temperaments. Springer Verlag, Berlin–New York.

[51] Parncutt, R.: Harmony: a psychoacoustical approach. Springer Series in Information Sciences, 19. Springer Verlag, Berlin–New York, 1989, 206 pp. MR 91c:00015.

[52] Penrose, R.—Rindler, W.: Spinors and spacetime. Vol. 1. Cambridge Monographs on Mathematical Physics. Cambridge University Press, Cambridge New York, 1984, 458 pp. MR 86h:83002.

[53] Pippard, A. B.: The physics of vibration. Vol. 2. (The simple vibrator in quantum mechanics). Cambridge University Press, Cambridge New York, 1983, 208 pp. MR 84m:81005b.

[54] Pierce, J. R.: Attaining Consonance in Arbitrary Scales. Journal of the Acoustical Society of America 40(1966), 249.

[55] Popivanov, P.—Tersian, S. A. (eds.): Conference on Differential Equations and Applications (in Russian). Technical University, Ruse, 1991, 509 pp.

[56] Reiss, J. W.: A shorter music encyclopedia (in Polish). PWM, Warsaw, 1960.

[57] Roederer, J. G.: The physics and psychophysics of music. Springer Verlag, New York, 1995, 221 pp.

[58] Riordan, J.: An Introduction to Combinatorial mathematics. Wiley, New York 1967.

[59] Saks, K. J.: Boethius and the Judgement of the Ears: A Hidden Challenge in Medieval and Renaissance Music. In: The Second Sense. Studies in Hearing and Musical Judgement from Antiquity to the Seventeenth Century (Ed. Charles Burnett). London, 1991.

[60] Sethares, W. A.: Tuning, Timbre, Spectrum, Scale. Springer-Verlag, Berlin, 1997.

[61] Šledzinski, S. (ed.): A Shorter Music Encyclopaedia (in Polish). PWM, Warsaw, 1981.

[62] Sikorski, K.: Harmony (in Polish). PWM, Warsaw, 1952.

[63] Smith, J. D. H.—Romanowska, A.: Post-modern Algebra. Wiley, New York, 1999.

[64] Sudarjana, P. J.—Surjodiningrat, W. –Susanto, A.: Tone Measurements of Outstanding Javanese Gamelans in Yogyakarta and Surakarta (Second Edition). Yogyakarta, Gadjah Mada University Press, 1993.

[65] Tyukhtin, V. S.—Urmantsev, Yu. A.—Larin, Yu. S.—Dalin, V. Ya.—Sharapov, I. P.—Malikov, A. V.—Koptsik, V. A.— Zamorzaev, A. M.—Didyk, Yu. K.—Petukhov, S. V.—Artemev, Yu. I.: System. Symmetry. Harmony (in Russian). Mysl, Moscow, 1988, 318 pp. MR 89f:00045.

[66] Ville, J.: Théorie et applications de la notion de signal analytique, C& T, Laboratoire de Télécommunications de la Société Alsacienne de Construction Mécanique, 2éme A. No. 1 (1948).

[67] Wilkinson, S. R.: Tuning In. Leonard Books, 1988.

[68] Zalewski, M.: Theoretical Harmony (in Polish). PWSM, Warsaw, 1973.

Papers

[69] Aprahamian Mateescu, M.: Classes of melodies and matrix grammars (in Romanian). Stud. Cerc. Mat. 31 (1979), no. 4, 389–397. MR 81b:68086.

[70] R. Balian: Un principe d'incertutude fort en théorie du signal ou en mécanique quantique, C. R. Acad. Sci. Paris, Sér. II, 292 (1981), pp. 1357–1361.

[71] Balzano, G. J.: The group-theoretic description of 12-fold and microtonal pitch systems. Computer music journal 5(1980), 66–84.

[72] Bakhmutova, I. V.—Gusev, V. D.—Titkova, T. N.: Search for and classification of imperfect repetitions of song melodies (in Russian). Tartu Riikl. Ul. Toimetised (1988), No. 827, 20–32. MR 90f:92045.

[73] Baker A.: Linear forms in the logarithms of algebraic numbers (IV). Mathematica 15(1968), 204–216.

[74] Barbour, J. M.: Music and Ternary continued fractions. Amer. Math. Monthly 55(1948), p. 545.

[75] Batens, D.: Some remarks on the structural similarity between music and logic. Comm. Cognition 19 (1986), no. 2, 135–151.

[76] Bazelow, A. R.: Integer matrices with fixed row and column sums. in Collection: Second International Conference on Combinatorial Mathematics (New York, 1978), pp. 593–594. Ann. New York Acad. Sci., 319, New York Acad. Sci., New York, 1979. MR 82a:05004.

[77] Bazelow, A. R.—Brickle, F.: A combinatorial problem in music theory Babbitt's partition problem. I. in Collection: Second International Conference on Combinatorial Mathematics (New York, 1978), pp. 47–63. Ann. New York Acad. Sci., 319, New York Acad. Sci., New York, 1979. MR 81k:05023a.

[78] Bazelow, A. R.—Brickle, F.: A combinatorial problem in music theory Babbitt's partition problem. II. Ars Combin. 9 (1980), 289–306. MR 81k:05023b.

[79] Benmansour, S.—Seke, B.: A model of propagation of information in a finite set (in French) Cahiers Math. 1987, no. 1, 31–37. MR 89f:94009.

[80] Beran, J.—Mazzola, G.: Visualising the relationship between time series by hierarchical smoothing. Journal of computational and graphical statistics 8(1999), pp. 213 – 238.

[81] Bertoni, A.—Haus, G.—Mauri, G.—Torelli, M.: Analysis and compacting of musical texts. J. Cybernet. 8 (1978), no. 34, 257– 272. MR 80m:68077.

[82] Binmore, K. G.—Samuelson, L.—Vaughan, R.: Musical chairs: modeling noisy evolution. Games Econom. Behav. 11 (1995), no. 1, 1–35.

[83] Boomsliter, P.—Creel, W.: The long pattern hypothesis in harmony and hearing. Journal of Music Theory 5(1961), pp. 2–30.

[84] Boroda, M. G.—Polikarpov, A. A.: The Zipf Mandelbrot law and units of different levels of text organization (in Russian). Tartu Riikl. Ul. Toimetised (1984), No. 689, 35–60. MR 86c:92030.

[85] Burridge, R.—Kappraff, J.—Morshedi, Ch.: The sitar string, a vibrating string with a onesided inelastic constraint. SIAM J. Appl. Math. 42 (1982), no. 6, 1231–1251. MR 83j:73030.

[86] Čanak, M.: The connection between the Riemann boundary value problem and a problem of mathematical music theory (in German). Mat. Vesnik 38 (1986), no. 4, 399–405. MR 89h:35172.

[87] Čanak, M.: Application of some results of mathematical theory of music in function and making of music instruments. I. The stringed instruments. Math. Semesterber. 38 (1991), no. 2, 252–258. MR 92j:00023.

[88] Canright, D.: Superparticular pentatonics. 1/1 (The Journal of the Just Intonation Network), 9(1995), no. 1, p. 10.

[89] Chowning, J. M.: The Synthesis of Complex Audio Spectra by Means of Frequency Modulation. J. Audio Engineering Society 21(1973), 526–534.

[90] Chemillier, M.: The free monoid and music, I. Do musicians need mathematics? (in French). RAIRO Inform. Theor. Appl. 21(1987), no. 3, 341–371. MR 89b:00045a.

[91] Chemillier, M.: The free monoid and music, II (in French). RAIRO Inform. Theor. Appl. 21(1987), no. 4, 379–417. MR 89b:00045b.

[92] Clough, J.—Myerson, G.: Musical scales and the generalized circle of fifths. Amer. Math. Monthly 93 (1986), no. 9, 695–701. MR 88a:05019.

[93] Combes, J. M.—Grossmann, A.—Tchamitchian, Ph. (eds.):Wavelets. Proc. of the International Conference held in Marseille, December 14–18, 1987, "Inverse Problems and Theoretical Imaging." Springer Verlag, Berlin–New York, 1989, 315 pp. MR 90g:42062.

[94] Dabby, D. S.: Musical variations from a chaotic mapping. Chaos 6 (1996), no. 2, 95–107.

[95] Daubechies, I.: Wavelet transforms and orthonormal wavelet bases. in Collection:Different perspectives on wavelets (San Antonio, TX, 1993), 1–33. Proc. Sympos. Appl. Math., 47, Amer. Math. Soc., Providence, 1993.

[96] Daubechies, I.: Time-frequency localization operators: A geometric phase space approach, IEEE Trans. Information Theory 34(1988), 605–612.

[97] Denckla, B. F.: Dynamic intonation in synthesizer performance, M.Sc. thesis, Massachusetts Institute of Technology, 1997, 61 pp.

[98] Detlovs, V. K.: Harmony and the melodic line. In Collection: Algebra and discrete mathematics: the theoretical foundations of software (in Russian), 54, 67–87, 161–162. Latv. Gos. Univ., Riga, 1986. MR 89b:00043.

[99] De Veaux, R. D.: Mixtures of linear regressions. Comput. Statist. DataAnal. 8 (1989), no. 3, 227–245. MR 90m:62158.

[100] Dickson, T. J.—Rogers, D. G.: Problems in graph theory. V. Magic valuations. Southeast Asian Bull. Math. 3 (1979), no. 1, 40–43. MR 81j:05092.

[101] Douthett, J.—Entringer, R.—Mullhaupt, A.: Musical scale construction: the continued fraction compromise. Utilitas Math. 42 (1992), 97–113.

[102] Ebcioglu, K.: An expert system for harmonizing chorales in the style of J. S. Bach. J. Logic Programming 8 (1990), no. 12, 145–185. MR 91b:00019.

[103] Ebeling, W.—Nicolis, G.: Word frequency and entropy of symbolic sequences: a dynamical perspective. Chaos Solitons Fractals 2(1992), no. 6, 635–650.

[104] Erlich, P.: Tuning, tonality, and 22-tone Equal Temperament. Xenharmonikôn 17(1998), 12–40.

[105] Fill, J. A.—Izenman, A. J.: Invariance properties of Schoenberg's tone row system. J. Austral. Math. Soc. Ser. B 21(1979/80), no. 3, 268–282. MR 83h:20004.

[106] Fill, J. A.—Izenman, A. J.: The structure of RI invariant twelvetone rows. J. Austral. Math. Soc. Ser. B 21(1979/80), no. 4, 402–417. MR 81g:20002.

[107] P. Flandrin: Some aspects of non-stationary signal processing with emphasis on time-frequency and time-scale methods. in: J. M. Combe, A. Grossman, and Ph. Tchamitchian (eds.): Wavelets, Springer-Verlag, Berlin, 1989, pp. 68–98.

[108] Fokker, A.: Equal temperament and the 31-keyed organ. Scientific Monthly 81(1955), 161–166.

[109] Fripertinger, H.: Enumeration in musical theory. in Collection: Seminaire Lotharingien de Combinatoire (Thurnau, 1991), 29–42. Publ. Inst. Rech. Math. Av., 476, Univ. Louis Pasteur, Strasbourg, 1992.

[110] Foks, H.: Elements of mathematical music theory (in Polish). Master's Thesis, Warsaw Technical University, 1994.

[111] Fripertinger, H.: Finite group actions on sets of functions. Burnside's lemma. Construction of representatives. Application in music theory (in Deutsch). Bayreuth. Math. Schr. 45(1993), 19–132.

[112] Gagliardo, E.: Enneadecaphonic music. A new system of harmonic tones. Atti Accad. Ligure Sci. Lett. 37 (1980), 536–546 (1981). MR 82j:00003.

[113] Gagliardo, E.: Musical styles guided by mathematical generalizations of tonality (in Italian). Rend. Sem. Mat. Fis. Milano 56(1986), 39–49(1988). MR 89k:00001.

[114] Gamer, C.–Wilson, R. J.: Musical block designs. Ars Combin. 16 (1983), A, 217–225. MR 85d:05096.

[115] Garmendia Rodríguez, M. J.—Navarro González, J. A.: Mathematical Theory of Musical Scales, Extracta Math., 11(1996), 369–374.

[116] Gale, D.: Tone perception and decomposition of periodic functions. Amer. Math. Monthly 86 (1979), no. 1, 36–42. MR 80c:92004.

[117] Gottlieb, H. P. W.: Harmonic properties of the annular membrane. J. Acoust. Soc. Amer. 66 (1979), no. 3, 647–650. MR 80f:73035.

[118] Gibiat, V.: Phase space representations of acoustical musical signals. J. Sound Vibration 123 (1988), no. 3, 529–536. MR 89i:58085.

[119] Gleijeses, B.: Musical projectivities (in Italian). Boll. Un. Mat. Ital. A (5) 18 (1981), no. 2, 275–282. MR 83f:51004.

[120] Goguen, J. A.: Complexity of hierarchically organized systems and the structure of musical experiences. Internat. J. Gen. Systems 3 (1976/77), no. 4, 233–251. MR 58 33185.

[121] Gombich, E. H.: Moment and movement in art , pp.115–212. in Collection:ed. P. T. Landsberg: The enigma of time. Adam Hilger, Ltd., Bristol–Heyden & Son, Inc., Philadelphia, 1984, 248 pp. MR 86g:83002.

[122] Halsley, G. D.—Hewitt, E.: More on the superparticular ratios in music. Amer. Math. Monthly 79(1972), 1 096–1 100. MR 47 1744.

[123] Halsley, G. D.—Hewitt, E.: A group theoretical method in the music theory (in German). Jahresber. Deutsch. Math. Verein. 80 (1978), no. 4, 151–207. MR 80g:20063.

[124] Haluška, J. : Diatonic scales summary, in: Proc. of the 7th IFSA World Congress (Prague, June 25–29, 1997), vol.IV., 320–322.

[125] Haluška, J. : On fuzzy coding of information in music, Busefal 69 (1997), 37–42.

[126] Haluška, J.: On two algorithms in music acoustics, Extracta Math. 12(1997), no. 3, 243–250.

[127] Haluška, J.: Uncertainty and tuning in music, Tatra Mt. Math. Publ. 12(1997), 113–129.

[128] Haluška, J.: On numbers 256/243, 25/24, 16/15. Tatra Mt. Math. Publ. 14(1998), 145–151.

[129] Haluška, J.: Comma 32 805/32 768 The Int. Journal of Uncertainty, Fuzziness and Knowledge - Based Systems, 6(1998), 295–305.

[130] Haluška, J.—Riečan, B: On a lattice of tetrachords. Tatra Mt. Math. Publ. 16(1999), 283–294.

[131] Haluška, J.: Searching the frontier of the Pythagorean System. Tatra Mt. Math. Publ. 16(1999), 273–282.

[132] Haluška, J.: *Uncertainty measures and well-tempered systems.* General Systems **31**(2002), 73–96.

[133] Haluška, J.: Equal Temperament and Pythagorean tuning: a geometrical interpretation in the plane. Fuzzy Sets and Systems 114/2(2000), 261–269.

[134] Halušková, E.—Studenovská, D.: Partial monounary algebras with common quasi-endomorphism, Czechoslovak Math. J. 42(117)(1992), 59–72.

[135] Halušková, E.: Comma sequences. (in: Haluška, J. (ed.): Ambiguity and Music. Polygrafia, Bratislava 2000), pp. 1–5.

[136] Hajdu, G.: Low Energy and Equal Spacing; the Multifactorial Evolution of Tuning Systems, Interface, 22, (1993), 319–333.

[137] Hellegouarch, Y.: Scales. C. R. Math. Rep. Acad. Sci. Canada 4 (1982), no. 5, 277–281. MR 83k:10092.

[138] Herlinger, J. W.: Marchetto's Division of the Whole Tone. J. of the Amer. Musicological Soc. 34(1981), 193–216.

[139] Hsu, K. J.—Hsu, A. J.: Fractal geometry of music. Proc. Nat. Acad. Sci. U. S. A. 87(1990), no. 3, 938–941. MR 91f:92041.

[140] Igoshev, L. A.: Encoded music (in Russian). In Collection: Number and thought, No. 9 (Russian), 142–152. Znanie, Moscow, 1986.

[141] Kaisler, D.: Six american composers on nonstandard tunings. Perspective of New Music 29(1991), no. 1, 176–211.

[142] Kent, J. T.: Ternary continued fractions and the evenlytempered musical scale. CWI Newslett. 13(1986), 21–33. MR 88h:00014.

[143] Kirkpatrick, S.—Gelatt, C. D.—Vecchi, M. P: Optimization by Simulated Annealing, Science 220(1983), No. 4598, May.

[144] Kondo, K.: Galois field theoretical approach to the dislocational and disclinational aspects of the musical lattice. Internat. J. Engrg. Sci. 30 (1992), no. 10, 1475–1482.

[145] Lefebvre, V. A.: A rational equation for attractive proportions. Math. Psychology 36(1992), 100–128.

[146] Lehmer, D. H.: On a problem of Størmer. Illinois J. Math. 8(1964), 57–79.

[147] Leiss, E. L.: The scordatura problem. Proc. Twentieth Southeastern Conference on Combinatorics, Graph Theory, and Computing, Boca Raton, 1989. Congr. Numer. 74 (1990), 138–144. MR 91b:00021.

[148] Liebermann, P.—Liebermann, R.: Symmetry in question and answer sequences in music. Comput. Math. Appl. 19 (1990), no. 7, 59–66. MR 91c:00013.

[149] Mason, S.—Saffle, M.: Selfsimilarity, FASS curves, and algorithms for musical structures. Symmetry Cult. Sci. 6(1995), no. 3, 465–467.

[150] Mathews, M. V.—Pierce, J. R.: Harmony and Nonharmonic Partials. Journal of the Acoustical Society of America 68(1980), 1252–1257.

[151] Mathews, M. V.—Pierce, J. R.—Reeves, A.—Roberts, L. A.: Theoretical and Experimental Explorations of the Bohlen–Pierce Scale. Journal of the Acoustical Society of America 84(1988), 1214-1222.

[152] Mazzola, G.—Wieser, H. G.—Brunner, V.—Muzzulini, D.: A symmetry oriented mathematical model of classical counterpoint and related neurophysiological investigations by depth EEG. Comput. Math. Appl. 17(1989), no. 46, 539–594. MR 90e:92104.

[153] McLaren, B.: A brief history of microtonality in the 20th century. Xenharmonicôn 17(1998), 57–110.

[154] Morales M., R. O.—Morales, M. E.: Analysis by induction of musical form and style using firstorder logic (in Spanish). (in Collection: XXVIth National Congress of the Mexican Mathematical Society (Spanish), Morelia, 1993), 437–449. Aportaciones Mat. Comun., 14, Soc. Mat. Mexicana, Mexico City, 1994.

[155] Mugavero, A. C.: Graphs for musical analysis. in Collection: Advances in graph theory, 275–283. Vishwa, Gulbarga, 1991.

[156] Mugavero, A. C.: Random walks through musical graphs. Congr. Numer. 89 (1992), 153–159.

[157] Muses, C.: Feedback and musical resonance. Kybernetes 23 (1994), no. 67, 111–122.

[158] Newland, D. E.: Harmonic and musical wavelets. Proc. Roy. Soc. London Ser. A 444(1994), no. 1922, 605–620.

[159] Nirenberg, L.: The use of topological, functional analytic, and variational methods in nonlinear problems. In: Proc. Third Sem. Funct. Anal. Appl., Bari, 1978, II. Confer. Sem. Mat. Univ. Bari (1979), No. 163168, 391–398, (1980). MR 81m:58026.

[160] Noll T.: Tone apperception and Weber–Fechner's law, 45 pp. In: Proc. 2nd Int. Conf. "Understanding and creating music," Caserta, November 21–25, 2002, CD-ROM ISBN 88-900456-2-0.

[161] Plomp R.—Levelt, W. J. M.: Tonal Consonance and Critical Bandwidth, Journal of the Acoustical Society of America 38(1965), 548–560.

[162] Polansky, L.: Morphological metrics, Journal of New Music Research (25)1996, no. 4.

[163] Praetorius, M.: Syntagmatis Musici Tomus Secundus. De Organographia. Wolfenbüttel 1619. Facsimile ed. by W. Gurlitt. Kassel 1958, 51980. (= Documenta Musicologica Band 1, XIV).

[164] Reiner, D. L.: Enumeration in music theory. Amer. Math. Monthly 92 (1985), no. 1, 51–54. MR 86c:05021.

[165] Romanowska, A.: An attempt to investigate the algebraic structure of a tone system (in Polish). Doctoral Thesis, Warsaw Technical University, 1973.

[166] Romanowska, A.: Algebraic structure of the tone system. Demonstratio Math. 7(1974), 525–542.

[167] Romanowska, A.: Über algebraische Aspekte der Theorie der Harmonie. Unpublished manuscript, 1975.

[168] Romanowska, A.: On some algebras connected with the theory of harmony, Colloq. Math. 41(1979), 181–185.

[169] Rothenberg, D.: A model for pattern perception with musical applications. (I: Pitch structures as orderpreserving maps; II: The information content of pitch structures; III. The graph embedding of pitch structures) Math. Systems Theory 11 (1977/78), 199–234, 353–372, 73–101. MR 58 32139a,b,c.

[170] Slaymaker, F. H.: Chords from Tones Having Stretched Partials. Journal of the Acoustical Society of America 47(1968), 1469–1571.

[171] Sidiskis, B.: Intervals as musical notations that transmit information on natural sound sequences (in Russian). Mat. Metody v Social. Nauk. Trudy Sem. Processy Optimal. Upravlenija. II. Sekcija (1973), Vyp. 2, 113–127. MR 57 15539.

[172] Silver A. L. S.: Musimatics or the nun's fiddle. Amer. Math. Monthly 78(1971), 351–357

[173] Stembridge, Ch.: Italian organ music to Frescobaldi. In: Nicholas Thistlethwaite and Geoffrey Webber: The Cambridge Companion to the Organ. Cambridge 1998. pp. 148–163.

[174] Størmer C.: Quelques théorèmes sur l'équations de Pell $x^2 - Dy^2 = \pm 1$ et leurs applications. Skrifter Videnskabs-selskabet (Christiania) I, Mat.-Naturv. Kl. 2(1897), p. 48.

[175] Størmer C.: Sur une équation indéterminée, C. R. Acad. Sci. Paris 127(1898), 752–754.

[176] Tagliavini, L. F.: Notes on Tuning Methods in Fifteenth-Century Italy. In: Fenner Douglass et. al. (eds.): Charles Brenton Fisk. Organ Builder. Volume One. (Essays in his Honor). Easthampton (Mass.) 1986. pp. 191–199.

[177] Tangyan, A. S.: A model of correlative perception and its applications to certain problems of pattern recognition (in Russian). Mat. Model. 2 (1990), no. 8, 90–111. MR 91j:00005.

[178] Tarber, J. A.: A just scale for music. Jour. Acoust. Soc. Amer. 18(1946), p. 167.

[179] Teaney, D. T.—Moruzzi, V. L.—Mintzer, F. C.: The tempered Fourier transform. J. Acoust. Soc. Amer. 67(1980), no. 6, 2063–2067. MR 81c:44010.

[180] Thirring, W.: Über vollkommene Tonsysteme (in German). Annalen der Physik 47(1990), 245–250.

[181] Vidyamurthy, G.—Murty, M. N.: Musical harmony a fuzzy entropic characterization. Fuzzy Sets and Systems 48 (1992), no. 2, 195–200.

[182] Voss, R. F.: Random fractals: selfaffinity in noise, music, mountains, and clouds. Phys. D 38 (1989), no. 13, 362–371. MR 91c:00021.

[183] Vuza, D.: Mathematical aspects of the modal theory of Anatol Vieru. IV (in French). Rev. Roumaine Math. Pures Appl. 28(1983), no. 8, 757–773. MR 85f:20006b.

[184] Vuza, D.: Mathematical aspects of the modal theory of Anatol Vieru. V, (in French). Rev. Roumaine Math. Pures Appl. 31(1986), no. 5, 399–413. MR 88a:20065.

[185] Wille, R.: Mathematik and Musiktheorie, (in Schnitzler, G. (ed.): Musik und Zahl, Verlag für systematische Musik-Wissenschaft, Bonn-Bad Godesberg, 1976, pp. 233–264.

[186] Wille, R.: Mathematische Sprache in der Musiktheorie, Preprint 505, TH Darmstadt, 1979.

[187] Wille, R.: Mathematical language in music theory. (in German) (in Collection: Yearbook: surveys of mathematics 1980, Bibliographisches Inst., Mannheim, 1980), pp. 167–184. MR 83h:00002.

[188] Wille, R.: Symmetry attempt at a definition (in German). (in Collection: Symmetry of discrete mathematical structures and their symmetry groups, Res. Exp. Math., 15, Heldermann, Berlin, 1991.), pp. 119–149. MR 92b:51001.

[189] Young, F. J.: The natural frequencies of musical horns. Acustica 10 (1960), 91–97. MR 22 2259.

[190] Zadeh, L.: Towards a theory of fuzzy information granulation and its centrality in human reasoning and fuzzy logic. Fuzzy sets and systems 90(1997), 111–127.

[191] Zaripov, R. H.: An algorithmic description of a process of musical composition. Original: Dokl. Akad. Nauk SSSR 132, 1283–1286 (in Russian); Soviet Physics. Dokl. 5 (1960), 479–482. MR 22 10219.

Web

[192] Op de Coul, M.:
      http://www.xs4all.nl/~ huygensf

[193] Frazer, P. A.: The development of musical Tuning systems.
      http://www.midicode.com/tunings/index.shtml

[194] Keenan, D.:
      http://www.uq.net.au/~ zzdkeena/

[195] Monzo, J.:
      http://www.ixpress.com/interval/dict/index.html

[196] Schulter, M.: Pythagorean System and Medieval Polyphony.
      http://www.medieval.org/emfaq/harmony/pyth.html

Appendix A

Extended list of tone systems

A.1 Tetrachords (2 periods)

ARCHY(CHRO)-7; Archytas' Chromatic;
28/27, 9/8, 4/3, 3/2, 14/9, 27/16, 2

ARCHY(DIAT)-7; Archytas' Diatonic, also Lyra tuning;
28/27, 32/27, 4/3, 3/2, 14/9, 16/9, 2

ARCHY(ENH)-7; Archytas' Enharmonic;
28/27, 16/15, 4/3, 3/2, 14/9, 8/5, 2

ARCHY(ENH/P)-7; Permutation of Archytas Enharmonic with the 36/35 first;
36/35, 16/15, 4/3, 3/2, 54/35, 8/5, 2

ARISTO(CHRO)-7; Dorian Mode, Neo-Chromatic tetrachord, 6 + 18 + 6 parts;
100, 400, 500, 700, 800, 1 100, 1 200

ARISTO(ARCH/ENH)-7; Aristoxenos Archaic Enharmonic, 4 + 3 + 23 parts, similar to Archytas' enharmonic;
200/3, 500/3, 500, 700, 2 300/3, 2 450/3, 1 200

ARISTO(CHRO/NO.2)-7; Dorian Mode, a 1:2 Chromatic, 8 + 18 + 4 parts;
400/3, 1 300/3, 500, 700, 2 500/3, 3 400/3, 1 200

ARISTO(CHRO/NO.3)-7; Aristoxenos 3 Chromatic, 7 + 7 + 16 parts;
445/416, 230/201, 295/221, 442/295, 928/579, 1159/676, , 2

ARISTO(CHRO/NO.4)-7; Aristoxenos Chromatic, 5.5 + 5.5 + 19 parts;
275/3, 550/3, 500, 700, 2 375/3, 2 650/3, 1 200

ARISTO(CHRO/ENH)-7; Aristoxenos Ch/Enh, 3 + 9 + 18 parts;
50, 200, 500, 700, 770, 900, 1 200

ARISTO(CHRO/INV)-7; Aristo's Inverted Chromatic, Dorian Mode 18 + 6 + 6 parts;
300, 400, 500, 700, 1 000, 1 100, 1 200

ARISTO(CHRO/REJ)-7; Aristoxenos Rejected Chromatic, 6 + 3 + 21 parts;
100, 150, 500, 700, 800, 850, 1 200

ARISTO(DIAT/NO.1)-7;
Phrygian octave species on E, 12 + 6 + 12 parts;
200, 300, 500, 700, 900, 1 000, 1 200

ARISTO(DIAT/NO.2)-7; Aristoxenos 2 Diatonic, 7 + 11 + 12 parts;
500/3, 300, 500, 700, 2 450/3, 1 000, 1 200

ARISTO(DIAT/NO.3)-7; Aristoxenos Diat 3, 9.5 + 9.5 + 11 parts;
475/3, 950/3, 500, 700, 2 575, 3 050/3, 1 200

ARISTO(DIAT/NO.4)-7; Aristoxenos Diatonic, 8 + 8 + 14 parts;
400/3, 800/3, 500, 700, 2 500/3, 2 900/3, 1 200

ARISTO(DIAT/OR)-7; Aristoxenos Redup. Diatonic, 14 + 2 + 14 parts;
700/3, 800/3, 500, 700, 2 800/3, 2 900/3, 1 200

ARISTO(DIAT/INV)-7; Lydian octave species on E, Major Mode, 12 + 12 + 6 parts;
200, 400, 500, 700, 900, 1 100, 1 200

ARISTO(DIAT/RED/NO.1)-7; Aristoxenos Redup. Diatonic, Dorian Mode, 14 + 14 + 2 parts;
700/3, 1 400/3, 500, 700, 2 800/3, 5 000/3, 1 200

ARISTO(DIAT/RED/NO.2)-7; Aristoxenos 2 Redup. Diatonic 2, 4 + 13 + 13 parts;
200/3, 850/3, 500, 700, 2 300/3, 2 950/3, 1 200

ARISTO(DIAT/RED/NO.3)-7; Aristoxenos 3 Redup. Diatonic, 8 + 11 + 11 parts;
400/3, 950/3, 500, 700, 2 500/3, 3 050/3, 1 200

ARISTO(ENH/NO.1)-7; Aristoxenos Enharmonion, Dorian Mode;
50, 100, 500, 700, 750, 800, 1 200

ARISTO(ENH/NO.2)-7; Aristoxenos 2 Enharmonic, 3.5 + 3.5 + 23 parts;
58.333, 500/3, 500, 700, 758.333, 2 450/3, 1 200

ARISTO(ENH/NO.3)-7; Aristoxenos Enharmonic, 2.5 + 2.5 + 25 parts;
125/3, 250/3, 500, 700, 2 225/3, 2 350/3, 1 200

ARISTO(HEM/CHRO/NO.1)-7; Chromatic Hemiolion, Dorian Mode;
75, 150, 500, 700, 775, 850, 1 200

ARISTO(HEM/CHRO/NO.2)-7; Aristoxenos Chromatic, 4.5 + 7.5 + 18 parts;
75, 200, 500, 700, 775, 900, 1 200

ARISTO(HEM/CHRO/NO.3)-7; Dorian mode of Aristoxenos' Hemiolic Chromatic according to Ptolemy's interpret;
80/77, 40/37, 4/3, 3/2, 120/77, 60/37, 2

ARISTO(HYP/ENH/NO.2)-7; Aristoxenos 2nd Hyperenharmonic, 37.5 + 37.5 + 425;
75/2, 75, 500, 700, 1 475/2, 775, 1 200

ARISTO(HYP/ENH/NO.3)-7; Aristoxenos 3 Hyperenharmonic, 1.5 + 1.5 + 27 parts;
25, 50, 500, 700, 725, 750, 1 200

ARISTO(HYP/ENH/NO.4)-7; Aristoxenos 4 Hyperenharmonic, 2 + 2 + 26 parts;
100/3, 200/3, 500, 700, 2 200/3, 2 300/3, 1 200

ARISTO(HYP/ENH/NO.5)-7;
Aristoxenos Hyperenharmonic, 23 + 23 + 454;
23, 46, 500, 700, 723, 746, 1 200

ARISTO(INT/DIAT)-7;
Dorian mode of Aristoxenos Intense Diatonic according to Ptolemy;
20/19, 20/17, 4/3, 3/2, 30/19, 30/17, , 2

ARISTO(P/ENH/NO.2)-7;
Permuted Aristoxenos Enharmonion, 3 + 24 + 3 parts;
50, 450, 500, 700, 750, 1150, 1 200

ARISTO(P/ENH/NO.3)-7;
Permuted Aristoxenos Enharmonion, 24 + 3 + 3 parts;
400, 450, 500, 700, 1 100, 1 150, 1 200

ARISTO(P/SCHRO/NO.2)-7;
Aristoxenos 2 Chromatic, 6.5 + 6.5 + 17 parts;
108.333, 216.667, 500, 700, 808.333, 916.667, 1 200

ARISTO(SOFT/CHRO/NO.1)-7;
Aristoxenos Chromatic Malakon, Dorian Mode;
200/3, 400/3, 500, 700, 2 300/3, 2 500/3, 1 200

ARISTO(SOFT/CHRO/NO.2)-7; Aristoxenos S. Chromatic, 6 + 16.5 + 9.5 parts;
100, 375, 500, 700, 800, 1 075, 1 200

ARISTO(SOFT/CHRO/NO.3)-7;
Aristoxenos Chromatic Malakon, 9.5 + 16.5 + 6 parts;
125, 400, 500, 700, 825, 1 100, 1 200

ARISTO(SOFT/CHRO/NO.4)-7;
Aristoxenos S. Chromatic, 6 + 7.5 + 16.5 parts;
100, 225, 500, 700, 800, 925, 1 200

ARISTO(SOFT/CHRO/NO.5)-7;
Dorian mode of Aristoxenos' Soft Chromatic according to Ptolemy's interpretati;
30/29, 15/14, 4/3, 3/2, 45/29, 45/28, 2

ARISTO(SOFT/DIAT/NO.1)-7;
Aristoxenos Diatonon Malakon, Dorian Mode;
100, 250, 500, 700, 800, 950, 1 200

ARISTO(SOFT/DIAT/NO.2)-7; Dorian Mode, 6 + 15 + 9 parts;
100, 350, 500, 700, 800, 1 050, 1 200

ARISTO(SOFT/DIAT/NO.3)-7; Dorian Mode, 9 + 15 + 6 parts;
150, 400, 500, 700, 850, 1 000, 1 200

ARISTO(SOFT/DIAT/NO.4)-7; Dorian Mode, 9 + 6 + 15 parts;
150, 250, 500, 700, 850, 950, 1 200

ARISTO(SOFT/DIAT/NO.5)-7; Dorian Mode, 15 + 6 + 9 parts;
250, 350, 500, 700, 950, 1 050, 1 200

ARISTO(SOFT/DIAT/NO.6)-7; Dorian Mode, 15 + 9 + 6 parts;
250, 400, 500, 700, 950, 1 100, 1 200

ARISTO(SOFT/DIAT/NO.7)-7; Dorian mode of Aristoxenos Soft Diatonic according to Ptolemy;
20/19, 8/7, 4/3, 3/2, 30/19, 12/7, 2

ARISTO(SYN/CHRO)-7;
Aristoxenos Chromatic Syntonon, Dorian Mode;
100, 200, 500, 700, 800, 900, 1 200

ARISTO(UN/C HRO)-7;
Aristoxenos Unnamed Chromatic, Dorian Mode;
200/3, 200, 500, 700, 2 300/3, 900, 1 200

ARISTO(UN/CHRO2)-7; Dorian Mode, a 1:2 Chromatic, 8 + 4 + 18 parts;
400/3, 200, 500, 700, 2 500/3, 900, 1 200

ARISTO(UN/CHRO3)-7; Dorian Mode, a 1:2 Chromatic, 18 + 4 + 8 parts;
300, 1 100/3, 500, 700, 1 000, 3 200/3, 1 200

ARISTO(UN/CHRO4)-7; Dorian Mode, a 1:2 Chromatic, 18 + 8 + 4 parts;
300, 1 300/3, 500, 700, 1 000, 3 400/3, 1 200

ATHANASOPOULOS(CHRO/NO.1)-7; Athanasopoulos's Byzantine Liturgical mode Chromatic;
150, 400, 500, 700, 850, 1 100, 1 200

ATHANASOPOULOS(CHRO/NO.2)-7; Athanasopoulos's Byzantine Liturgical mode 2nd Chromatic;
100, 400, 500, 700, 800, 1 100, 1 200

DORIAN(DIES2)-7; Dorian Diatonic, 2 part Diesis;
100/3, 300, 500, 700, 2 200/3, 1 000, 1 200

DORIAN(DIES5)-7; Dorian Diatonic, 5 part Diesis;
250/3, 300, 500, 700, 2 350/3, 1 000, 1 200

DORIAN(DIAT/ENH/NO.1)-7; Diat. + Enharm. Diesis, Dorian Mode;
50, 300, 500, 700, 750, 1 000, 1 200

DORIAN(DIAT/ENH/NO.2)-7; Diat. + Enharm. Diesis, Dorian Mode 3 + 12 + 15 parts;
50, 250, 500, 700, 750, 950, 1 200

DORIAN(DIAT/ENH/NO.3)-7; Diat. + Enharm. Diesis, Dorian Mode, 15 + 3 + 12 parts;
250, 300, 500, 700, 950, 1 000, 1 200

DORIAN(DIAT/ENH/NO.4)-7; Diat. + Enharm. Diesis, Dorian Mode, 15 + 12 + 3 parts;
250, 450, 500, 700, 950, 1150, 1 200

DORIAN(DIAT/ENH/NO.5)-7; Dorian Mode, 12 + 15 + 3 parts;
200, 450, 500, 700, 900, 1 150, 1 200

DORIAN(DIAT/ENH/NO.6)-7; Dorian Mode, 12 + 3 + 15 parts;
200, 250, 500, 700, 900, 950, 1 200

DORIAN(DIAT/RED11)-7; Dorian mode of a diatonic genus with reduplicated 11/10;
11/10, 121/100, 4/3, 3/2, 33/20, 363/200, 2

DORIAN(DIAT/HEM/CHRO)-7; Diat. + Hem. Chrom. Diesis, Another genus of Aristoxenos, Dorian Mode;
75, 300, 500, 700, 775, 1 000, 1 200

DORIAN(DIAT/SOFT/CHRO)-7; Diat. + Soft Chrom. Diesis, Another genus of Aristoxenos, Dorian Mode;
200/3, 300, 500, 700, 2 300/3, 1 000, 1 200

DIATONIC(SOFT/NO.1)-7; Soft Diatonic genus 5 + 10 + 15 parts;
250/3, 250, 500, 700, 2 350/3, 950, 1 200

DORIAN(DIA/SOFT/NO.2)-7; Soft Diatonic genus with equally divided Pyknon, Dorian Mode;
125, 250, 500, 700, 825, 950, 1 200

DIDY(CHRO)-7; Didymus Chromatic;
16/15, 10/9, 4/3, 3/2, 8/5, 5/3, 2

DIDY(CHRO/NO.1)-7; Permuted Didymus Chromatic;
16/15, 32/25, 4/3, 3/2, 8/5, 48/25, , 2

DIDY(CHRO/NO.2)-7; Didymus's Chromatic, 6/5 x 25/24 x 16/15;
6/5, 5/4, 4/3, 3/2, 9/5, 15/8, 2

DIDY(CHRO/NO.3)-7; Didymus's Chromatic, 25/24 x 16/15 x 6/5;
25/24, 10/9, 4/3, 3/2, 25/16, 5/3, 2

DIDY(DIAT)-7; Didymus Diatonic;
16/15, 32/27, 4/3, 3/2, 8/5, 16/9, 2

DIDY(ENH/NO.2)-7; Permuted Didymus Enharmonic;
256/243, 16/15, 4/3, 3/2, 128/81, 8/5, 2

DIDY(ENH)-7; Dorian mode of Didymos's Enharmonic;
32/31, 16/15, 4/3, 3/2, 48/31, 8/5, 2

ERATOSTHENES(CHRO)-7; Dorian mode of Eratosthenes's Chromatic. same as Ptol. Intense Chromatic;
20/19, 10/9, 4/3, 3/2, 30/19, 5/3, 2

ERATOSTHENES(DIAT)-7; Dorian mode of Eratosthenes's Diatonic, Pythagorean;
256/243, 32/27, 4/3, 3/2, 128/81, 16/9, 2

ERATOSTHENES(ENH)-7; Dorian mode of Eratosthenes's Enharmonic;
40/39, 20/19, 4/3, 3/2, 20/13, 30/19, 2

GOLDSECTION(DIAT)-7; Diatonic scale with ratio between whole and half tone the Golden Section;
192.429, 384.858, 503.785, 696.215, 888.644, 1 081.072, 1 200.000

IASTI-7; Iasti-aiolika, kithara tuning: tonic diatonic and ditonic diatonic;
28/27, 32/27, 4/3, 3/2, 27/16, 16/9, 2

IASTIA-7; Iastia or Lydia, kithara tuning: intense diatonic and tonic diatonic;
28/27, 32/27, 4/3, 3/2, 8/5, 9/5, 2

PHRYGIAN-7; Old Phrygian(?), also McClain's 7-tone scale;
10/9, 6/5, 4/3, 3/2, 5/3 9/5, 2

PTO(AIOLIKA)-7; Ptolemy's kithara tuning, mixture of Tonic Diatonic and Di-tone Diatonic;
28/27, 32/27, 4/3, 3/2, 27/16, 16/9, 2

PTO(CHRO)-7; Ptolemy Soft Chromatic;
28/27, 10/9, 4/3, 3/2, 14/9, 5/3, 2

PTO(DIAT/NO.1)-7; Ptolemy's Diatonon Ditoniaion;
28/27, 32/27, 4/3, 3/2, 14/9, 16/9, 2

PTO(DIAT/NO.2)-7; Dorian mode of a permutation of Ptolemy's Tonic Diatonic; 7
28/27, 7/6, 4/3, 3/2, 14/9, 7/4, 2

PTO(DIAT/NO.3)-7;
Dorian mode of the remaining permutation of Ptolemy's Intense Diatonic;
9/8, 6/5, 4/3, 3/2, 27/16,9/5, 2

PTO(DIAT/NO.4)-7; Permuted Ptolemy's diatonic;
8/7, 32/27, 4/3, 3/2, 12/7, 16/9, 2

PTO(DOR)-7;
Dorian mode of Ptolemy's Intense Diatonic (Diatonon Syntonon);
16/15, 6/5, 4/3, 3/2, 8/5, 9/5, 2

PTO(ENH)-7; Dorian mode of Ptolemy's Enharmonic;
46/45, 16/15, 4/3, 3/2, 23/15, 8/5, 2

PTO(HOM)-7; Dorian mode of Ptolemy's Diatonon Homalon;
12/11, 6/5, 4/3, 3/2, 18/11, 9/5, 2

PTO(ONT/CHRO)-7; Dorian mode of Ptolemy's Intense Chromatic;
22/21, 8/7, 4/3, 3/2, 11/7 12/7, 2

PTO(MAL/DIAT)-7; Ptolemy soft diatonic;
21/20, 7/6, 4/3, 3/2, 63/40, 7/4, 2

PTO(MAL/DIAT/NO.2)-7; Permuted Ptolemy soft diatonic;
10/9, 7/6, 4/3, 3/2, 5/3, 7/4, 2

PTO(MAL/DIAT/NO.3)-7; Permuted Ptolemy soft diatonic;
8/7, 6/5, 4/3, 3/2, 12/7, 9/5, 2

QUINTILIANUS(CHRO)-7; Aristides Quintilianus' Chromatic genus;
18/17, 9/8, 4/3, 3/2, 27/17, 27/16, 2

A.2   Pentatonics

DIAPHONIC-5; D5-tone Diaphonic Cycle;
8/7, 4/3, 3/2, 12/7, 2

DIM/TET/B-5; A pentatonic form on the 9/7;
9/8, 9/7, 14/9, 7/4, 2

SCHLESINGER(DIV-FIFTH/NO.1)-5; Divided Fifth No. 1, From Schlesinger;
24/23, 12/11, 4/3, 3/2, 2

SCHLESINGER(DIV-FIFTH/NO.2)-5; Divided Fifth No. 2, From Schlesinger;
16/15, 8/7, 4/3, 3/2, 2

SCHLESINGER(DIV-FIFTH/NO.3)-5; Divided Fifth No. 3, From Schlesinger;
28/27, 7/6, 4/3, 3/2, 2

SCHLESINGER(DIV-FIFTH/NO.4)-5; Divided Fifth No. 4, From Schlesinger;
21/20, 7/6, 21/16, 3/2, 2

SCHLESINGER(DIV-FIFTH/NO.5)-5; Divided Fifth No. 5, From Schlesinger;
11/10, 11/9, 11/8, 11/7, 2

HEXANY15-5; 1.3.5.15 2)4 hexany (1.15 tonic) degenerate, symmetrical pentatonic;
5/4, 4/3, 3/2, 8/5, 2

HO-MAI-5; Ho Mai Nhi (Nam Hue) dan tranh scale;
11/10, 4/3, 3/2, 33/20, 2

GOLDEN-5; Golden pentatonic;
5/4, 21/16, 3/2, 13/8, 2

ISFAHAN(NO.2)-5; Isfahan (IG No. 2, DF No. 8), from Rouanet;
13/12, 7/6, 5/4, 4/3, 2

ISLAM(NO.2)-5; Islamic Genus No. 1 (DFNo. 7), From Rouanet;
13/12, 7/6, 91/72, 4/3, 2

KOREA-5; According to L. Harrison, called "the Delightful" in Korea;
9/8, 4/3, 3/2, 9/5, 2

HARRISON(PE/NO.1)-5; From L. Harrison, a Pelog style pentatonic;
16/15, 6/5, 3/2, 8/5, 2

HARRISON(PE/NO.2)-5; From L. Harrison, a Pelog style pentatonic;
12/11, 6/5, 3/2, 8/5, 2

HARRISON(PE/NO.3)-5; From L. Harrison, a Pelog style pentatonic;
28/27, 4/3, 3/2, 14/9, 2

HARRISON(PE/NO.4)-5; From L. Harrison, a Pelog style pentatonic;
16/15, 6/5, 3/2, 15/8, 2

HARRISON(MINOR/NO.1)-5; From L. Harrison, a symmetrical pentatonic with minor thirds;
6/5, 4/3, 3/2, 5/3, 2

HARRISON(MINOR/NO.2)-5; A minor pentatonic;
8/7, 4/3, 8/5, 16/9, 2

HARRISON(MIXED/NO.1)-5; A "Mixed type" pentatonic, from L. Harrison;
12/11, 6/5, 3/2, 13/8, 2

HARRISON(MIXED/NO.2)-5; A "Mixed type" pentatonic, from L. Harrison;
6/5, 4/3, 3/2, 15/8, 2

HARRISON(MIXED/NO.3)-5; A "Mixed type" pentatonic, from L. Harrison;
6/5, 9/7, 3/2, 8/5, 2

HARRISON(MIXED/NO.4)-5; A "Mixed type" pentatonic, from L. Harrison;
15/14, 5/4, 3/2, 12/7, 2

MONTFORD-5; Montford's Spondeion, a mixed septimal and undecimal pentatonic;
28/27, 4/3, 3/2, 18/11, 2

NEUTR/PENT(NO.1)-5; Quasi-Neutral Pentatonic 1, 15/13 x 52/45 in each trichord, after Dudon;
52/45, 4/3, 3/2, 26/15, 2

NEUTR/PENT(NO.2)-5; Quasi-Neutral Pentatonic 2, 15/13 x 52/45 in each trichord, after Dudon;
15/13, 4/3, 3/2, 45/26, 2

OLYMPOS-5; Scale of ancient Greek flutist Olympos, 6th century BC as reported by Partch;
16/15, 4/3, 64/45, 16/9, 2

PRIME-5; What L. Harrison calls "the Prime Pentatonic," a widely used scale;
9/8, 5/4, 3/2, 5/3, 2

FARRISON/DUDON(SEPT/SLE/NO.2)-5; Septimal Slendro 2, From L. Harrison, J. Dudon's APTOS;
9/8, 21/16, 3/2, 12/7, 2

MILLS(SEPT/SLE/NO.3)-5; Septimal Slendro 3, Dudon, Harrison, called "MILLS" after Mills Gamelan;
9/8, 9/7, 3/2, 12/7, 2

NAT/HARRISON/DUDON(SEPT/SLE/NO.4)-5; Septimal Slendro 4, from L. Harrison, Jacques Dudon, called "NAT";
9/8, 21/16, 3/2, 7/4, 2

HARRISON/DUDON(SEPT/SLE/NO.5)-5; Septimal Slendro 5, from Jacques Dudon;
7/6, 21/16, 49/32, 343/192, 2

SEPT/SLE(NO.1)-5; A Slendro type pentatonic which is based on intervals of 7; from L. Harrison;
8/7, 9/7, 3/2, 12/7, 2

SEPT/SLE(NO.2)-5; A Slendro type pentatonic which is based on intervals of 7, no. 2;
7/6, 4/3, 3/2, 7/4, 2

SEPT/SLE(NO.4)-5; A Slendro type pentatonic which is based on intervals of 7, no. 4;
9/8, 4/3, 3/2, 12/7, 2

SLEN/PEL/JC-12; Slendro/JC PEL S1c, P1 c♯, S2 d, e♭, P2 e, S3 f,P3 f♯, S4 g, a♭, P4 a, S5 b♭, P5 b;
1/1, 8/7, 8/7, 16/15, 64/49, 4/3, 3/2, 3/2, 3/2, 12/7, 8/5, 2

SCHMIDT(SLEN/PEL/NO.1)-12; D. Schmidt (Pelog white, Slendro black);
9/8, 7/6, 5/4, 4/3, 11/8, 3/2, 3/2, 7/4, 7/4, 15/8, 2

SCHMIDT(SLEN/PEL/NO.2)-12; D. Schmidt with 13,17,19,21,27;
17/16, 9/8, 19/16, 5/4, 21/16, 11/8, 3/2, 13/8, 27/16, 7/4, 15/8, 2

DUDON(SLE/A1)-5; Slendro A1, J. Dudon: Seven-Limit Slendro Mutations, 1/1 8:2 Jan 1994;
8/7, 4/3, 3/2, 7/4, 2

DUDON(SLE/A2)-5; Slendro A2, J. Dudon: Seven-Limit Slendro Mutations, 1/1 8:2 Jan 1994;
8/7, 64/49, 32/21, 12/7, 2

DUDON(SLE/M)-5; Dudon's Slendro M from "Seven-Limit Slendro Mutations," 1/1 8:2 Jan 1994;
8/7, 4/3, 3/2, 12/7, 2

DUDON(SLE/S/NO.1)-5; Dudon's Slendro S1 from "Seven-Limit Slendro Mutations," 1/1 8:2 Jan 1994;
8/7, 4/3, 32/21, 7/4, 2

DUDON(SLE/S/NO.2)-5; Dudon's Slendro S2;
8/7, 64/49, 32/21, 256/147, 2

UDAN-MAS(SLE)-5; Slendro Udan Mas (approx.);
7/6, 47/35, 20/13, 16/9, 2

TRANH(NO.1)-5; B. D. Tranh scale;
10/9, 4/3, 3/2, 5/3, 2

TRANH(NO.2)-5; B. D. Tranh scale;
10/9, 20/17, 3/2, 5/3, 2

WINNINGTON/INGRAM-5; Winnington-Ingram's Spondeion;
12/11, 4/3, 3/2, 18/11, 2

A.3   Heptatonics

ARCHY(ENH/3)-7; Complex 9 of p. 113 based on Archytas's Enharmonic;
28/27, 16/15, 9/7, 4/3, 48/35, 12/7, 2

ARCHY(ENH/T)-7; Complex 6 of p. 113 based on Archytas's Enharmonic;
36/35, 28/27, 16/15, 9/7, 4/3, 27/14, 2

ARCHY(ENH/T/2)-7; Complex 5 of p. 113 based on Archytas's Enharmonic;
28/27, 16/15, 5/4, 4/3, 15/8, 35/18, 2

ARCHY(ENH/T/3)-7; Complex 1 of p. 113 based on Archytas's Enharmonic;
28/27, 16/15, 784/729, 448/405, 4/3, 112/81, 2

ARCHY(ENH/T/4)-7; Complex 8 of p. 113 based on Archytas's Enharmonic;
28/27, 16/15, 5/4, 35/27, 4/3, 5/3, 2

ARCHY(ENH/T/5)-7; Complex 10 of p. 113 based on Archytas's Enharmonic;
245/243, 28/27, 16/15, 35/27, 4/3, 35/18, 2

ARCHY(ENH/6)-7; Complex 2 of p. 113 based on Archytas's Enharmonic;
28/27, 16/15, 448/405, 256/225, 4/3, 64/45, 2

ARCHY(ENH/7)-7; Complex 11 of p. 113 based on Archytas's Enharmonic;
36/35, 28/27, 16/15, 192/175, 4/3, 48/35, 2

BAGPIPEA-7; Highland Bagpipe, from Acustica4: 231 (1954) J.M.A Lenihan and S. McNeill;
9/8, 5/4, 27/20, 3/2, 5/3, 9/5, 2

BASTARD-7; 1)7 7-any from 1.3.5.7.9.11.13 and Schlesinger's "Bastard" Hypodorian Harmonia;
8/7, 16/13, 4/3, 16/11, 8/5, 16/9, 2

BYZ-PALACE-7; Byzantine Palace mode;
18/17, 9/7, 4/3, 3/2, 18/11, 9/5, 2

CHINESE(BRONZE)-7; A scale found on an ancient Chinese bronze instrument from the 3rd century BC;
8/7, 6/5, 5/4, 4/3, 3/2, 5/3, 2

CHRO15-7; Tonos-15 Chromatic;
15/14, 15/13, 15/11, 3/2, 30/19, 5/3, 2

CHRO15-INV)-7; Inverted Chromatic Tonos-15 Harmonia;
6/5, 19/15, 4/3, 22/15, 26/15, 28/15, 2

CHRO15-INV/NO.2)-7; A harmonic form of the Chromatic Tonos-15 inverted;
16/15, 17/15, 4/3, 22/15, 23/15, 8/5, 2

CHRO17-7; Tonos-17 Chromatic;
17/16, 17/15, 17/12, 17/11, 34/21, 17/10, 2

CHRO17/CON/)-7; Conjunct Tonos-17 Chromatic;
17/16, 17/15, 17/12, 34/23, 17/11, 17/9, 2

CHRO19-7; Tonos-19 Chromatic;
19/18, 19/17, 19/14, 19/13, 38/25, 19/12, 2

CHRO19/CON/)-7; Conjunct Tonos-19 Chromatic;
19/18, 19/17, 19/14, 38/27, 19/13, 19/11, 2

CHRO21-7; Tonos-21 Chromatic;
21/20, 21/19, 21/16, 3/2, 14/9, 21/13, 2

CHRO21/INV)-7; Inverted Chromatic Tonos-21 Harmonia;
26/21, 9/7, 4/3, 32/21, 38/21, 40/21, 2

CHRO21-INV/NO.2)-7; Inverted harmonic form of the Chromatic Tonos-21;
16/15, 8/7, 4/3, 32/21, 34/21, 12/7, 2

CHRO23-7; Tonos-23 Chromatic;
23/22, 23/21, 23/18, 23/16, 23/15, 23/14, 2

CHRO23-COM-7; Conjunct Tonos-23 Chromatic;
23/22, 23/21, 23/18, 23/17, 23/16, 23/13, 2

CHRO25-7; Tonos-25 Chromatic;
50/47, 25/22, 25/18, 25/16, 5/3, 25/14, 2

CHRO25-COM-7; Conjunct Tonos-25 Chromatic;
50/47, 25/22, 25/18, 25/17, 25/16, 25/13, 2

CHRO27-7; Tonos-27 Chromatic;
18/17, 9/8, 27/20, 3/2, 27/17, 27/16, 2

CHRO27-INV)-7; Inverted Chromatic Tonos-27 Harmonia;
32/27, 34/27, 4/3, 40/27, 16/9, 17/9, 2

CHRO27-INV/NO.2)-7; Inverted harmonic form of the Chromatic Tonos-27;
28/27, 29/27, 4/3, 40/27, 14/9, 5/3, 2

CHRO29-7; Tonos-29 Chromatic;
29/28, 29/27, 29/22, 29/20, 29/19, 29/18, 2

CHRO29/CON)-7; Conjunct Tonos-29 Chromatic;
29/28, 29/27, 29/22, 29/21, 29/20, 29/16, 2

CHRO33-7; Tonos-33 Chromatic. A variant is 66-63-60-48;
33/31, 33/29, 11/8, 3/2, 11/7, 33/20, 2

CHRO33/CON/)-7; Conjunct Tonos-33 Chromatic;
33/31, 33/29, 11/8, 33/23, 3/2, 11/6, 2

DIAPHONIC-7; 7-tone Diaphonic Cycle, disjunctive form on 4/3 and 3/2;
12/11, 6/5, 4/3, 16/11, 8/5, 16/9, 2

DIAT13-7; This genus is from K.S's diatonic Hypodorian harmonia;
16/15, 16/13, 4/3, 3/2, 8/5, 24/13, 2

DIM/TET(A)-7; A heptatonic form on the 9/7;
27/25, 27/23, 9/7, 14/9, 42/25, 42/23, 2

DORIAN(CHRO/NO.2)-7; Schlesinger's Dorian Harmonia, the chromatic genus;
22/21, 11/10, 11/8, 11/7, 44/27, 22/13, 2

DORIAN(CHRO/INV)-7; A harmonic form of Schlesinger's Chromatic Dorian inverted;
24/23, 12/11, 14/11, 16/11, 17/11, 18/11, 2

DORIAN(DIAT/CON)-7; A Dorian Diatonic with its own trite synemmenon replacing paramese;
11/10, 11/9, 11/8, 22/15, 11/7, 11/6, 2

DORIAN(ENH/NO.2)-7; Schlesinger's Dorian Harmonia, the enharmonic genus;
44/43, 22/21, 11/8, 11/7, 44/27, 22/13, 2

DORIAN(ENH/INV)-7; A harmonic form of Schlesinger's Dorian enharmonic inverted;
48/47, 24/23, 14/11, 16/11, 3/2, 17/11, 2

DORIAN(PENTA)-7; Schlesinger's Dorian Harmonia, the pentachromatic genus;
55/53, 11/10, 11/8, 11/7, 55/34, 22/13, 2

DORIAN(TRI/CHROM/NO.1)-7;
Schlesinger's Dorian Harmonia in the first trichromatic genus;
33/32, 33/31, 11/8, 11/7, 66/41, 33/20, 2

DORIAN(TRI/CHROM/NO.2)-7; Schlesinger's Dorian Harmonia in the second trichromatic genus;
33/32, 11/10, 11/8, 11/7, 66/41, 22/13, 2

DUDON(DIAT)-7; Dudon Neutral Diatonic, Obtained by taking mediants of each term of the major and minor modes;
9/8, 27/22, 59/44, 3/2, 18/11, 81/44, 2

ENH15-7; Tonos-15 Enharmonic;
30/29, 15/14, 15/11, 3/2, 20/13, 30/19, 2

ENH15(INV/NO.1)-7; Inverted Enharmonic Tonos-15 Harmonia;
19/15, 13/10, 4/3, 22/15, 28/15, 29/15, 2

ENH15(INV/NO.2)-7; Inverted harmonic form of the enharmonic Tonos-15;
31/30, 16/15, 4/3, 22/15, 3/2, 23/15, 2

ENH17-7; Tonos-17 Enharmonic;
34/33, 17/16, 17/12, 17/11, 68/43, 34/21, 2

ENH17(CON)-7; Conjunct Tonos-17 Enharmonic;
34/33, 17/16, 17/12, 68/47, 34/23, 17/9, 2

ENH19-7; Tonos-19 Enharmonic;
38/37, 19/18, 19/14, 19/13, 76/51, 38/25, 2

ENH19(CON)-7; Conjunct Tonos-19 Enharmonic;
38/37, 19/18, 19/14, 76/55, 38/27, 19/11, 2

ENH2-7; 1:2 Enharmonic. New genus 2 + 4 + 24 parts;
100/3, 100.000, 500, 700, 2 200/3, 800.000, 1 200

ENH21-7; Tonos-21 Enharmonic;
42/41, 21/20, 21/16, 3/2, 84/55, 14/9, 2

ENH21(INV/NO.1)-7; Inverted Enharmonic Tonos-21 Harmonia;
9/7, 55/42, 4/3, 32/21, 40/21, 41/21, 2

ENH21(INV/NO.2)-7; Inverted harmonic form of the enharmonic Tonos-21;
32/31, 16/15, 4/3, 32/21, 11/7, 34/21, 2

ENH23-7; Tonos-23 Enharmonic;
46/45, 23/22, 23/18, 23/16, 46/31, 23/15, 2

ENH23(CON)-7; Conjunct Tonos-23 Enharmonic;
46/45, 23/22, 23/18, 46/35, 23/17, 23/13, 2

ENH25-7; Tonos-25 Enharmonic;
100/97, 50/47, 25/18, 25/16, 50/31, 5/3, 2

ENH25(CON)-7; Conjunct Tonos-25 Enharmonic;
100/97, 50/47, 25/18, 10/7, 25/17, 25/13, 2

ENH27-7; Tonos-27 Enharmonic;
36/35, 18/17, 27/20, 3/2, 54/35, 27/17, 2

ENH27(INV/NO.1)-7; Inverted Enharmonic Tonos-27 Harmonia;
34/27, 35/27, 4/3, 40/27, 17/9, 35/18, 2

ENH27(INV/NO.2)-7; Inverted harmonic form of the enharmonic Tonos-27;
56/55, 28/27, 4/3, 40/27, 41/27, 14/9, 2

ENH29-7; Tonos-29 Enharmonic;
58/57, 29/28, 29/22, 29/20, 58/39, 29/19, 2

ENH29(CON)-7; Conjunct Tonos-29 Enharmonic;
58/57, 29/28, 29/22, 58/43, 29/21, 29/16, 2

ENH33-7; Tonos-33 Enharmonic;
33/32, 33/31, 11/8, 3/2, 66/43, 11/7, 2

ENH33(CON)-7; Conjunct Tonos-33 Enharmonic;
33/32, 33/31, 11/8, 66/47, 33/23, 11/6, 2

ENH/INV/CON)-7; Inverted Enharmonic Conjunct Phrygian Harmonia;
13/12, 17/12, 35/24, 3/2, 23/12, 47/24, 2

ENHMOD-7; Enharmonic After Wilson's Purvi Modulations, See page 111;
9/8, 7/6, 4/3, 3/2, 14/9, 8/5, 2

EPIMORE/ENH)-7; New Epimoric Enharmonic, Dorian mode of the 4th new Enharmonic on Hofmann's list;
76/75, 16/15, 4/3, 3/2, 38/25, 8/5, 2

FARABI-7; Al Farabi's Chromatic cca 700 AD;
9/8, 27/20, 729/512, 3/2, 9/5, 19/10, 2

FARABI.BLUE-7; Another tuning from Al Farabi, ccca 700 AD;
9/8, 45/32, 131/90, 3/2, 15/8, 31/16, 2

FARNSWORTH-7; Farnsworth's scale;
9/8, 5/4, 21/16, 3/2, 27/16, 15/8, 2

HARM/DOREN/INV/NO.1-7; 1st Inverted Schlesinger's Enharmonic Dorian Harmonia;
27/22, 5/4, 14/11, 16/11, 21/11, 43/22, 2

HARM/DOR/INV/NO.1-7; 1st Inverted Schlesinger's Chromatic Dorian Harmonia;
13/11, 27/22, 14/11, 16/11, 20/11, 21/11, 2

HARM/LYD/ENH/INV/NO.1-7;
1st Inverted Schlesinger's Enharmonic Lydian Harmonia;
17/13, 35/26, 18/13, 20/13, 25/13, 51/26, 2

HARM/MIXOl/CHROM/INV/NO.1-7;
1st Inverted Schlesinger's Chromatic Mixolydian Harmonia;
9/7, 19/14, 10/7, 11/7, 13/7, 27/14, 2

HARM/MIXOl/ENH/INV/NO.1-7;
1st Inverted Schlesinger's Enharmonic Mixolydian Harmonia;
19/14, 39/28, 10/7, 11/7, 27/14, 55/28, 2

HARM1C15-7; Harm1C-15-Harmonia;
6/5, 19/15, 4/3, 22/15, 26/15, 28/15, 2

HARM1C(DORIAN)-7; Harm1C-Dorian;
3/11, 27/22, 14/11, 16/11, 20/11, 21/11, 2

HARM1C/MIX-7; Harm1C-Con Mixolydian;
8/7, 10/7, 3/2, 11/7, 13/7, 27/14, 2

HARM1C/MIXOLYDIAN-7; Harm1C-Mixolydian;
15/14, 8/7, 10/7, 11/7, 23/14, 12/7, 2

HARM1C(CON/PHRYGIAN)-7; Harm1C-ConPhryg;
13/12, 4/3, 17/12, 3/2, 11/6, 23/12, 2

HARM1C(PHRYGIAN)-7; Harm1C-Phrygian;
7/6, 5/4, 4/3, 3/2, 11/6, 23/12, 2

HARM1E15-7; Harm1E-15-Harmonia;
19/15, 13/10, 4/3, 22/15, 28/15, 29/15, 2

HARM4-7; Fourth octave of the harmonic overtone series;
9/8, 5/4, 11/8, 3/2, 13/8, 7/4, 15/8

HARM(BASTARD)-7; 6)7 7-any from 1.3.5.7.9.11.13 and inversion of "Bastard" Hypodorian Harmonia;
9/8, 5/4, 11/8, 3/2, 13/8, 7/4, 2

HARMD15)-7; HarmD-15-Harmonia;
16/15, 6/5, 4/3, 22/15, 8/5, 26/15, 2

HARMD/CONMIX-7; HarmD-ConMixolydian;
8/7, 9/7, 3/2, 11/7, 12/7, 13/7, 2

HARMD/MIX-7; HarmD-Mixolydian;
8/7, 9/7, 10/7, 11/7, 12/7, 13/7, 2

HEM/CHRO-7; Hemiolic Chromatic genus has the strong or 1:2 division of the 12/11 pyknon;
34/33, 12/11, 4/3, 3/2, 17/11, 18/11, 2

HEM/CHRO11-7;
11'al Hemiolic Chromatic genus with a CI of 11/9, Winnington-Ingram;
24/23, 12/11, 4/3, 3/2, 36/23, 18/11, 2

HEM/CHRO13-7; 13'al Hemiolic Chromatic or neutral-third genus has a CI of 16/13;
26/25, 13/12, 4/3, 3/2, 39/25, 13/8, 2

HHIDJAZI-7; Medieval Arabic scale;
65 536/59 049, 32/27, 4/3, 262 144/177 147, 32 768/19 683, 16/9, 2

HHOSAINI-7; Medieval Arabic scale;
65 536/59 049, 32/27, 4/3, 262 144/177 147, 27/16, 16/9, 2

HIGGS-7; From Greg Higgs announcement of the formation of an Internet Tuning list;
3/2, 8/5, 21/13, 34/21, 13/8, 5/3, 2

HYPOD(CHRO/NO.2)-7; Schlesinger's Chromatic Hypodorian Harmonia;
16/15, 8/7, 4/3, 16/11, 32/21, 8/5, 2

HYPOD(CHROM/OENH)-7;
Schlesinger's Hypodorian Harmonia in a mixed chromatic-enharmonic genus;
32/31, 16/15, 4/3, 16/11, 32/21, 8/5, 2

HYPOD(CHROM/INV)-7; A harmonic form of Schlesinger's Chromatic Hypodorian Inverted;
17/16, 9/8, 11/8, 3/2, 25/16, 13/8, 2

HYPOD(DIAT/CON)-7; A Hypodorian Diatonic with its own trite synemmenon replacing paramese;
16/15, 16/13, 4/3, 32/23, 8/5, 16/9, 2

HYPOD(ENH/INV)-7; Inverted Schlesinger's Enharmonic Hypodorian Harmonia;
21/16, 43/32, 11/8, 3/2, 15/8, 31/16, 2

HYPOD(ENH/INV/NO.2)-7; A harmonic form of Schlesinger's Hypodorian enharmonic inverted;
33/32, 17/16, 11/8, 3/2, 49/32, 25/16, 2

HYPOD(INV)-7; Inverted Schlesinger's Chromatic Hypodorian Harmonia;
5/4, 21/16, 11/8, 3/2, 7/4, 15/8, 2

HYPOL(CHROM/INV/NO.2)-7; Harmonic form of Schlesinger's Chromatic Hypolydian inverted;
21/20, 11/10, 13/10, 7/5, 3/2, 8/5, 2

HYPOL(CHROM/INV/NO.3)-7; A harmonic form of Schlesinger's Chromatic Hypolydian inverted;
21/20, 11/10, 13/10, 3/2, 8/5, 17/10, 2

HYPOL(DIAT/CON)-7; A Hypolydian Diatonic with its own trite synemmenon replacing paramese;
10/9, 5/4, 4/3, 20/13, 5/3, 20/11, 2

HYPOL(ENH/INV/NO.2)-7; A harmonic form of Schlesinger's Hypolydian enharmonic inverted;
41/40, 21/20, 13/10, 7/5, 29/20, 3/2, 2

HYPOL(ENH/INV/NO.3)-7; A harmonic form of Schlesinger's Hypolydian enharmonic inverted;
41/40, 21/20, 13/10, 3/2, 31/20, 8/5, 2

HYPOP(CHROM/ENH)-7;
Schlesinger's Hypophrygian Harmonia in a mixed chromatic-enharmonic genus;
36/35, 18/17, 18/13, 3/2, 36/23, 18/11, 2

HYPOP(CHROM/INV)-7; Inverted Schlesinger's Chromatic Hypophrygian Harmonia;
11/9, 23/18, 4/3, 13/9, 16/9, 17/9, 2

HYPOP(CHROM/INV/NO.2)-7;
A harmonic form of Schlesinger's Chromatic Hypophrygian inverted;
19/18, 10/9, 4/3, 13/9, 14/9, 5/3, 2

HYPOP(DIAT/CON)-7; A Hypophrygian Diatonic with its own trite synemmenon replacing paramese;
9/8, 6/5, 18/13, 36/25, 18/11, 9/5, 2

HYPOP(ENH/INV)-7; Inverted Schlesinger's Enharmonic Hypophrygian Harmonia;
23/18, 47/36, 4/3, 13/9, 17/9, 35/18, 2

HYPOP(ENH/INV/NO.2)-7; A harmonic form of Schlesinger's Hypophrygian enharmonic inverted;
37/36, 19/18, 4/3, 13/9, 3/2, 14/9, 2

IONIC-7; Ancient greek Ionic;
9/8, 5/4, 4/3, 3/2, 5/3, 9/5, 2

ISLAM-7; Medieval Islamic scale of Zalzal;
9/8, 81/64, 4/3, 40/27, 130/81, 16/9, 2

RING(K/NO.2)-7; Double-tie circular mirroring of 6:7:8;
8/7, 7/6, 4/3, 3/2, 12/7, 7/4, 2

RING(K/NO.3)-7; Double-tie circular mirroring of 3:5:7;
7/6, 6/5, 7/5, 10/7, 5/3, 12/7, 2

RING(K/NO.4)-7; Double-tie circular mirroring of 4:5:7;
8/7, 5/4, 7/5, 10/7, 8/5, 7/4, 2

RING(K/NO.5)-7; Double-tie circular mirroring of 5:6:7;
7/6, 6/5, 7/5, 10/7, 5/3, 12/7, 2

RING(K/NO.6)-7; Double-tie circular mirroring of 6:7:9;
7/6, 9/7, 4/3, 3/2, 14/9, 12/7, 2

RING(K/NO.7)-7; Double-tie circular mirroring of 5:7:9;
10/9, 9/7, 7/5, 10/7, 14/9, 9/5, 2

HARRISON(MID)-7; L. Harrison mid mode;
9/8, 6/5, 4/3, 3/2, 5/3, 7/4, 2

HARRISON(MID/NO.2)-7; L. Harrison mid mode 2;
9/8, 6/5, 4/3, 3/2, 12/7, 9/5, 2

LIU-MIN-7; Linus Liu's Harmonic Minor;
10/9, 6/5, 4/3, 40/27, 8/5, 50/27, 2

LIU-PENT-7; Linus Liu's "pentatonic scale";
9/8, 81/64, 27/20, 3/2, 27/16, 243/128, 81/40

LUTE-7; Scale on the "Scholar's Lute";
8/7, 6/5, 5/4, 4/3, 3/2, 5/3, 2

LYDIAN(CHRO/NO.2)-7; Schlesinger's Lydian Harmonia, the chromatic genus;
26/25, 13/12, 13/10, 13/9, 26/17, 13/8, 2

LYDIAN(CHRO/INV)-7; A harmonic form of Schlesinger's Chromatic Lydian inverted;
27/26, 14/13, 18/13, 20/13, 21/13, 22/13, 2

LYDIAN(DIAT/CON)-7; A Lydian Diatonic with its own trite synemmenon replacing paramese;
13/12, 13/11, 13/10, 26/19, 13/8, 13/7, 2

LYDIAN(ENH/NO.2)-7; Schlesinger's Lydian Harmonia, the enharmonic genus;
52/51, 26/25, 13/10, 13/9, 52/35, 26/17, 2

LYDIAN(ENH/INV)-7; A harmonic form of Schlesinger's Enharmonic Lydian inverted;
53/52, 27/26, 18/13, 20/13, 41/26, 21/13, 2

LYDIAN(PENTA)-7; Schlesinger's Lydian Harmonia, the pentachromatic genus;
65/63, 13/12, 13/10, 13/9, 65/43, 13/8, 2

LYDIAN(TRI/NO.1)-7; Schlesinger's Lydian Harmonia, the first trichromatic genus;
39/38, 39/37, 13/10, 13/9, 3/2, 39/25, 2

LYDIAN(TRI/NO.2)-7;
Schlesinger's Lydian Harmonia in the second trichromatic genus;
39/38, 13/12, 13/10, 13/9, 3/2, 13/8, 2

MALAKA-7; Malaka, lyra tuning: soft or intense chromatic and tonic diatonic;
28/27, 10/9, 4/3, 3/2, 14/9, 16/9, 2

MALCOLM(MID-EAST)-7; Malcolm's Mid-East;
9/8, 5/4, 11/8, 3/2, 7/4, 15/8, 2

METABOLIKA-7; Metabolika, lyra tuning: soft diatonic and tonic diatonic;
21/20, 7/6, 4/3, 3/2, 14/9, 16/9, 2

ARCHY(MID/ENH/NO.2)-7;
Permutation of Archytas' Enharmonic with the 5/4 medially and 28/27 first;
28/27, 35/27, 4/3, 3/2, 14/9, 35/18, 2

MIXOL(/CHRO/NO.2)-7;
Schlesinger's Mixolydian Harmonia in the chromatic genus;
28/27, 14/13, 14/11, 7/5, 28/19, 14/9, 2

MIXOL(CHRO/INV)-7;
A harmonic form of Schlesinger's Chromatic Mixolydian inverted;
16/15, 8/7, 10/7, 11/7, 23/14, 12/7, 2

MIXOL(DIATCON)-7; A Mixolydian Diatonic with its own trite synemmenon
replacing paramese;
14/13, 7/6, 14/11, 3/2, 14/9, 7/4, 2

MIXOL(DIAT/INV)-7; A Mixolydian Diatonic with its own trite synemmenon
replacing paramese;
8/7, 9/7, 4/3, 11/7, 12/7 13/7, 2

MIXOL(ENH/NO.2)-7; Schlesinger's Mixolydian Harmonia in the enharmonic
genus;
56/55, 28/27, 14/11, 7/5, 56/39, 28/19, 2

MIXOL(ENH/INV)-7; A harmonic form of Schlesinger's Mixolydian inverted;
31/30, 16/15, 10/7, 11/7, 45/28, 23/14, 2

MIXOL(PENTA)-7; Schlesinger's Mixolydian Harmonia in the pentachromatic
genus;
35/34, 14/13, 14/11, 7/5, 35/24, 14/9, 2

MIXOL(TRI/NO.1)-7; Schlesinger's Mixolydian Harmonia in the first trichromatic genus;
42/41, 21/20, 14/11, 7/5, 42/29, 3/2, 2

MIXOL(TRI/NO.2)-7; Schlesinger's Mixolydian Harmonia in the second trichromatic genus;
42/41, 14/13, 14/11, 7/5, 42/29, 14/9, 2

SCHMIDT(PEL)-7; Modern Pelog designed by D. Schmidt and used by Berkeley
Gamelan;
11/10, 6/5, 7/5, 3/2, 8/5 9/5, 2

PHRYG(CHROM/CON)-7; Inverted Conjunct Chromatic Phrygian;
13/12, 4/3, 17/12, 3/2, 11/6, 23/12, 2

PHRYG(CHROM/CON/NO.2)-7; Harmonic Conjunct Chromatic Phrygian;
13/12, 9/8, 7/6, 3/2, 19/12 5/3, 2

PHRYG(CHROM/INV)-7; Inverted Schlesinger's Chromatic Phrygian;
25/24, 13/12, 4/3, 3/2, 19/12, 5/3, 2

PHRYG(DIAT/CON)-7; A Phrygian Diatonic with its own trite synemmenon
replacing paramese;
12/11, 6/5, 4/3, 24/17, 12/7, 24/13, 2

PHRYG(ENH)-7; Schlesinger's Phrygian Harmonia in the enharmonic genus;
48/47, 24/23, 4/3, 3/2, 48/31, 8/5, 2

PHRYG(ENH/CON)-7; Harmonic Conjunct Enharmonic Phrygian;
13/12, 53/48, 9/8, 3/2, 37/24, 19/12, 2

PHRYG(ENH/INV)-7; Inverted Schlesinger's Enharmonic Phrygian Harmonia;
5/4, 31/24, 4/3, 3/2, 23/12, 47/24, 2

PHRYG(ENH/INV/NO.2)-7;
Inverted harmonic form of Schlesinger's Enharmonic Phrygian;
49/48, 25/24, 4/3, 3/2, 37/24, 19/12, 2

PHRYG(INV)-7; Inverted Schlesinger's Chromatic Phrygian Harmonia;
7/6, 5/4 4/3, 3/2, 11/6, 23/12, 2

PHRYG(INV/CON)-7; Inverted Conjunct Phrygian Harmonia with 17, the local Trite Synemmenon;
13/12, 7/6, 17/12, 3/2, 5/3, 11/6, 2

PHRYG(PENTA)-7;
Schlesinger's Phrygian Harmonia, the pentachromatic genus;
30/29, 12/11, 4/3, 3/2, 30/19, 12/7, 2

PHRYG(TRI/NO.1)-7; Schlesinger's Phrygian Harmonia, the chromatic genus;
24/23, 12/11, 4/3, 3/2, 8/5, 12/7, 2

PHRYG(TRI/NO.2)-7; Schlesinger's Phrygian Harmonia, the second trichromatic genus;
36/35, 12/11, 4/3, 3/2, 36/23, 12/7, 2

PHRYG(TRI/NO.3)-7;
Schlesinger's Phrygian Harmonia, the first trichromatic genus;
36/35, 18/17, 4/3, 3/2, 36/23, 18/11, 2

POOLE-7; Poole's double diatonic or dichordal scale;
9/8, 5/4, 4/3, 3/2, 5/3, 7/4, 2

PORTUGUESE(BAG/NO.1)-7; Portugese bagpipe tuning;
14/13, 81/68, 32/25, 36/25, 128/81, 7/4, 2

PS/DORIAN-7; Complex 4 of p. 115 based on Archytas's Enharmonic;
28/27, 16/15, 4/3, 3/2, 15/8, 27/14, 2

PS/ENH)-7; Dorian mode of an Enharmonic genus found in Ptolemy's Harmonics;
56/55, 16/15, 4/3, 3/2, 84/55, 8/5, 2

PS/HYPOD-7; Complex 7 of p. 115 based on Archytas's Enharmonic;
9/8, 45/32, 81/56, 3/2, 14/9, 8/5, 2

PS/HYPOD/NO.1-7; Complex 7 of p. 115 based on Archytas's Enharmonic;
9/8, 45/32, 81/56, 3/2, 14/9, 8/5, 2

PS/HYPOD/NO.2-7; Complex 8 of p. 115 based on Archytas's Enharmonic;
9/8, 7/6, 6/5, 3/2, 15/8, 27/14, 2

PS/MIXOL-7; Complex 3 of p. 115 based on Archytas's Enharmonic;
28/27, 16/15, 4/3, 5/3, 12/7, 16/9, 2

PTO-7; Intense Diatonic Systonon;
9/8, 5/4, 4/3, 3/2, 5/3, 15/8, 2

*PTO(IASTIA)-7; Ptolemy's Iastia or Lydia tuning, mixture of Tonic Diatonic &
Intense Diatonic;*
28/27, 32/27, 4/3, 3/2, 8/5, 9/5, 2

*PTO(MALAKA)-7;
Ptolemy's Malaka lyra tuning, a mixture of Intense Chrom. & Tonic Diatonic;*
22/21, 8/7, 4/3, 3/2, 14/9, 16/9, 2

*PTO(MALAKA/NO.2)-7; Malaka lyra, mixture of his Soft Chromatic and Tonic
Diatonic;*
28/27, 10/9, 4/3, 3/2, 14/9, 16/9, 2

*PTO(METABOLIKA)-7; Metabolika lyra tuning, mixture of Soft Diatonic &
Tonic Diatonic;*
21/20, 7/6, 4/3, 3/2, 14/9, 16/9, 2

*RAMEAU(MIN)-9; Rameau's minor diatonic system on E (asc. 4-6-8-9, desc.
9-7-5-4);*
9/8, 6/5, 27/20, 3/2, 8/5, 27/16, 9/5, 15/8, 2

RAST-7; Medieval arabic scale;
9/8, 8 192/6 561, 4/3, 3/2, 32 768/19 683, 16/9, 2

*RAT(DOR/ENH)-7; Rationalized Schlesinger's Dorian Harmonia in the enhar-
monic genus;*
44/43, 22/21, 11/8, 11/7, 8/5, 44/27, 2

*RAT(HYPOD/ENH)-7; 1+1 rationalized enharmonic genus derived from K.S.'s
'Bastard' Hypodorian;*
32/31, 16/15, 4/3, 16/11, 64/43, 32/21, 2

*RAT(HYPOD/ENH/NO.2)-7; 1+2 rationalized enharmonic genus derived from
K.S.'s 'Bastard' Hypodorian;*
32/31, 32/29, 4/3, 16/11, 64/43, 64/41, 2

*RAT(HYPOD/ENH/NO.3)-7; 1+3 Rationalized enharmonic genus derived from
K.S.'s 'Bastard' Hypodorian;*
32/31, 8/7, 4/3, 16/11, 64/43, 8/5, 2

*RAT(HYPOD/HEX)-7; 1+1 Rationalized hexachromatic/hexenharmonic genus
derived from K. S. 'Bastard';*
48/47, 24/23, 4/3, 16/11, 96/65, 3/2, 2

*RAT(HYPOD/HEX/NO.2)-7; 1+2 RAT. hexachromatic/hexenharmonic genus
derived from K.S.'s 'Bastard' Hypodorian;*
48/47, 16/15, 4/3, 16/11, 96/65, 32/21, 2

*RAT(HYPOD/HEX/NO.3)-7; 1+3 RAT. hexachromatic/hexenharmonic genus
from K.S.'s 'Bastard' Hypodorian;*
48/47, 12/11, 4/3, 16/11, 96/65, 48/31, 2

*RAT(HYPOD/HEX/NO.4)-7; 1+4 RAT. hexachromatic/hexenharmonic genus
from K.S.'s 'Bastard' Hypodorian;*
48/47, 48/43, 4/3, 16/11, 96/65, 96/61, 2

RAT(HYPOD/HEX/NO.5)-7; 1+5 RAT. hexachromatic/hexenharmonic genus from K.S.'s 'Bastard' Hypodorian;
48/47, 8/7, 4/3, 16/11, 96/65, 8/5, 2

RAT(HYPOD/HEX/NO.6)-7; 2+3 Rationalized hexachromatic/hexenharmonic genus from K.S.'s 'Bastard' Hypodorian;
24/23, 48/43, 4/3, 16/11 3/2, 96/61, 2

RAT(HYPOD/PENTA)-7; 1+1 Rationalized pentachromatic/pentenharmonic genus derived from K.S.'s 'Bastard' Hypodorian;
40/39, 20/19, 4/3, 16/11, 40/27, 80/53, 2

RAT(HYPOD/PENTA/NO.2)-7; 1+2 Rationalized pentachromatic/pentenharmonic genus from K.S.'s 'Bastard' Hypodorian;
40/39, 40/37, 4/3, 16/11 40/27, 20/13, 2

RAT(HYPOD/PENTA/NO.3)-7; 1+3 Rationalized pentachromatic/pentenharmonic genus from from K.S.'s 'Bastard' Hypodorian;
40/39, 10/9, 4/3, 16/11 40/27, 80/51, 2

RAT(HYPOD/PENTA/NO.4)-7; 1+4 Rationalized pentachromatic/pentenharmonic genus from 'Bastard' Hypodorian;
40/39, 8/7, 4/3, 16/11, 40/27, 8/5, 2

RAT(HYPOD/PENTA/NO.5)-7; 2+3 Rationalized pentachromatic/pentenharmonic genus from 'Bastard' Hypodorian;
20/19, 10/9, 4/3, 16/11, 80/53, 80/51 2/1

RAT(HYPOD/PENTA/NO.6)-7; 2+3 Rationalized pentachromatic/pentenharmonic genus from 'Bastard' Hypodorian;
40/39, 8/7, 4/3, 16/11, 80/53, 8/5, 2

RAT(HYPODTRI)-7; Rationalized first (1+1) trichromatic genus derived from K.S.'s 'Bastard' hyp;
24/23, 12/11, 4/3, 16/11, 3/2, 48/31, 2

RAT(HYPOD/TRI/NO.2)-7; Rationalized second (1+2) trichromatic genus derived from K.S.'s 'Bastard' Hypodorian;
24/23, 8/7, 4/3, 16/11, 3/2, 8/5, 2

RAT(HYPOP/CHROM)-7; Rationalized Schlesinger's Hypophrygian Harmonia in the chromatic genus;
18/17, 9/8, 18/13, 3/2, 36/23, 18/11, 2

RAT(HYPOP/ENH)-7; Rationalized Schlesinger's Hypophrygian Harmonia in the enharmonic genus;
36/35, 18/17, 18/13, 3/2, 72/47, 36/23, 2

RAT(HYPOP/PENTA)-7; Rationalized Schlesinger's Hypophrygian Harmonia in the pentachromatic genus;
45/43, 9/8, 18/13, 3/2, 45/29, 18/11, 2

RAT(HYPOP/TRI)-7; Rationalized Schlesinger's Hypophrygian Harmonia in the first trichromatic genus;
27/26, 27/25, 18/13, 3/2, 54/35, 27/17, 2

RAT(HYPOP/TRI/NO.2)-7; Rationalized Schlesinger's Hypophrygian Harmonia in second trichromatic genus;
27/26, 9/8, 18/13, 3/2, 54/35, 18/11, 2

SCOTTISH(BAG/NO.1)-7; Scottish bagpipe tuning;
10/9, 5/4, 15/11, 40/27, 5/3, 11/6, 2

SCOTTISH(BAG/NO.2)-7; Scottish bagpipe tuning 2;
10/9, 11/9, 4/3, 3/2, 18/11, 9/5, 2

SCOTTISH(BAG/NO.3)-7; Scottish bagpipe tuning 3;
9/8, 5/4, 11/8, 3/2, 27/16, 11/6, 2

SCOTTISH(BAG/NO.4)-7; Scottish Bagpipe Ellis/Land;
197, 341, 495, 703, 853, 1009, 1 200

SERRE(ENH)-7; Dorian mode of the Serre's Enharmonic;
64/63, 16/15, 4/3, 3/2, 32/21, 8/5, 2

SUB/HARM/1C/CON/M-7; Subharm1C-ConMixolydian;
7/6, 28/23, 14/11, 14/9, 28/17, 7/4, 2

SUB/HARM/1C/CON/P-7; Subharm1C-ConPhryg;
6/5, 24/19, 4/3, 8/5, 12/7, 24/13, 2

SUB/HARM/1C/MIX-7; Subharm1C-Mixolydian;
7/6, 28/23, 14/11, 7/5, 7/4, 28/15, 2

SUB/HARM/1C/PHRYG-7; Subharm1C-Phrygian;
6/5, 24/19, 4/3, 3/2, 24/13, 48/25, 2

SUB/HARM/1E/CON/M-7; Subharm1E-ConMixolydian;
28/23, 56/45, 14/11, 28/17, 56/33, 7/4, 2

SUB/HARM/1E/CON/P-7; Subharm1E-ConPhrygian;
24/19, 48/37, 4/3, 12/7, 16/9, 24/13, 2

SUB/HARM/1E/MIX-7; Subharm1E-Mixolydian;
28/23, 56/45, 14/11, 7/5, 28/15, 56/29, 2

SUB/HARM/1E/PHRYG-7; Subharm1E-Phrygian;
24/19, 48/37, 4/3, 3/2, 48/25, 96/49, 2

SUB/HARM2C/15-7; Subharm2C-15-Harmonia;
5/4, 30/23, 15/11, 3/2, 30/17, 15/8, 2

SUB/HARM2E/15-7; Subharm2E-15-Harmonia;
30/23, 4/3, 15/11, 3/2, 15/8, 60/31, 2

SUPER-7; Most equal superparticular 7-tone scale;
9/8, 5/4, 11/8, 3/2, 5/3, 11/6, 2

TRISUB1-7; Sub-(6-7-8) Tritriadic;
8/7, 4/3, 3/2, 32/21, 12/7, 16/9, 2

TRITRIAD-7; Tritriadic scale of the 10:12:15 triad, natural minor mode;
9/8, 6/5, 4/3, 3/2, 8/5, 9/5, 2

TRITRIAD10-7; Tritriadic scale of the 10:14:15 triad;
21/20, 9/8, 4/3, 7/5, 3/2, 28/15, 2

TRITRIAD11-7; Tritriadic scale of the 11:13:15 triad;
13/11, 15/11, 22/15, 195/121, 26/15, 225/121, 2

TRITRIAD13-7; Tritriadic scale of the 10:13:15 triad;
9/8, 13/10, 4/3, 3/2, 26/15, 39/20, 2

TRITRIAD14-7; 14.18.21 Tritriadic. Primary triads 1/1, 9/7, 3/2, secondary are 1/1, 7/6, 3/2;
9/8, 9/7, 4/3, 3/2, 12/7, 27/14, 2

TRITRIAD18-7; Tritriadic scale of the 18:22:27 triad;
9/8, 11/9, 4/3, 3/2, 44/27, 11/6, 2

TRITRIAD22-7; Tritriadic scale of the 22:27:33 triad;
9/8, 27/22, 4/3, 3/2, 18/11, 81/44, 2

TRITRIAD26-7; Tritriadic scale of the 26:30:39 triad;
9/8, 15/13, 4/3, 3/2, 20/13, 45/26, 2

TRITRIAD32-7; Tritriadic scale of the 26:32:39 triad;
9/8, 16/13, 4/3, 3/2, 64/39, 24/13, 2

TRITRIAD6-7; Tritriadic scale of the 6:7:9 triad;
9/8, 7/6, 4/3, 3/2, 14/9, 7/4, 2

TRITRIAD64-7; Tritriadic scale of the 64:81:96 triad;
9/8, 81/64, 4/3, 3/2, 27/16, 243/128, 2

TRITRIAD7-7; Tritriadic scale of the 7:9:11 triad;
99/98, 121/98, 14/11, 9/7, 11/7, 18/11, 2

TRITRIAD9-7; Tritriadic scale of the 9:11:13 triad;
169/162, 11/9, 18/13, 13/9, 22/13, 143/81, 2

VONG-7; Vong Co Dan Tranh scale;
11/10, 36/29, 27/20, 3/2, 33/20, 51/28, 2

ZALZAL(NO.1)-7; Modern Arabic scale;
9/8, 27/22, 4/3, 3/2, 18/11, 16/9, 2

ZALZAL(NO.2)-7; Zalzal's Scale, a medieval Islamic with Ditone Diatonic & 10/9 x 13/12 x 72/65;
9/8, 81/64, 4/3, 40/27, 130/81, 16/9, 2

ZENKOULEH-7; Medieval Arabic scale;
9/8, 8 192/6 561, 4/3, 262 144/177 147, 32 768/19 683, 16/9, 2

A.4   6, 8, 9, 10, 11 tones per octave

AUG/TET(J)-6; 9/8 C.I. comprised of 11:10:9:8 subharmonic series on 1 and 8:9:10:11 on 16/11;
11/10, 11/9, 11/8, 16/11, 18/11, 20/11

AUG/TET(K)-6; 9/8 C.I. This is the converse form of Aug/TET(J);
9/8, 5/4, 11/8, 16/11, 8/5, 16/9

AUG/TET(L)-6; 9/8 C.I. This is the harmonic form of Aug/TET(I);
9/8, 5/4, 11/8, 16/11, 18/11, 20/11

CHRO(STRONG)-6; Strong 32/27 chromatic;
14/13, 16/13, 4/3, 56/39, 3/2, 2

CLUSTER(6/A)-6; Six-tone triadic cluster 4:5:6;
5/4, 4/3, 3/2, 5/3, 15/8, 2

CLUSTER(6/B)-6; Six-tone triadic cluster 4:6:5;
6/5, 5/4, 3/2, 8/5, 15/8, 2

CLUSTER(6/C)-6; Six-tone triadic cluster 3:4:5;
10/9, 6/5, 4/3, 8/5, 5/3, 2

CLUSTER(6/D)-6; Six-tone triadic cluster 3:5:4;
10/9, 5/4, 4/3, 3/2, 5/3, 2

CLUSTER(6/E)-6; Six-tone triadic cluster 5:6:8;
6/5, 5/4, 3/2, 8/5, 48/25, 2

CLUSTER(6/F)-6; Six-tone triadic cluster 5:8:6;
6/5, 4/3, 8/5, 5/3, 48/25, 2

CLUSTER(6/G)-6; Six-tone triadic cluster 4:5:7;
35/32, 8/7, 5/4, 10/7, 7/4, 2

CLUSTER(6/H)-6; Six-tone triadic cluster 4:7:5;
35/32, 5/4, 7/5, 8/5, 7/4, 2

CLUSTER(6/I)-6; Six-tone triadic cluster 5:6:7;
6/5, 7/5, 10/7, 42/25, 12/7, 2

CLUSTER(6/J)-6; Six-tone triadic cluster 5:7:6;
7/6, 6/5, 7/5, 5/3, 42/25, 2

HEXANY1-6; Two out of 1 3 5 7 hexany;
35/32, 5/4, 21/16, 3/2, 7/4, 15/8

HEXANY49-6; 1.3.21.49 2)4 hexany (1.21 tonic);
8/7, 7/6, 3/2, 49/32, 7/4, 2

HEXANY(TETR)-6; Complex 12 of p. 115, a hexany based on Archytas's enharmonic;
36/35, 16/15, 9/7, 4/3, 48/35, 2

HEXANY(TRANS)-6; Complex 1 of p. 115, a hexany based on Archytas's enharmonic;
28/27, 16/15, 35/27, 4/3, 112/81, 2

HEXANY(TRANS/NO.2)-6; Complex 2 of p. 115, a hexany based on Archytas's enharmonic;
28/27, 16/15, 4/3, 48/35, 64/45, 2

HEXANY(TRANS/NO.3)-6; Complex 9 of p. 115, a hexany based on Archytas's enharmonic;
28/27, 16/15, 5/4, 9/7, 4/3, 2

JOYOUS-6; L. Harrison's Joyous 6;
9/8, 5/4, 3/2, 5/3, 15/8, 2

AUG/TET(A)-8; Linear Division of the 11/8, duplicated on the 16/11;
44/41, 22/19, 44/35, 11/8, 16/11, 64/41, 32/19, 64/35

AUG/TET(B)-8; harmonic mean division of 11/8;
88/85, 44/41, 22/19, 11/8, 16/11, 96/85, 64/41, 32/19

AUG/TET(C)-8; 11/10 C.I.;
15/14, 15/13, 5/4, 11/8, 16/11, 120/77, 240/143, 20/11

AUG/TET(D)-8; 11/9 C.I.;
27/26, 27/25, 9/8, 11/8, 16/11, 216/143, 432/275, 18/11

AUG/TET(E)-8; 5/4 C.I.;
33/32, 33/31, 11/10, 11/8, 16/11, 3/2, 48/31, 8/5

AUG/TET(F)-8; 5/4 C.I. again;
99/98, 33/32, 11/10, 11/8, 16/11, 72/49, 3/2, 8/5

AUG/TET(G)-8; 9/8 C.I.;
33/31, 33/29, 11/9, 11/8, 16/11, 48/31, 48/29, 16/9

AUG/TET(H)-8; 9/8 C.I. A gapped version of this scale is called AugTETI;
33/31, 11/10, 11/9, 11/8, 16/11, 48/31, 8/5, 16/9

NAM(ARTIFICIAL)-9; Artificial Nam system;
11/10, 17/14, 36/29, 4/3, 27/20, 3/2, 33/20, 38/21, 2

AUG/TET(M)-8; Linear Division of the 7/5, duplicated on the 10/7;
14/13, 7/6, 14/11, 7/5, 10/7, 20/13, 5/3, 20/11

*BOETHIUS(ENH)-8; Boethius's enharmonic, with a CI of 81/64 and added
16/9;*
512/499, 256/243, 4/3, 3/2, 768/499, 16/9, 128/81, 2

CHRO31-8; Tonos-31 chromatic. Tone 24 alternates with 23 as MESE or A;
31/29, 31/27, 31/24, 31/23, 31/22, 31/21, 31/20, 2

CHRO(31/CON)-8; Conjunct Tonos-31 chromatic;
31/29, 31/27, 31/24, 31/23, 31/22, 31/21, 31/18, 2

CLUSTER(8/A)-8; Eight-tone triadic cluster 4:5:6;
9/8, 5/4, 4/3, 45/32, 3/2, 5/3, 15/8, 2

CLUSTER(8/B)-8; Eight-tone triadic cluster 4:6:5;
75/64, 6/5, 5/4, 3/2, 25/16, 8/5, 15/8, 2

CLUSTER(8/C)-8; Eight-tone triadic cluster 3:4:5;
10/9, 6/5, 4/3, 25/18, 8/5, 5/3, 50/27, 2

CLUSTER(8/D)-8; Eight-tone triadic cluster 3:5:4;
10/9, 5/4, 4/3, 40/27, 3/2, 5/3, 16/9, 2

CLUSTER(8/E)-8; Eight-tone triadic cluster 5:6:8;
6/5, 5/4, 32/25, 3/2, 192/125, 8/5, 48/25, 2

CLUSTER(8/F)-8; Eight-tone triadic cluster 5:8:6;
144/125, 6/5, 4/3, 36/25, 8/5, 5/3, 48/25, 2

CLUSTER(8/G)-8; Eight-tone triadic cluster 4:5:7;
35/32, 8/7, 5/4, 10/7, 49/32, 7/4, 245/128, 2

CLUSTER(8/H)-8; Eight-tone triadic cluster 4:7:5;
35/32, 5/4, 175/128, 7/5, 25/16, 8/5, 7/4, 2

CLUSTER(8/I)-8; Eight-tone triadic cluster 5:6:7;
147/125, 6/5, 7/5, 10/7, 42/25, 12/7, 49/25, 2

CLUSTER(8/J)-8; Eight-tone triadic cluster 5:7:6;
126/125, 7/6, 6/5, 7/5, 36/25, 5/3, 42/25, 2

DIAMOND(TETR)-8; Tetrachord modular diamond based on Archytas's enharmonic;
28/27, 16/15, 5/4, 9/7, 35/27, 4/3, 48/35, 2

DIAT(15)-8; Tonos-15 diatonic and its own trite synemmenon B_b;
15/13, 5/4, 15/11, 10/7, 3/2, 5/3, 15/8, 2

DIAT(15/INV)-8; Inverted Tonos-15 harmonia, a harmonic series from 15 from 30;
16/15, 6/5, 4/3, 7/5, 22/15, 8/5, 26/15, 2

DIAT(17)-8; Tonos-17 diatonic and its own trite synemmenon B_b;
17/15, 17/13, 17/12, 34/23, 17/11, 17/10, 17/9, 2

DIAT(19)-8; Tonos-19 diatonic and its own trite synemmenon B_b;
19/18, 19/16, 19/14, 38/27, 19/13, 19/12, 19/11, 2

DIAT(21)-8; Tonos-21 diatonic and its own trite synemmenon B_b;
21/19, 7/6, 21/16, 7/5, 3/2, 21/13, 7/4, 2

DIAT(21/INV)-8; Inverted Tonos-21 harmonia, a harmonic series from 21 from 42;
8/7, 26/21, 4/3, 10/7, 32/21, 12/7, 38/21, 2

DIAT(23)-8; Tonos-23 diatonic and its own trite synemmenon B_b;
23/21, 23/20, 23/18, 23/17, 23/16, 23/14, 23/13, 2

DIAT(25)-8; Tonos-25 diatonic and its own trite synemmenon B_b;
25/22, 5/4, 25/18, 25/17, 25/16, 25/14, 25/13, 2

DIAT(27)-8; Tonos-27 diatonic and its own trite synemmenon B_b;
9/8, 9/7, 27/20, 27/19, 3/2, 27/16, 27/14, 2

DIAT(27/INV)-8; Inverted Tonos-27 harmonia, a harmonic series from 27 from 54;
28/27, 32/27, 4/3, 13/9, 40/27, 14/9, 16/9, 2

DIAT(29)-8; Tonos-29 diatonic and its own trite synemmenon B_b;
29/26, 29/24, 29/22, 29/21, 29/20, 29/18, 29/16, 2

DIAT(31)-8; Tonos-31 diatonic. The disjunctive and conjunctive diatonic forms are the same;
31/28, 31/26, 31/24, 31/23, 31/22, 31/20, 31/18, 2

DIAT(33)-8; Tonos-33 diatonic. The conjunctive form is 23 (B_b instead of B) 20 18 33/2;
11/10, 11/9, 11/8, 33/23, 3/2, 33/20, 11/6, 2

DORIAN(DIAT/2)-8;
Schlesinger's Dorian harmonia, a subharmonic series through 13 from 22;
11/10, 11/9, 11/8, 22/15, 11/7, 22/13, 11/6, 2

DORIAN(INV)-8; Inverted Schlesinger's Dorian harmonia, a harmonic series from 11 from 22;
12/11, 13/11, 14/11, 15/11, 16/11, 18/11, 20/11, 2

ENH(31)-8; Tonos-31 enharmonic. Tone 24 alternates with 23 as MESE or A;
31/30, 31/29, 31/24, 31/23, 31/22, 62/43, 31/21, 2

ENH(31/CON)-8; Conjunct Tonos-31 enharmonic;
31/30, 31/29, 31/24, 31/23, 62/45, 31/22, 31/18, 2

EULER(GM)-8; Euler's Genus Musicum, Octony based on Archytas's enharmonic;
28/27, 16/15, 48/405, 4/3, 112/81, 64/45, 1792/1215, 2

HARM(1C/HYPOD)-8; HarmC-hypodorian;
5/4, 21/16, 11/8, 23/16, 3/2, 7/4, 15/8, 2

HARM(1C/HYPOL)-8; HarmC-hypolydian;
21/20, 11/10, 13/10, 7/5, 3/2, 8/5, 17/10, 2

HARM1C(LYDIAN)-8; Harm1C-Lydian;
27/26, 14/13, 18/13, 19/13, 20/13, 21/13, 22/13, 2

HARMD(DOR)-8; HarmD-Dorian;
12/11, 13/11, 14/11, 15/11, 16/11, 18/11, 20/11, 2

HARMD(HYPOL)-8; HarmD-hypolydian;
11/10, 6/5, 13/10, 7/5, 3/2, 8/5, 9/5, 2

HARME(HYPOD)-8; HarmE-hypodorian;
21/16, 43/32, 11/8, 23/16, 3/2, 15/8, 31/16, 2

HARME(HYPOL)-8; HarmE-hypolydian;
43/40, 21/20, 13/10, 7/5, 3/2, 31/20, 8/5, 2

HARRISON-8; Harrison 8-tone from Serenade for Guitar;
16/15, 6/5, 5/4, 45/32, 3/2, 5/3, 16/9, 2

HYPOD(DIAT/NO.2)-8;
Schlesinger's hypodorian harmonia, a subharmonic series through 13 from 16;
16/15, 16/13, 4/3, 32/23, 16/11, 8/5, 16/9, 2

HYPOL/CHROM-8; Schlesinger's hypolydian harmonia in the chromatic genus;
20/19, 10/9, 4/3, 10/7, 20/13, 8/5, 5/3, 2

HYPOL/CHROM/INV-8; Inverted Schlesinger's chromatic hypolydian harmonia;
6/5, 5/4, 13/10, 7/5, 3/2, 9/5, 19/10, 2

HYPOL/DIAT-8;
Schlesinger's hypolydian harmonia, a subharmonic series through 13 from 20;
10/9, 5/4, 4/3, 10/7, 20/13, 5/3, 20/11, 2

HYPOL/DIAT/INV-8; Inverted Schlesinger's hypolydian harmonia, a harmonic series from 10 from 20;
11/10, 6/5, 13/10, 7/5, 3/2, 8/5, 9/5, 2

HYPOL/ENH-8; Schlesinger's hypolydian harmonia in the enharmonic genus;
40/39, 20/19, 4/3, 10/7, 20/13, 8/5, 5/3, 2

HYPOL/ENH/INV-8; Inverted Schlesinger's enharmonic hypolydian harmonia;
5/4, 51/40, 13/10, 7/5, 3/2, 19/10, 39/20, 2

HYPOL/PENTA-8;
Schlesinger's hypolydian harmonia in the pentachromatic genus;
25/24, 10/9, 4/3, 10/7, 20/13, 100/63, 5/3, 2

HYPOL/TRI-8;
Schlesinger's hypolydian harmonia in the first trichromatic genus;
30/29, 15/14, 4/3, 10/7, 20/13, 30/19, 60/37, 2

HYPOL/TRI2-8; Schlesinger's hypolydian harmonia in the second trichromatic genus;
30/29, 10/9, 4/3, 10/7, 20/13, 30/19, 5/3, 8/1

HYPOP-DIAT2-8; Schlesinger's hypophrygian harmonia;
9/8, 6/5, 18/13, 36/25, 3/2, 18/11, 9/5, 2

HYPOP/DIAT/INV-8; Inverted Schlesinger's hypophrygian harmonia, a harmonic series from 9 from 18;
10/9, 11/9, 4/3, 25/18, 13/9, 5/3, 16/9, 2

HARMD/HYPOD-9; HarmD-hypodorian;
9/8, 5/4, 11/8, 23/16, 3/2, 13/8, 7/4, 15/8, 2

HARM/MEAN-9; Harm. Mean 9-tonic 8/7 is HM of 1/1 and 4/3, etc.;
32/31, 16/15, 8/7, 4/3, 3/2, 48/31, 8/5, 12/7, 2

HARMC/HYPOP-9; HarmC-hypophrygian;
11/9, 23/18, 4/3, 25/18, 13/9, 14/9, 16/9, 17/9, 2

HARM(D/HYPOD)-9; Harm D-hypodorian;
9/8, 5/4, 11/8, 23/16, 3/2, 13/8, 7/4, 15/8, 2

HARMD/HYPOP-9; HarmD-hypophrygian;
10/9, 11/9, 4/3, 25/18, 13/9, 14/9, 5/3, 16/9, 2

HARMD/LYD-9; HarmD-Lydian;
14/13, 15/13, 16/13, 18/13, 19/13, 20/13, 22/13, 24/13, 2

HARME/HYPOP-9; HarmE-hypophrygian;
23/18, 47/36, 4/3, 25/18, 13/9, 14/9, 17/9, 35/18, 2

GILSON-10; Gilson's 10-tone JI;
75/64, 6/5, 5/4, 32/25, 3/2, 25/16, 8/5, 15/8, 48/25, 2

DIAPHONIC-10; 10-tone diaphonic cycle;
18/17, 9/8, 6/5, 9/7, 18/13, 3/2, 8/5, 12/7, 24/13, 2

DEKANY(NO.1)-10; Dekany 1.3.5.7.11;
55/48, 7/6, 5/4, 11/8, 35/24, 77/48, 5/3, 7/4, 11/6, 2

DEKANY(NO.2)-10; Dekany 1.3.5.7;
16/15, 8/7, 6/5, 4/3, 48/35, 32/21, 8/5, 12/7, 16/9, 2

ITER-FIFTH-10; Iterated 3/2 Scale, IE=3/2, PD=3, SD=2;
207.987, 311.980, 381.309, 415.973, 467.970, 571.963, 623.960, 658.624, 675.957, 3/2

HARM9-10; 6/7/8/9 harmonics, first 9 overtones of 5th through 9th harmonics;
9/8, 7/6, 5/4, 4/3, 49/36, 3/2, 14/9, 7/4, 16/9, 2

TEMES-10; Temes' 5-tone ϕ scale, 2 cycles;
273.000, 366.910, 466.181, 560.090, 833.090, 1 100.729, 1 200.000, 1 299.271, 1 393.181, 1 666.181

HEXANIC-11; Composed of 1.3.5.45, 1.3.5.75, 1.3.5.9, and 1.3.5.25 hexanies;
16/15, 9/8, 6/5, 5/4, 4/3, 3/2, 25/16, 8/5, 15/8, 48/25, 2

KANZELMEYER-11;
Bruce Kanzelmeyer, 11 harmonics from 16 to 32. Base 388.3614815 Hz;
17/16, 19/16, 5/4, 11/8, 23/16, 3/2, 13/8, 7/4, 29/16, 31/16, 2

HYPOD/DIAT/INV-9; Inverted Schlesinger's hypodorian harmonia, a harmonic series from 8 from 16;
9/8, 5/4, 11/8, 23/16, 3/2, 13/8, 7/4, 15/8, 2

LIU-MEL-9; Linus Liu's Melodic Minor, use 5 and 7 descending and 6 and 8 ascending;
10/9, 6/5, 4/3, 3/2, 81/50, 5/3, 9/5, 50/27, 2

LYDIAN/DIAT-INV)-8; Inverted Schlesinger's Lydian harmonia, a harmonic series from 13 from 26;
14/13, 16/13, 18/13, 19/13, 20/13, 22/13, 24/13, 2

LYDIAN/DIAT-INV)-8; Inverted Schlesinger's Lydian harmonia, a harmonic series from 13 from 26;
14/13, 16/13, 18/13, 19/13, 20/13, 22/13, 24/13, 2

MCCLAIN-8; McClain's 8-tone scale, c.f. p. 51 of The Myth of Invariance;
9/8, 5/4, 45/32, 3/2, 25/16, 27/16, 15/8, 2

MIN3/30/29-9; 30/29 × 29/28 × 28/27 + 6/5;
30/29, 15/14, 10/9, 4/3, 3/2, 45/29, 45/28, 5/3, 2

MISCA-9; 21/20 × 20/19 × 19/18 = 7/6. 7/6 × 8/7 = 4/3;
21/20, 21/19, 7/6, 4/3, 3/2, 63/40, 63/38, 7/4, 2

MISCB-9; 33/32 × 32/31 × 31/27 = 11/9. 11/9 × 12/11 = 4/3;
33/32, 33/31, 11/9, 4/3, 3/2, 99/64, 99/62, 11/6 2/1

MISCC-9; 96/91 × 91/86 × 86/54 = 32/27. 32/27 × 9/8 = 4/3;
96/91, 48/43, 32/27, 4/3, 3/2, 144/91, 72/43, 16/9, 2

MISCD-9; 27/26 × 26/25 × 25/24 = 9/8. 9/8 × 32/27 = 4/3;
27/26, 27/25, 9/8, 4/3, 3/2, 81/52, 81/50, 27/16, 2

MISCE-9; 15/14 × 14/13 × 13/12 = 5/4. 5/4 × 16/15 = 4/3;
15/14, 15/13, 5/4, 4/3, 3/2 45/28, 45/26, 15/8, 2

MISCF-9; SupraEnh1;
28/27, 16/15, 4/3, 81/56, 3/2, 14/9, 8/5, 27/14, 2

MISCG-9; SupraEnh 2;
28/27, 16/15, 9/7, 4/3, 3/2, 14/9 8/5, 27/14, 2

MISCH-9; SupraEnh 3;
28/27, 16/15, 9/7, 4/3, 3/2, 14/9, 15/8, 27/14, 2

MIXOL/DIAT/NO.2-8;
Schlesinger's Mixolydian harmonia, a subharmonic series though 13 from 28;
14/13, 7/6, 14/11, 4/3, 7/5, 14/9, 7/4, 2

MIXOL/DIATINV/NO.2-8; Inverted Schlesinger's Mixolydian harmonia, a harmonic series from 14 from 28;
8/7, 9/7, 4/3, 10/7, 11/7, 12/7, 13/7, 2

OCTANY(NO.1)-8; Octany from 1.3.5.7.9.11.13.15, 1.3 tonic;
9/8, 5/4, 11/8, 3/2, 13/8, 7/4, 15/8, 2

OCTANY(NO.7)-8; Octany from 1.3.5.7.9.11.13.15, 1.3.5.7.9.11.13 tonic;
15/14, 15/13, 5/4, 15/11, 3/2, 5/3, 15/8, 2

OCTONY(MIN)-8; Octony on harmonic minor, from Palmer on an album of Turkish music;
9/8, 6/5, 5/4, 4/3, 3/2, 8/5, 15/8, 2

OCTONY(ROT)-8; Rotated Octony on harmonic minor;
5/4, 4/3, 3/2, 25/16, 8/5, 5/3, 15/8, 2

OCTONY(TRANS/NO.1)-8; Complex 10 of p. 115, an Octony based on Archytas's enharmonic;
28/27, 16/15, 5/4, 4/3, 25/16, 45/28, 5/3, 2

OCTONY(TRANS/NO.2)-8; Complex 6 of p. 115 based on Archytas's enharmonic, an Octony;
28/27, 16/15, 135/112, 243/196, 9/7, 4/3, 27/14, 2

OCTONY(TRANS/NO.3)-8; Complex 5 of p. 115 based on Archytas's enharmonic, an Octony;
28/27, 16/15, 75/64, 135/112, 5/4, 4/3, 15/8 2/1

OCTONY(TRANS/NO.4)-8; Complex 11 of p. 115, an Octony based on Archytas's enharmonic;
28/27, 16/15, 9/7, 4/3, 45/28, 81/49, 12/7, 2

OCTONY(TRANS/NO.5)-8; Complex 15 of p. 115, an Octony based on Archytas's enharmonic;
28/27, 16/15, 175/144, 5/4, 35/27, 4/3, 35/18, 2

OCTONY(TRANS/NO.6)-8; Complex 14 of p. 115, an Octony based on Archytas's enharmonic;
36/35, 28/27, 16/15, 9/7, 324/245, 4/3, 48/35, 2

PHRYG(DIAT-8;
Schlesinger's Phrygian harmonia, a subharmonic series through 13 from 24;
12/11, 6/5, 4/3, 24/17, 3/2, 12/7, 24/13, 2

PHRYG(DIAT-INV)-8; Inverted Schlesinger's Phrygian harmonia, a harmonic series from 12 from 24;
13/12, 7/6, 4/3, 17/12, 3/2, 5/3, 11/6, 2

PORTUGUESE(BAG/NO.2)-10; Portugese bagpipe tuning 2;
21/20, 14/13, 32/27, 17/14, 21/16, 64/45, 3/2, 25/16, 59/32, 2

PROG/ENNEA/NO.1-9; Progressive Enneatonic, appr. 50+100+150+200; in each half (500;);
36/35, 12/11, 19/16, 4/3, 3/2, 17/11, 18/11, 16/9, 2

PROG/ENNEA/NO.2-9; Progressive Enneatonic, appr. 50+100+200+150; in each half (500;);
34/33, 12/11, 27/22, 4/3, 3/2, 17/11, 18/11, 81/44, 2

PROG/ENNEA/NO.3-9; Progressive Enneatonic, appr. 50+100+150+200; in each half (500;);
34/33, 12/11, 32/27, 4/3, 3/2, 17/11, 18/11, 16/9, 2

RAT(HYPOL/ENH)-8;
Rationalized Schlesinger's hypolydian harmonia in the enharmonic genus;
40/39, 20/19, 4/3, 10/7, 20/13, 80/51, 8/5, 2

SCHOLZ-8; Simple Tune No. 1 Carter Scholz;
28/27, 8/7, 7/6, 4/3, 3/2, 14/9, 7/4, 2

SUB/HARM/1C/DOR-8; Subharm1C-Dorian;
11/9, 22/17, 11/8, 22/15, 11/7, 11/6, 23/11, 2

SUB/HARM/1C-LYD-8; Subharm1C-Lydian;
13/11, 26/21, 13/10, 19/13, 13/9, 13/7, 52/27, 2

SUB/HARM/1E/DOR-8; Subharm1E-Dorian;
22/17, 4/3, 11/8, 22/15, 11/7, 44/23, 88/45, 2

SUB/HARM/1E/LYD-8; Subharm1E-Lydian;
26/21, 52/41, 13/10, 19/13, 13/9, 52/27, 104/53, 2

SUB/HARM2C-HYPOD-8; SHarm2C-hypodorian;
16/13, 32/25, 4/3, 32/23, 16/11, 16/9, 32/17, 2

SUB/HARM2C/HYPOL-8; SHarm2C-hypolydian;
20/17, 5/4, 4/3, 10/7, 20/13, 20/11, 40/21, 2

SUB/HARM2C/HYPOP-8; SHarm2C-hypophrygian;
9/7, 4/3, 18/13, 36/25, 3/2, 9/5, 36/19, 2

SUB/HARM2E/HYPOD-8; SHarm2E-hypodorian;
32/25, 64/49, 4/3, 32/23, 16/11, 32/17, 64/33, 2

SUB/HARM2E/HYPOL-8; SHarm2E-hypolydian;
5/4, 40/31, 4/3, 10/7, 20/13, 40/21, 80/41, 2

SUB/HARM/NO.2/E/HYPOP-8; SHarm2E-hypophrygian;
4/3, 72/53, 18/13, 36/25, 3/2, 36/19, 72/37, 2

SIX-STAR(A)-8; Harmonic six-star, group A, from Fokker;
16/15, 6/5, 32/25, 4/3, 3/2, 8/5, 48/25, 2

SIX-STAR(B)-8; Harmonic six-star, group B, from Fokker;
28/25, 8/7, 32/25, 7/5, 8/5 7/4, 64/35, 2

SIX-STAR(C)-8; Harmonic six-star, group C on E♭, from Fokker;
8/7, 7/6, 4/3, 32/21, 14/9, 7/4, 16/9, 2

SOLEMN-6; Solemn 6;
6/5, 4/3, 3/2, 8/5, 9/5, 2

SPONDEION-6; A subharmonic six-tone series, notated as a whole-tone scale;
11/10, 11/9, 11/8, 11/7, 11/6, 2

SUB-8; Subharmonic series 1/16 - 1/8;
16/15, 8/7, 16/13, 4/3, 16/11, 8/5, 16/9, 2

TRICHORD-11; Trichordal Undecatonic;
9/8, 7/6, 5/4, 21/16, 4/3, 3/2, 5/3, 27/16, 7/4, 15/8, 2

YASSER-6; Yasser Hexad, 6 of 19 as whole tone scale;
189.474, 378.947, 568.421, 757.895, 947.368, 1 200

A.5 Dodekatonics

AGRICOLA-12; Agricola's Monochord;
135/128, 9/8, 1215/1 024, 81/64, 4/3, 45/32, 3/2, 405/256, 27/16, 16/9, 243/128, 2

Multiple Archytas;
28/27, 16/15, 5/4, 9/7, 4/3, 112/81, 3/2, 14/9, 8/5, 15/8, 27/14, 2

ARIEL(NO.1)-12; Ariel 1;
27/25, 9/8, 6/5, 5/4, 4/3, 25/18, 3/2, 8/5, 5/3, 9/5, 15/8, 2

ARIEL(NO.2)-12; Ariel 2;
16/15, 10/9, 6/5, 5/4, 4/3, 25/18, 3/2, 8/5, 5/3, 9/5, 15/8, 2

ARIEL(NO.3)-12; Ariel's 12-tone JI scale;
16/15, 10/9, 32/27, 100/81, 4/3, 25/18, 3/2, 8/5, 5/3, 16/9, 50/27, 2

AWRAAMOFF-12; Awraamoff Septimal Just;
9/8, 8/7, 6/5, 5/4, 21/16, 4/3, 3/2, 8/5, 12/7, 7/4, 15/8, 2

BAGPIPE-12; Bagpipe Tuning;
117/115, 146/131, 196/169, 89/73, 141/106, 81/59, 150/101, 125/82, 139/84, 205/116, 11/6, 2

BOOMSLITER-12; Boomsliter–Creel basic set of their referential tuning;
9/8, 7/6, 6/5, 5/4, 4/3, 7/5, 3/2, 8/5, 5/3, 7/4, 9/5, 2

BULGARIAN(BAG)-12; Bulgarian bagpipe tuning;
66, 202, 316, 399, 509, 640, 706, 803, 910, 1011, 1 092, 1 200

BURT(NO.1)-12; W. Burt's 13diatsub No. 01;
26/25, 13/12, 26/23, 13/11, 13/10, 26/19, 13/9, 27/17, 13/8, 26/15, 13/7, 2

BURT(NO.2)-12; W. Burt's 13enhsub No. 02;
104/103, 52/51, 104/101, 26/25, 13/10, 104/79, 4/3, 104/77, 26/19, 13/11, 13/7, 2

BURT(NO.3)-12; W. Burt's 13enhharm No. 03;
14/13, 33/26, 19/13, 77/52, 3/2, 79/52, 20/13, 25/13, 101/52, 51/26, 103/52, 2

BURT(NO.04)-12; W. Burt's 13diatharm No. 04, see his post 3/30/94 in Tuning Digest No. 57;
14/13, 15/13, 16/13, 17/13, 18/13, 19/13, 20/13, 22/13, 23/13, 24/13, 25/13, 2

BUTR(NO.05)-12; W. Burt's 17diatsub No. 05;
17/16, 17/15, 17/14, 17/13, 17/16, 34/23, 17/11, 34/21, 17/10, 34/19, 17/9, 2

BURT(NO.06)-12; W. Burt's 17enhsub No. 06;
68/67, 34/33, 68/65, 17/16, 17/12, 34/23, 17/11, 136/87, 68/43, 8/5, 34/21, 2

BURT(NO.07)-12; W. Burt's 17enhharm No. 07;
21/17, 5/4, 43/34, 87/68, 22/17, 23/17, 24/17, 32/17, 65/34, 33/17, 67/34, 2

BURT(NO.08)-12; W. Burt's 17diatharm No. 08;
18/17, 19/17, 20/17, 21/17, 22/17, 23/17, 24/17, 26/17, 28/17, 30/17, 32/17, 2

BURT(NO.09)-12; W. Burt's 19diatsub No. 09;
38/37, 19/18, 19/17, 19/16, 19/14, 38/27, 19/13, 38/25, 19/12, 38/23, 19/11, 2

BURT(NO.10)-12; W. Burt's 19enhsub No. 10;
76/75, 38/37, 76/73, 19/18, 19/14, 38/27, 19/13, 152/103, 76/51, 152/101, 38/25, 2

BURT(NO.11)-12; W. Burt's 19enhharm No. 11;
25/19, 101/76, 51/38, 103/76, 26/19, 27/19, 28/19, 36/19, 73/38, 37/19, 75/38, 2

BURT(NO.12)-12; W. Burt's 19diatharm No. 12;
22/19, 23/19, 24/19, 25/19, 26/19, 27/19, 28/19, 32/19, 34/19, 36/19, 37/19, 2

BURT(NO.13)-12; W. Burt's 23diatsub No. 13;
23/22, 23/21, 46/41, 23/20, 23/18, 23/17, 23/16, 23/15, 23/14, 46/27, 23/13, 2

BURT(NO.14)-12; W. Burt's 23enhsub No. 14;
92/91, 46/45, 92/89, 23/22, 23/18, 23/17, 23/16, 92/63, 46/31, 92/61, 23/15, 2

BURT(NO.15)-12; W. Burt's 23enhharm No. 15;
30/23, 61/46, 31/23, 63/46, 32/23, 34/23, 36/23, 44/23, 89/46, 45/23, 91/46, 2

BURT(NO.16)-12; W. Burt's 23diatharm No. 16;
26/23, 27/23, 28/23, 30/23, 32/23, 34/23, 36/23, 40/23, 41/23, 42/23, 44/23, 2

CANRIGHT-12; David Canright's piano tuning for "Canon for Seven Hands";
28/27, 9/8, 7/6, 5/4, 4/3, 45/32, 3/2, 14/9, 5/3, 7/4, 15/8, 2

CARLOS(HARM)-12; Carlos Harmonic;
17/16, 9/8, 19/16, 5/4, 21/16, 11/8, 3/2, 13/8, 27/16, 7/4, 15/8, 2

CARLOS(SUPER)-12; Carlos Super Just;
17/16, 9/8, 6/5, 5/4, 4/3, 11/8, 3/2, 13/8, 5/3, 7/4, 15/8, 2

COLONNA(NO.1)-12; Colonna 1;
25/24, 10/9, 85/72, 5/4, 4/3, 25/18, 3/2, 55/36, 5/3, 85/48, 15/8, 2

COLONNA(NO.2)-12; Colonna 2;
25/24, 9/8, 6/5, 5/4, 4/3, 7/5, 3/2, 8/5, 5/3, 9/5, 11/6, 2

CRUCIFORM-12; Cruciform Lattice;
9/8, 75/64, 6/5, 5/4, 4/3, 45/32, 3/2, 25/16, 8/5, 5/3, 15/8, 2

DIAPHONIC(NO.1)-12; 12-tone Diaphonic Cycle, conjunctive form on 3/2 and 4/3;
21/20, 21/19, 7/6, 21/17, 21/16, 7/5, 3/2, 30/19, 5/3, 30/17, 15/8, 2

DIAPHONIC(NO.2)-12; 2nd 12-tone Diaphonic Cycle, conjunctive form on 10/7 and 7/5;
21/20, 21/19, 7/6, 21/17, 21/16, 7/5, 28/19, 14/9, 28/17, 7/4, 28/15, 2

RING(K/DOUBLE/NO.1)-12; Double-tie circular mirroring of 4:5:6:7;
21/20, 7/6, 6/5, 49/40, 5/4, 7/5, 3/2, 42/25, 12/7, 7/4, 9/5, 2

RING(K/DOUBLE/NO.2)-12; Double-tie circular mirroring of 3:5:7:9;
21/20, 7/6, 63/50, 9/7, 27/20, 7/5, 3/2, 14/9, 49/30, 5/3, 9/5, 2

DODECENY-12; Degenerate eikosany 3:6 from 1.3.5.9.15.45 tonic 1.3.15;
135/128, 9/8, 75/64, 6/5, 5/4, 4/3, 45/32, 3/2, 5/3, 27/16, 15/8, 2

SCHLESINGER(DORIAN)-12; Schlesinger's Dorian Piano Tuning (Sub 22);
22/21, 11/10, 22/19, 11/9, 22/17, 11/8, 22/15, 11/7, 22/13, 44/25, 11/6, 2

DOWLAND-12; Dowland lute tuning;
33/31, 9/8, 33/28, 264/211, 4/3, 24/17, 3/2, 99/62, 27/16, 99/56, 396/211, 2

DUNCAN-12; D. Duncan's Superparticular Scale;
17/16, 9/8, 6/5, 5/4, 4/3, 7/5, 3/2, 8/5, 5/3, 7/4, 15/8, 2

DUODENE-12; Ellis's Duodene : genus [33355];
16/15, 9/8, 6/5, 5/4, 4/3, 45/32, 3/2, 8/5, 5/3, 9/5, 15/8, 2

DUODENE(14-18-21)-12; 14-18-21 Duodene;
28/27, 9/8, 7/6, 9/7, 4/3, 81/56, 3/2, 14/9, 12/7, 7/4, 27/14, 2

DUODENE(3-11-9)-12; 3-11/9 Duodene;
12/11, 9/8, 11/9, 27/22, 4/3, 11/8, 3/2, 44/27, 18/11, 11/6, 81/44, 2

DUODENE(3-7)-12; 3-7 Duodene;
9/8, 8/7, 7/6, 9/7, 21/16, 4/3, 3/2, 32/21, 12/7, 7/4, 63/32, 2

DUODENE(6-7-9)-12; 6-7-9 Duodene;
9/8, 8/7, 7/6, 9/7, 21/16, 4/3, 3/2, 14/9, 12/7, 7/4, 27/14, 2

DUODENE(ROT)-12; Ellis's Duodene rotated : genus [33555];
1125/1 024, 9/8, 75/64, 5/4, 45/32, 375/256, 3/2, 25/16, 225/128, 15/8, 125/64, 2

DUODENE(SKEW)-12; Rotated 6/5x3/2 duodene;
27/25, 10/9, 6/5, 5/4, 4/3, 36/25, 3/2, 8/5, 5/3, 9/5, 48/25, 2

EFG55577-12; Genus [55577];
35/32, 125/112, 8/7, 5/4, 175/128, 10/7, 25/16, 875/512, 7/4, 25/14, 125/64, 2

EFG55777-12; Genus [55777];
35/32, 8/7, 49/40, 5/4, 7/5, 10/7, 49/32, 8/5, 7/4, 64/35, 245/128, 2

RING(K/E/NO.1)-12; Single-tie circular mirroring of 3:4:5;
9/8, 6/5, 5/4, 27/20, 45/32, 36/25, 25/16, 8/5, 5/3, 9/5, 15/8, 2

RING(K/E/NO.2)-12; Single-tie circular mirroring of 6:7:8;
9/8, 8/7, 7/6, 9/7, 21/16, 72/49, 49/32, 12/7, 7/4, 27/14, 63/32, 2

RING(K/E/NO.3)-12; Single-tie circular mirroring of 4:5:7;
50/49, 8/7, 400/343, 5/4, 125/98, 64/49, 25/16, 8/5, 80/49, 7/4, 25/14, 2

RING(K/E/NO.4)-12; Single-tie circular mirroring of 4:5:6;
16/15, 6/5, 32/25, 4/3, 36/25, 3/2, 192/125, 5/3, 128/75, 16/9, 48/25, 2

RING(K/E/NO.5)-12; Single-tie circular mirroring of 3:5:7;
126/125, 36/35, 7/6, 216/175, 7/5, 10/7, 36/25, 72/49, 42/25, 12/7, 49/25, 2

RING(K/E/NO.)6-12; Single-tie circular mirroring of 6:7:9;
54/49, 8/7, 9/7, 4/3, 72/49, 3/2, 14/9, 81/49, 16/9, 648/343, 96/49, 2

RING(K/E/NO.7)-12; Single-tie circular mirroring of 5:7:9;
50/49, 10/9, 500/441, 100/81, 9/7, 450/343, 14/9, 100/63, 81/49, 9/5, 90/49, 2

ELLIS(HARM)-12; Ellis's Just Harmonium;
16/15, 9/8, 6/5, 5/4, 4/3, 27/20, 3/2, 8/5, 5/3, 9/5, 15/8, 2

ETHIOPIAN-12; Ethiopian Tunings from [17];
15/14, 32/29, 97/83, 26/21, 41/31, 55/39, 53/36, 19/12, 21/13, 310/117, 37/20, 2

GANASSI-12; Ganassi;
20/19, 10/9, 20/17, 5/4, 4/3, 24/17, 3/2, 30/19, 5/3, 30/17, 15/8, 2

GILSON(NO.1)-12; Gilson septimal I;
8/7, 6/5, 5/4, 3/2, 10/7, 3/2, 25/16, 25/16, 25/14, 15/8, 2

GILSON(NO.2)-12; Gilson septimal II;
15/14, 8/7, 6/5, 9/7, 10/7, 10/7, 3/2, 8/5, 9/5, 9/5, 2

HARM(1-23)-12; Harmonics 1 to 23;
17/16, 9/8, 19/16, 5/4, 21/16, 11/8, 23/16, 3/2, 13/8, 7/4, 15/8, 2

HARM/D/PHRYG(12; With 5 extra tones;
25/24, 13/12, 9/8, 7/6, 4/3, 5/4, 3/2, 19/12, 5/3, 7/4, 11/6, 2

HARM-12; See pp. 17 and 466–468 [26], lower 4 oct. Instr. designed & tuned by Ellis;
10/9, 9/8, 6/5, 5/4, 4/3, 3/2, 8/5, 5/3, 7/4, 9/5, 15/8, 2

ELLIS(HARM/UP)-12; Upper 2 octaves of Ellis's Harmonical;
17/16, 9/8, 19/16, 5/4, 11/8, 7/4, 3/2, 25/16, 13/8, 29/16, 15/8, 2

HEXANY(NO.1)-12; Hexany Cluster 1;
9/8, 144/125, 6/5, 5/4, 4/3, 27/20, 36/25, 3/2, 8/5, 9/5, 48/25, 2

HEXANY(NO.2)-12; Hexany Cluster 2;
25/24, 9/8, 6/5, 5/4, 125/96, 4/3, 25/18, 3/2, 25/16, 5/3, 15/8, 2

HEXANY(NO.3)-12; Hexany Cluster 3;
25/24, 10/9, 6/5, 5/4, 4/3, 3/2, 8/5, 5/3, 9/5, 15/8, 48/25, 2

HEXANY(NO.4)-12; Hexany Cluster 4;
25/24, 9/8, 6/5, 5/4, 4/3, 36/25, 3/2, 8/5, 5/3, 9/5, 15/8, 2

HEXANY(NO.5)-12; Hexany Cluster 5;
9/8, 6/5, 5/4, 4/3, 3/2, 25/16, 8/5, 5/3, 9/5, 15/8, 48/25, 2

354 APPENDIX A

HEXANY(N.6)-12; Hexany Cluster 6;
25/24, 10/9, 9/8, 6/5, 5/4, 4/3, 3/2, 25/16, 8/5, 5/3, 15/8, 2

HEXANY(NO.7)-12; Hexany Cluster 7;
25/24, 6/5, 5/4, 4/3, 25/18, 3/2, 25/16, 8/5, 5/3, 9/5, 15/8, 2

HEXANY(NO.8)-12; Hexany Cluster 8;
25/24, 6/5, 5/4, 125/96, 4/3, 3/2, 25/16, 8/5, 5/3, 15/8, 48/25, 2

HEXANY(S)-12; Hexany S 13579;
35/32, 9/8, 5/4, 21/16, 45/32, 3/2, 105/64, 27/16, 7/4, 15/8, 63/32, 2

HEXANY(S2)-12; Hexany S 1371113;
77/64, 13/8, 7/4, 33/32, 91/64, 3/2, 231/128, 39/32, 11/8, 21/16, 143/128, 2

HYPOL(CHRO)-12; Hypolydian Chromatic Tonos;
20/19, 40/37, 10/9, 4/3, 10/7, 40/27, 20/13, 8/5, 80/49, 5/3, 40/23, 2

HYPO(DIAT)-12; Hypolydian Diatonic Tonos;
10/9, 20/17, 5/4, 4/3, 10/7, 40/27, 20/13, 5/3, 40/23, 20/11, 40/21, 2

HYPO(ENH)-12; Hypolydian Enharmonic Tonos;
40/39, 80/77, 20/19, 4/3, 10/7, 40/27, 20/13, 80/51, 160/101, 8/5, 16/9, 2

HYPOD(CHRO)-12; Hypodorian Chromatic Tonos;
16/15, 32/29, 8/7, 16/13, 4/3, 32/23, 16/11, 32/21, 64/41, 8/5, 16/9, 2

HYPOD(DIAT)-12; Hypodorian Diatonic Tonos;
16/15, 8/7, 16/13, 32/25, 4/3, 32/23, 16/11, 8/5, 32/19, 16/9, 32/17, 2

HYPOD(ENH)-12; Hypodorian Enharmonic Tonos;
32/31, 64/61, 16/15, 32/27, 4/3, 32/23, 16/11, 64/43, 128/85, 32/21, 64/37, 2

HYPOP(CHRO)-12; Hypophrygian Chromatic Tonos;
18/17, 12/11, 9/8, 9/7, 18/13, 36/25, 3/2, 36/23, 8/5, 18/11, 9/5, 2

HYPOP(DIAT)-12; Hypophrygian Diatonic Tonos;
9/8, 36/31, 6/5, 9/7, 18/13, 36/25, 3/2, 18/11, 12/7, 9/5, 36/19, 2

HYPOP(ENH)-12; Hypophrygian Enharmonic Tonos;
36/35, 24/23, 18/17, 6/5, 18/13, 36/25, 3/2, 72/47, 48/31, 36/23, 9/5, 2

JI-12; Basic JI with 7-limit tritone;
16/15, 9/8, 6/5, 5/4, 4/3, 7/5, 3/2, 8/5, 5/3, 9/5, 15/8, 2

JOHNSTON-12; Ben Johnston's combined otonal-utonal scale;
135/128, 9/8, 135/112, 5/4, 11/8, 45/32, 3/2, 135/88, 27/16, 7/4, 15/8, 2

JORGENSEN-12; Jorgensen's 5 &7 temperament;
51.429, 171.429, 291.429, 342.857, 514.286, 531.429,
685.714, 771.429, 857.143, 1011.429, 1028.571, 1 200.000

KILROY-12; Kilroy;
9/8, 6/5, 5/4, 4/3, 45/32, 3/2, 8/5, 5/3, 27/16, 16/9, 15/8, 2

KLONARIS-12; Scale by Johnny Klonaris;
17/16, 9/8, 19/16, 5/4, 21/16, 11/8, 3/2, 25/16, 13/8, 7/4, 15/8, 2

LAMONTE-12; LaMonte Young, Tuning of For Guitar(1958). See 1/1 March 1992;
16/15, 10/9, 6/5, 5/4, 4/3, 45/32, 3/2, 8/5, 5/3, 9/5, 15/8, 2

LEFTPISTOL-12; Left Pistol;
135/128, 16/15, 9/8, 5/4, 4/3, 45/32, 3/2, 8/5, 5/3, 27/16, 15/8, 2

LORINA-12; Lorina;
28/27, 28/25, 7/6, 6/5, 4/3, 4/3, 28/19, 14/9, 7/4, 7/4, 16/9, 2

M-REINHARD-12; M. Reinhard's Harmonic-13 scale. 1/1=440Hz;
14/13, 13/12, 16/13, 13/10, 18/13, 13/9, 20/13, 13/8, 22/13, 13/7, 208/105, 2

MAJOR/CLUS-12; Major Mode Cluster, c.f. [7];
135/128, 10/9, 9/8, 5/4, 4/3, 45/32, 3/2, 5/3, 27/16, 16/9, 15/8, 2

MAJOR/WING-12; Major Wing with 7 major and 6 minor triads;
25/24, 9/8, 6/5, 5/4, 4/3, 3/2, 25/16, 8/5, 5/3, 9/5, 15/8, 2

MARPURG(NO.1)-12; Marpurg I;
25/24, 9/8, 6/5, 5/4, 4/3, 45/32, 3/2, 25/16, 5/3, 9/5, 15/8, 2

MARPURG(NO.3)-12; Marpurg III;
25/24, 9/8, 6/5, 5/4, 4/3, 45/32 3/2, 25/16, 27/16, 16/9, 15/8, 2

MARPURG(NO.4)-12; Marpurg IV, also Yamaha Pure Minor;
25/24, 10/9, 6/5, 5/4 4/3, 25/18, 3/2, 25/16, 5/3 9/5, 15/8, 2

MC-CLAIN-12; McClain's 12-tone scale, c.f. page 119 of The Myth of Invariance;
135/128, 9/8, 75/64, 5/4, 81/64, 45/32, 3/2, 25/16, 27/16, 15/8, 125/64, 2

MINOR/CLUS-12; Minor Mode Cluster, c.f. [7];
16/15, 9/8, 6/5, 4/3, 27/20, 64/45, 3/2, 8/5, 27/16, 16/9, 9/5, 2

MINOR/DUO-12; Minor Duodene;
10/9, 9/8, 6/5, 5/4, 4/3, 27/20, 3/2, 8/5, 5/3, 9/5, 15/8, 2

MINOR-WING-12; Minor Wing with 7 minor and 6 major triads, c.f. [7];
9/8, 6/5, 5/4, 4/3, 36/25, 3/2, 8/5, 5/3, 9/5, 15/8, 48/25, 2

MONTVALLON-12; Montvallon;
135/128, 9/8, 6/5, 5/4, 4/3, 45/32, 3/2, 405/256, 5/3, 16/9, 15/8, 2

ODD(NO.1)-12;
25/24, 6/5, 5/4, 36/25, 3/2, 25/16, 8/5, 5/3, 9/5, 15/8, 48/25, 2

ODD(NO.)2-12;
10/9, 9/8, 75/64, 6/5, 5/4, 4/3, 25/18, 3/2, 5/3, 9/5, 15/8, 2

OLDANI-12; This scale by N. L. Oldani appeared in Interval 5(3), p. 10–11;
25/24, 9/8, 32/27, 5/4, 4/3, 45/32, 3/2, 25/16, 5/3, 16/9, 15/8, 2

PALACE-12; Palace mode;
18/17, 9/8, 8/7, 9/7, 4/3, 10/7, 3/2, 36/23, 18/11, 12/7, 9/5, 2

PAREJA-12; Ramis de Pareja;
135/128, 10/9, 32/27, 5/4, 4/3, 45/32, 3/2, 128/81, 5/3, 16/9, 15/8, 2

PARTSH/GREEK-12; Partch Greek scales from "Two Studies on Ancient Greek Scales" on black/white. The black notes are the 5-tone scale of Olympus (C♮ =

1/1). The white notes are the 7-tone enharmonic (C = 1/1).;
28/27, 9/8, 16/15, 4/3, 6/5, 3/2, 3/2, 14/9, 8/5, 8/5, 2

PEL/JC-12; Pelog/BH Slendro, c.f. [7];
8/7, 8/7 9/8, 64/49, 6/5, 3/2 3/2, 3/2, 12/7, 8/5, 2

PENTAGON(1)-12; Pentagonal scale 9/8, 3/2, 16/15, 4/3, 5/3;
27/25, 9/8, 6/5, 81/64, 729/512, 36/25, 243/160, 81/50, 27/16, 9/5, 243/128, 2

PENTAGON(3)-12; Pentagonal scale 7/4, 4/3, 15/8, 32/21, 6/5;
49/48, 35/32, 7/6, 1225/1024, 245/192 49/36, 25/18, 49/32, 5/3, 7/4, 175/96, 2

PERKIS-12; Perkis 60-30;
15/14, 10/9, 6/5, 5/4, 4/3, 10/7, 3/2, 30/19, 5/3, 12/7, 15/8, 2

PHRYGIAN(OLD)-12; Old Phrygian (?);
10/9, 6/5, 5/4, 4/3, 27/20, 40/27, 3/2, 8/5, 5/3, 16/9, 9/5, 2

PHRYGIAN(ENH)-12; Phrygian Enharmonic Tonos;
18/17, 9/8, 36/31, 72/61, 6/5, 4/3, 3/2, 72/47, 48/31, 36/2372/41, 2

PHRYGIAN(HARM)-12; Phrygian Harmonia-Aliquot 24 (flute tuning);
24/23, 12/11, 8/7, 6/5, 24/19, 4/3, 24/17, 3/2, 8/5, 12/7, 24/13, 2

PIANO(7-LIM)-12; Enhanced piano 7-limit;
135/128, 9/8, 7/6, 5/4, 4/3, 45/32 3/2, 14/9, 27/16, 7/4, 15/8, 2

PURE(MAJ)-12; Pure (just) C major, c.f. Wilkinson: Tuning In;
16/15, 9/8, 6/5, 5/4, 4/3, 45/32, 3/2, 8/5, 5/3, 16/9, 15/8, 2

PURE(MIN)-12; Pure (just) C minor, c.f. Wilkinson: Tuning In;
25/24, 10/9, 6/5, 5/4, 4/3, 45/32, 3/2, 8/5, 5/3, 16/9, 15/8, 2

PYRAMID-UP-12; This scale may also be called the "Wedding Cake";
9/8, 75/64, 5/4, 4/3, 45/32, 3/2, 25/16, 5/3, 27/16, 16/9, 15/8, 2

PYRAMID-DOWN-12; Upside-Down Wedding Cake (divorce cake);
16/15, 9/8, 6/5, 32/25, 4/3, 3/2, 8/5, 27/16, 16/9, 9/5, 48/25, 2

REINHARD-12; J. Reinhard;
18/17, 9/8, 45/38, 5/4, 4/3, 24/17, 3/2, 30/19, 5/3, 30/17, 15/8, 2

RILEY-12; Terry Riley's Harp of New Albion scale;
16/15, 9/8, 6/5, 5/4, 4/3, 64/45, 3/2, 8/5, 5/3, 16/9, 15/8, 2

ROBOT-12; Dead Robot (see lattice);
25/24, 16/15, 9/8, 75/64, 6/5, 5/4, 4/3, 45/32, 3/2, 5/3, 15/8, 2

ROBOT-LIVE-12; Live Robot;
9/8, 6/5, 5/4, 32/25, 4/3, 36/25, 3/2, 8/5, 128/75, 15/8, 48/25, 2

RSR-12; RSR 7-limit JI;
16/15, 8/7, 6/5, 5/4, 4/3, 7/5, 3/2, 8/5, 5/3, 9/5, 15/8, 2

SCHISMIC-12; Scale with major thirds flat by a schisma;
2187/2048, 9/8, 19683/16384, 8192/6561, 4/3, 1024/729,
3/2, 6561/4096, 32768/19683, 59049/32768, 4096/2187, 2

SEIKILOS-12; Seikilos Tuning;
28/27, 9/8, 7/6, 9/7, 4/3, 49/36, 3/2, 14/9, 27/16, 7/4, 27/14, 2

SHENG-12; Sheng scale on naturals starting on d, from [17];
141/134, 34/31, 55/46, 71/58, 4/3, 80/57, 117/80, 107/67, 63/38, 59/33, 63/34, 2

SIMONTON-12; Simonton Integral Ratio Scale, c.f. JASA: A new integral ratio scale;
17/16, 9/8, 19/16, 5/4, 4/3, 17/12, 3/2, 19/12, 5/3, 16/9, 17/9, 2

SLE(MAT)-12; Dudon's Slendro Matrix from "Seven-Limit Slendro Mutations," 1/1 8:2, Jan 1994;
8/7, 8/7, 64/49, 21/16, 4/3, 3/2, 32/21, 12/7, 256/147, 7/4, 2

SONGLINES-12; Songlines. DEM, Bill Thibault and Scott Gresham-Lancaster. 1992 ICMC;
7/6, 6/5, 5/4, 4/3, 7/5, 3/2, 8/5, 5/3, 7/4, 9/5, 11/6, 2

STANHOPE-12; Well temperament of Charles, third earl of Stanhope, 1806;
91.202, 196.741, 295.112, 5/4, 4/3, 589.247, 3/2, 793.157, 891.527, 16/9, 15/8, 2

STEL/HEX(NO.2)-12; Stellated two out of 1 3 5 9 hexany;
135/128, 9/8, 5/4, 81/64, 27/20, 45/32, 3/2, 25/16, 5/3, 27/16, 15/8, 2

SUB40-12; 12 of sub 40;
20/19, 10/9, 20/17, 5/4, 4/3, 10/7, 20/13, 8/5, 5/3, 20/11, 40/21, 2

SUB48-12; 12 of sub 48 (Leven);
16/15, 8/7, 6/5, 24/19, 4/3, 24/17, 3/2, 8/5, 12/7, 16/9, 48/25, 2

SUB50-12; 12 of sub 50;
25/24, 10/9, 25/21, 5/4, 25/19, 10/7, 25/17, 25/16, 5/3, 25/14, 50/27, 2

TAU-SIDE-12; Tau-on-Side;
25/24, 16/15, 9/8, 5/4, 4/3, 45/32, 3/2, 25/16, 8/5, 5/3, 15/8, 2

TANBUR-12; Sub-40 tanbur scale;
40/39, 20/19, 40/37, 10/9, 8/7, 320/273, 160/133, 320/259, 80/63, 64/49, 160/119, 2

TETRAGAM(DIAT/2)-12; Tetragam Dia2;
16/15, 10/9, 10/9, 5/4, 4/3, 64/45, 3/2, 8/5 5/3, 5/3, 7/4, 2

TETRAGAM(ENH)-12; Tetragam Enharm.;
28/27, 16/15, 16/15, 5/4, 4/3, 7/5, 3/2, 14/9, 8/5, 8/5, 7/4, 2

TETRAGAM/HEXGAM-12; Tetragam/Hexgam;
28/27, 9/8, 7/6, 5/4, 21/16, 35/24, 3/2, 14/9, 5/3, 7/4, 15/8, 2

TETRAGAM(PYTH)-12; Tetragam Pyth.;
256/243, 9/8, 9/8, 81/64, 4/3, 729/512, 3/2, 128/81, 27/16, 27/16, 16/9, 2

TETRAGAM-SP-12; Tetragam Septimal;
28/27, 28/27, 28/27, 9/7, 4/3, 7/5, 3/2, 14/9, 14/9, 14/9, 7/4, 2

TETRAGAM-UN-12; Tetragam Undecimal;
33/32, 12/11, 12/11, 11/9, 4/3, 11/8, 3/2, 99/64, 18/11, 18/11, 11/6, 2

TRI(1)-12; 12-tone Tritriadic of 7:9:11;
99/98, 81/77, 11/9, 121/98, 14/11, 9/7, 14/9, 11/7, 18/11, 81/49, 121/63, 2

TRIAPHONIC-12; 12-tone Triaphonic Cycle, conjunctive form on 4/3, 5/4 and 6/5;
20/19, 10/9, 20/17, 5/4, 4/3, 80/57, 40/27, 80/51, 5/3, 30/17, 15/8, 2

TT456DTDMT-12; TT456DTDMT;
25/24, 9/8, 6/5, 5/4, 4/3, 25/16, 3/2 8/5, 5/3, 9/5, 15/8, 2

TT679DTDMMT-12; TT679DTDMMT;
9/8, 7/6, 9/7, 4/3, 49/36, 3/2, 14/9, 12/7, 7/4, 49/27, 27/14, 2

WIL-HELIX-12; Wilson's Helix Song, c.f. David Rosenthal, Helix Song, XH 7&8, 1979;
13/12, 9/8, 7/6, 5/4, 4/3, 11/8, 3/2, 13/8, 5/3, 7/4, 11/6, 2

YASSER(JI)-12; Yasser's JI Scale, 2 Yasser hexads, a 121/91 apart;
121/112, 9/8, 121/104, 5/4, 121/91, 11/8, 1089/728, 13/8, 605/364, 7/4, 1331/728, 2

YASSER(DIAT)-12; Yasser's Supra-Diatonic;
126.316, 189.474, 6/5, 378.947, 505.263, 631.579, 694.737, 821.053, 5/3, 1010.526, 1073.684, 2

YUGOBAG-12; Yugoslavian bagpipe tuning;
99, 202, 362, 463, 655, 754, 861, 949, 991, 1 047, 1 129, 1 200

A.6   More than 12 tones per octave

ANY-19; 2 out of 1/7 1/5 1/3 1 3 5 7 CPS;
16/15, 35/32, 8/7, 7/6, 6/5, 5/4, 21/16, 4/3, 7/5, 10/7, 3/2, 32/21, 8/5, 5/3, 12/7, 7/4, 64/35, 15/8, 2

ANY-21; 1.3.5.7.11.13 2)7 21-any, 1.3 tonic;
33/32, 13/12, 9/8, 55/48, 7/6, 39/32, 5/4, 21/16, 65/48, 11/8, 35/24, 143/96, 3/2, 77/48, 13/8, 5/3, 7/4, 11/6, 15/8, 91/48, 2

ANY-26; 6)8 28-any from 1.3.5.7.9.11.13, only 26 tones;
65/64, 15/14, 13/12, 195/176, 65/56, 13/11, 39/32, 5/4, 195/154, 13/10, 65/48, 15/11, 39/28, 13/9, 65/44, 3/2, 65/42, 13/8, 5/3, 195/112, 39/22, 65/36, 13/7, 15/8, 65/33, 2

ANY56-48; 3)8 56-any from 1.3.5.7.11.13, 1.3.5 tonic, only 48 tones;
65/64, 33/32, 1001/960, 21/20, 13/12, 35/32, 11/10, 143/128, 9/8, 91/80, 7/6, 143/120, 77/64, 39/32, 99/80, 5/4, 77/60, 13/10, 21/16, 429/320, 11/8, 7/5, 45/32, 91/64, 231/160, 117/80, 143/96, 3/2, 91/60, 99/64, 63/40, 77/48, 13/8, 33/20, 27/16, 273/160, 55/32, 7/4, 143/80, 9/5, 117/64, 11/6, 15/8, 91/48, 77/40, 39/20, 63/32, 2

ANY-70; 1.3.5.7.11.13.17.19, 4:8, 70-any, tonic 1.3.5.7;
323/320, 2 717/2 688, 143/140, 247/240, 3 553/3 360, 17/16, 13/12, 2431/2240, 209/192, 4 199/3 840, 11/10, 247/224, 187/168, 221/192, 323/280, 187/160, 19/16, 143/120, 2 717/2 240, 17/14, 209/168, 4 199/3 360, 2 431/1 920, 143/112, 247/192, 13/10, 209/160, 221/168, 3 553/2 688, 187/140, 323/240, 19/14, 11/8,

221/160, 2 717/1 920, 17/12, 323/224, 2 431/1 680, 247/168, 143/96, 209/140,
247/160, 187/120, 4 199/2 688, 11/7, 221/140, 19/12, 3 553/2 240, 2 717/1 680,
13/8, 187/112, 323/192, 17/10, 143/84, 46 189/26 880, 209/120, 247/140, 143/80,
2431/1344, 11/6, 221/120, 3 553/1 920, 13/7, 209/112, 4 199/2 240, 19/10, 323/168,
187/96, 221/112, 2

ARIEL-19; Ariel's 19-tone scale;
25/24, 16/15, 10/9, 125/108, 6/5, 5/4, 32/25, 4/3, 25/18, 36/25, 3/2, 25/16, 8/5,
5/3, 216/125, 9/5, 15/8, 48/25, 2

ARIEL-31; Ariel's 19-tone scale;
128/125, 25/24, 16/15, 625/576, 9/8, 144/125, 75/64, 6/5, 625/512, 5/4, 32/25,
125/96, 4/3, 512/375, 25/18, 36/25, 375/256, 3/2, 192/125, 25/16, 8/5,
1 024/625, 5/3, 128/75, 125/72, 16/9, 1 152/625, 15/8, 48/25, 125/64, 2

BARSTOW-18; Guitar scale for Partch's Barstow;
16/15, 11/10, 10/9, 9/8, 8/7, 6/5, 5/4, 4/3, 11/8, 10/7, 3/2, 8/5, 5/3, 12/7, 9/5,
11/6, 15/8, 2

BELET-13; Brian Belet, Proc. of the ICMC 1992, pp. 158–161;
16/15, 10/9, 9/8, 6/5, 5/4, 4/3, 11/8, 3/2, 8/5, 13/8, 7/4, 15/8, 2

BURT-19; W. Burt 19-tone Forks. Interval 5(3): pp. 13+23 Winter 1986-87;
28/27, 16/15, 10/9, 9/8, 6/5, 5/4, 9/7, 4/3, 7/5, 10/7, 3/2, 14/9, 8/5, 5/3, 16/9,
9/5, 15/8, 27/14, 2

CATLER-24; Catler 24-tone JI from "Over and Under the 13 Limit," 1/1 3(3);
33/32, 16/15, 9/8, 8/7, 7/6, 6/5, 128/105, 16/13, 5/4, 21/16, 4/3, 11/8, 45/32,
16/11, 3/2, 8/5, 13/8, 5/3, 27/16, 7/4, 16/9, 24/13, 15/8, 2

CHALMERS-19; 19-tone with more hexanies than Perrett's Tierce-Tone, c.f. [7];
21/20, 16/15, 9/8, 7/6, 6/5, 5/4, 21/16, 4/3, 7/5, 35/24, 3/2, 63/40, 8/5, 5/3,
7/4, 9/5, 28/15, 63/32, 2

CHORDAL-40; Chordal Notes S & H;
3/2, 5/4, 7/4, 9/4, 11/4, 13/4, 15/4, 15/4, 15/8, 17/8, 19/8, 19/16, 2, 4/3, 8/5,
8/7, 16/9, 16/11, 16/13, 16/15, 7/3, 7/2, 10/3, 8/3, 5/2, 12/5, 12/7, 11/9, 13/9,
17/10, 17/5, 9/7, 9/8, 16/9, 11/7, 7/6, 7/5, 10/7, 6/5, 9/5

RING(CK/NO.1)-13; Double-tie circular mirroring, common pivot of 4:5:6:7;
8/7, 7/6, 6/5, 5/4, 4/3, 7/5, 10/7, 3/2, 8/5, 5/3, 12/7, 7/4, 2

RING(CK/NO.2)-13; Double-tie circular mirroring, common pivot of 3:5:7:9;
10/9, 7/6, 6/5, 9/7, 4/3, 7/5, 10/7, 3/2, 14/9, 5/3, 12/7, 9/5, 2

CLUSTER-13; 13-tone 5-limit Tritriadic Cluster;
25/24, 9/8, 6/5, 5/4, 4/3, 36/25, 3/2, 25/16, 8/5, 5/3, 9/5, 15/8, 2

CONCERTINA-14; English Concertina, c.f. Helmholtz, p. 470. from Ellis;
25/24, 10/9, 9/8, 75/64, 5/4, 4/3, 45/32, 3/2, 25/16, 5/3, 27/16, 16/9, 15/8, 2

DARREG-19; This set of 19 ratios in 5-limit JI is for his megalyra family;
25/24, 16/15, 10/9, 9/8, 75/64, 6/5, 5/4, 4/3, 45/32, 64/45, 3/2, 25/16, 8/5,
5/3, 27/16, 225/128, 9/5, 15/8, 2

DAVID(11-LIM)-22; 11-limit system from Gary David, 1967;
33/32, 21/20, 12/11, 9/8, 7/6, 77/64, 5/4, 14/11, 21/16, 11/8, 7/5, 63/44, 3/2,
14/9, 77/48, 18/11, 27/16, 7/4, 11/6, 15/8, 21/11, 2

DIACYCLE(13)-23; Diacycle on 20/13, 13/10; there are also nodes at 3/2, 4/3;
13/9, 18/13;
40/39, 20/19, 40/37, 10/9, 8/7, 20/17, 40/33, 5/4, 40/31, 4/3, 40/29, 10/7,
40/27, 20/13, 30/19, 60/37, 5/3, 12/7, 30/17, 20/11, 15/8, 60/31, 2

DIAMOND(33-LIM)-59; 33-limit Diamond + 2nd ratios. See Novaro: , Sistema
Natural . . . , 1927;
33/32, 16/15, 15/14, 14/13, 13/12, 12/11, 11/10, 10/9, 9/8, 8/7, 15/13, 7/6,
13/11, 32/27, 6/5, 39/32, 11/9, 16/13, 5/4, 14/11, 9/7, 13/10, 21/16, 4/3, 15/11,
11/8, 18/13, 7/5, 45/32, 64/45, 10/7, 13/9, 16/11, 22/15, 3/2, 32/21, 20/13,
14/9, 11/7, 8/5, 13/8, 18/11, 64/39, 5/3, 27/16, 22/13, 12/7, 26/15, 7/4, 16/9,
9/5, 20/11, 11/6, 24/13, 13/7, 28/15, 15/8, 64/33, 2

DIAMOND(9-LIM)-19; 9-limit Diamond;
10/9, 9/8, 8/7, 7/6, 6/5, 5/4, 9/7, 4/3, 7/5, 10/7, 3/2, 14/9, 8/5, 5/3, 12/7, 7/4,
16/9, 9/5, 2

DIAMOND(OMD)-13; 13-tone Octave Modular Diamond, based on Archytas's
Enharmonic;
36/35, 28/27, 16/15, 5/4, 9/7, 4/3, 3/2, 14/9, 8/5, 15/8, 27/14, 35/18, 2

DORIAN/CHRO-24; Dorian Chromatic Tonos;
16/15, 8/7, 32/27, 64/53, 16/13, 4/3, 16/11, 32/21, 64/41, 8/5, 16/9, 2, 32/15,
16/7, 64/27, 128/53, 32/13, 8/3, 32/11, 64/21, 128/41, 16/5, 32/9, 4

DORIAN/DIAT-24; Dorian Diatonic Tonos;
16/15, 8/7, 16/13, 32/25, 4/3, 32/23, 16/11, 8/5, 32/19, 16/9, 32/17, 2, 32/15,
16/7, 32/13, 64/25, 8/3, 64/23, 32/11, 16/5, 64/19, 32/9, 64/17, 4

DORIAN/ENH)-24; Dorian Enharmonic Tonos;
16/15, 8/7, 64/55, 128/109, 32/27, 4/3, 16/11, 64/43, 128/85, 32/21, 16/9, 2,
32/15, 16/7, 128/55, 256/109, 64/27, 8/3, 32/11, 128/43, 256/85, 64/21, 32/9, 4

DORIAN/PIS-15; Diatonic PIS in the Dorian Tonos, a non-rep. 16-tone gamut;
8/7, 16/13, 4/3, 16/11, 8/5, 16/9, 2, 32/15, 16/7, 32/13, 8/3, 32/11, 16/5, 32/9, 4

EFG333333333337-24; Genus [333333333337];
137 781/131 072, 2 187/2 048, 567/512, 9/8, 1 240 029/1 048 576, 19 683/16 384,
5103/4 096, 81/64, 21/16, 177 147/131 072, 45 927/32 768, 729/512, 189/128, 3/2,
413 343/262 144, 6 561/4 096, 1 701/1 024, 27/16, 7/4, 59 049/32 768,
15 309/8 192, 243/128, 63/32, 2

EFG33335555-25; Genus bis-ultra-chromaticum [33335555];
25/24, 16/15, 10/9, 9/8, 256/225, 75/64, 6/5, 5/4, 32/25, 4/3, 25/18, 45/32,
64/45, 36/25, 3/2, 25/16, 8/5, 5/3, 128/75, 225/128, 16/9, 9/5, 15/8, 48/25, 2

EFG333555-16; Genus diatonico-hyperchromaticum [333555];
25/24, 16/15, 10/9, 75/64, 6/5, 5/4, 4/3, 25/18, 64/45, 3/2, 25/16, 8/5, 5/3,
16/9, 15/8, 2

EFG; Genus diatonico-enharmonicum [333557];
64/63, 16/15, 15/14, 10/9, 8/7, 6/5, 128/105, 5/4, 80/63, 4/3, 48/35, 64/45, 10/7, 3/2, 32/21, 8/5, 512/315, 5/3, 12/7, 16/9, 64/35, 15/8, 40/21, 2

EFG335577-27; Genus chromaticum septimis triplex [335577];
21/20, 16/15, 15/14, 35/32, 8/7, 7/6, 6/5, 128/105, 5/4, 21/16, 4/3, 48/35, 7/5, 10/7, 35/24, 3/2, 32/21, 8/5, 105/64, 5/3, 12/7, 7/4, 64/35, 28/15, 15/8, 40/21, 2

EFG35711-15; Genus [3 5 7 11];
33/32, 11/10, 8/7, 33/28, 6/5, 44/35, 48/35, 3/2, 11/7, 8/5, 33/20, 12/7, 64/35, 66/35, 2

EIKOSANY-20; 3:6, 1.3.5.7.9.11 Eikosany (1.3.5 tonic);
33/32, 21/20, 11/10, 9/8, 7/6, 99/80, 77/60, 21/16, 11/8, 7/5, 231/160, 3/2, 63/40, 77/48, 33/20, 7/4, 9/5, 11/6, 77/40, 2

EPIMORE-40; Epimore (Scholz);
4, 5, 6, 7, 8, 9, 10, 11, 12, 13, 14, 15, 16, 18, 20, 21, 22, 24, 25, 26, 27, 28, 32, 33, 35, 36, 39, 40, 42, 44, 45, 48, 49, 50, 54, 55, 56, 63, 64, 65

EXP/HEX-32; Expanded hexany 1.3.5.7.9.11;
2 079/2 048, 33/32, 135/128, 35/32, 9/8, 1 155/1 024, 297/256, 77/64, 315/256, 5/4, 10 395/8 192, 165/128, 21/16, 693/512, 11/8, 45/32, 1 485/1 024, 189/128, 3/2, 385/256, 99/64, 105/64, 27/16, 3 465/2 048, 55/32, 7/4, 231/128, 945/512, 15/8, 495/256, 63/32, 2

GARCIA-29; Linear 29-tone scale by Jose L. Garcia, 1988 15/13-52/45 alternating;
40/39, 27/26 16/15, 128/117, 9/8, 15/13, 32/27, 6/5, 16/13, 81/64, 135/104, 4/3, 160/117, 18/13, 64/45, 512/351, 3/2, 20/13, 81/52, 8/5, 64/39, 27/16, 45/26, 16/9, 9/5, 24/13, 256/135, 405/208, 2

GRADY-14; Kraig Grady, letter to L. Harrison, published in 1/1 7 (1) 1991 p. 5. Has 13 species of superparticular pentatonic scales and modes;
21/20, 9/8, 7/6, 5/4, 21/16, 4/3, 7/5, 3/2, 63/40, 27/16, 7/4, 15/8, 63/32, 2

HANSON-19; JI version of Hanson's 19 out of 53-TET scale;
25/24, 27/25, 9/8, 125/108, 6/5, 5/4, 125/96, 4/3, 25/18, 36/25, 3/2, 25/16, 8/5, 5/3, 125/72, 9/5, 15/8, 48/25, 2

HARM(10)-13; 6/7/8/9/10 harmonics;
35/32, 9/8, 5/4, 81/64, 21/16, 45/32, 3/2, 49/32, 25/16, 27/16, 7/4, 63/32, 2

HARM(15)-15; Fifth octave of the harmonic overtone series;
17/16, 9/8, 19/16, 5/4, 21/16, 11/8, 23/16, 3/2, 25/16, 13/8, 27/16, 7/4, 29/16, 15/8, 31/16

HARM(30-30)-59; First 30 subharmonics & harmonics;
16/15, 32/29, 8/7, 32/27, 16/13, 32/25, 4/3, 32/23, 32/21, 8/5, 32/19, 16/9, 32/17, 2, 32/15, 16/7, 32/13, 8/3, 32/11, 16/5, 32/9, 4 , 32/7, 16/3, 32/5, 8/1, 32/3, 16/1, 32/1, 33/1, 34/1, 35/1, 36/1, 37/1, 38/1, 39/1, 40/1, 41/1, 42/1, 43/1, 44/1, 45/1, 46/1, 47/1, 48/1, 49/1, 50/1, 51/1, 52/1, 53/1, 54/1, 55/1, 56/1, 57/1, 58/1, 59/1, 60/1, 61/1, 62/1 15/14, 10/9, 6/5, 5/4, 4/3, 10/7, 3/2, 30/19, 5/3, 12/7, 15/8, 2

HARM(7-LIM)-47; 7-limit harmonics;
2, 3, 4 , 5, 6, 7, 8, 9, 10, 12, 14, 15, 16, 18, 20, 21, 22, 24, 25, 28, 30, 32, 35, 36,
40, 42, 45, 48, 49, 50, 56, 60, 63, 64, 70, 72, 75, 80, 81, 84, 90, 96, 98, 100, 105,
112, 120

HARRISON-16;
Harrison 16-tone, L. Harrison 16tone scale based on Boethius's account of Pto-
lemy's superparticular divisions of the 4/3 as 16/15 × 5/4, 10/9 × 6/5, and 8/7
× 7/6. See 1/1 7(1)(1991), 4-5;
16/15, 10/9, 8/7, 7/6, 6/5, 5/4, 4/3, 17/12, 3/2, 8/5, 5/3, 12/7, 7/4, 9/5, 15/8, 2

DIAMOND(HEPT)-25; Inverted-Prime Heptatonic Diamond based on Archytas's
Enharmonic;
36/35, 28/27, 16/15, 9/8, 7/6, 6/5, 98/81, 56/45, 5/4, 32/25, 9/7, 4/3, 3/2, 14/9,
25/16, 8/5, 45/28, 81/49, 5/3, 12/7, 16/9, 15/8, 27/14, 35/18, 2

DIAMOND(HEPT/P)-27;
Heptatonic Diamond based on Archytas's Enharmonic, tones;
36/35, 28/27, 16/15, 9/8, 7/6, 6/5, 5/4, 9/7, 35/27, 4/3, 48/35, 112/81, 45/32,
64/45, 81/56, 35/24, 3/2, 54/35, 14/9, 8/5, 5/3, 12/7, 16/9, 15/8, 27/14, 35/18, 2

HERF-14; E. Sims: Reflections on This and That, 1991. Used by Herf in Ek-
melischer Gesang;
33/32, 17/16, 9/8, 19/16, 5/4, 21/16, 11/8, 23/16, 3/2, 13/8, 27/16, 7/4, 15/8, 2

HEXAGONAL-13; Star hexagonal 13-tone scale;
25/24, 16/15, 10/9, 6/5, 5/4, 4/3, 3/2, 8/5, 5/3, 9/5, 15/8, 48/25, 2

HEXAGONAL-37; Star hexagonal 37-tone scale;
25/24, 16/15, 27/25, 625/576, 10/9, 9/8, 256/225, 144/125, 75/64, 6/5, 100/81,
5/4, 32/25, 125/96, 4/3, 27/20, 25/18, 45/32, 64/45, 36/25, 40/27, 3/2, 192/125,
25/16, 8/5, 81/50, 5/3, 128/75, 125/72, 225/128, 16/9, 9/5, 1152/625, 50/27,
15/8, 48/25, 2

HYPODORIAN(PIS)-15; Diatonic perfect immutable system (PIS) in the hy-
podorian tonos;
12/11, 6/5, 4/3, 3/2, 8/5, 24/13, 2, 48/23, 24/11, 12/5, 8/3, 3/1, 24/7, 48/13, 4

HYPOLYDIAN/PIS-15; The Diatonic PIS in the Hypolydian Tonos;
14/13, 7/6, 14/11, 7/5, 14/9, 7/4, 28/15, 2, 28/13, 7/3, 28/11, 14/5, 28/9, 7/2, 4

HYPOPHRYG(PIS)-15;
The Diatonic PIS in the Hypophrygian Tonos;
13/12, 13/11, 13/10, 13/9, 13/8, 26/15, 2, 52/25, 13/6, 26/11, 13/5, 26/9, 13/4,
26/7, 4 *IIVV17-21; 17-limit IIVV;*

33/32, 17/16, 13/12, 9/8, 7/6, 39/32, 5/4, 21/16, 4/3, 11/8, 45/32, 17/12, 3/2,
51/32, 13/8, 5/3, 27/16, 7/4, 11/6, 15/8, 2

JUST(5-LIM)-31; A just 5-limit 31-tone scale;
128/125, 135/128, 27/25, 1125/1 024, 9/8, 144/125, 75/64, 6/5, 625/512, 5/4,
32/25, 675/512, 27/20, 5625/4 096, 45/32, 36/25, 375/256, 3/2, 192/125, 25/16,
8/5, 3 375/2 048, 27/16, 216/125, 225/128, 9/5, 1875/1 024, 15/8, 48/25, 125/64, 2

JUST(7-LIM)-31; A just 7-limit 31-tone scale;
128/125, 25/24, 16/15, 35/32, 9/8, 8/7, 7/6, 6/5, 128/105, 5/4, 32/25, 21/16, 4/3, 48/35, 7/5, 10/7, 35/24, 3/2, 32/21, 25/16, 8/5, 105/64, 5/3, 12/7, 7/4, 16/9, 64/35, 15/8, 48/25, 125/64, 2

JOHNSTON(JUST/ENH)-21; Johnston 21-tone just enharmonic scale;
25/24, 27/25, 9/8, 75/64, 6/5, 5/4, 32/25, 125/96, 4/3, 25/18, 36/25, 3/2, 25/16, 8/5, 5/3, 125/72, 9/5, 15/8, 48/25, 125/64, 2

JOHNSTON(JUST/ENH)-25; Johnston 25-tone just enharmonic scale;
25/24, 135/128, 16/15, 10/9, 9/8, 75/64, 6/5, 5/4, 81/64, 32/25, 4/3, 27/20, 45/32, 36/25, 3/2, 25/16, 8/5, 5/3, 27/16, 225/128, 16/9, 9/5, 15/8, 48/25, 2

KANZELMEYER-18;
Kanzelmeyer, 18 harmonics from 32 to 64. Base 388.3614815 Hz;
17/16, 37/32, 19/16, 5/4, 41/32, 43/32, 11/8, 23/16, 47/32, 3/2, 13/8, 53/32, 7/4, 29/16, 59/32, 61/32, 31/16, 2

KANZELMEYER-32;
Kanzelmeyer, 32 harmonics from 32 to 64. Base 388.3614815 Hz;
33/32, 17/16, 35/32, 9/8, 37/32, 19/16, 39/32, 5/4, 41/32, 21/16, 43/32, 11/8, 45/32, 23/16, 47/32, 3/2, 49/32, 25/16, 51/32, 13/8, 53/32, 27/16, 55/32, 7/4, 57/32, 29/16, 59/32, 15/8, 61/32, 31/16, 63/32, 2

LYDIAN/CHRO-24; Lydian Chromatic Tonos;
20/19, 10/9, 20/17, 40/33, 5/4, 10/7, 20/13, 8/5, 80/49, 5/3, 20/11, 2, 40/19, 20/9, 40/17, 80/33, 5/2, 20/7, 40/13, 16/5, 160/49, 10/3, 40/11, 4

LYDIAN/DIAT-24; Lydian Diatonic Tonos;
20/19, 10/9, 5/4, 4/3, 10/7, 40/27, 20/13, 5/3, 40/23, 20/11, 40/21, 2, 40/19, 20/9, 5/2, 8/3, 20/7, 88/27, 40/13, 10/3, 80/23, 40/11, 80/21, 4

SCHLESINGER(LYDIAN/DIAT)-26; Schlesinger's Lydian Harmonia, a subharmonic series through 13 from 26;
13/12, 13/11, 13/10, 26/19, 13/9, 13/8, 13/7, 2

LYDIAN/ENH)-24; Lydian Enharmonic Tonos;
20/19, 10/9, 8/7, 80/69, 20/17, 10/7, 20/13, 80/51, 160/101, 8/5, 20/11, 2, 40/19, 20/9, 16/7, 160/69, 40/17, 20/7, 40/13, 160/51, 320/101, 16/5, 40/11, 4

LYDIAN/PIS-15; The Diatonic PIS in the Lydian Tonos;
10/9, 5/4, 10/7, 20/13, 5/3, 20/11, 2, 40/19, 20/9, 5/2, 20/7, 40/13, 10/3, 40/11, 4

MANDELBAUM(5-LIM)-19; Mandelbaum's 5-limit 19-tone scale;
25/24, 27/25, 10/9, 125/108, 6/5, 5/4, 125/96, 4/3, 25/18, 36/25, 3/2, 125/81, 8/5, 5/3, 125/72, 9/5, 50/27, 48/25, 2

MANDELBAUM(7-LIM)-19; Mandelbaum's septimal 19-tone scale;
25/24, 15/14, 9/8, 7/6, 6/5, 5/4, 9/7, 4/3, 7/5, 36/25, 3/2, 14/9, 8/5, 5/3, 7/4, 9/5, 15/8, 27/14, 2

MARION(NO.1)-24; Marion's 7-limit Scale no. 1;
225/224, 25/24, 15/14, 35/32, 9/8, 7/6, 25/21, 5/4, 9/7, 21/16, 75/56, 45/32, 10/7, 35/24, 3/2, 25/16, 45/28, 5/3, 7/4, 25/14, 175/96, 15/8, 63/32, 2

MARION(NO.10)-25; Marion's 7-limit Scale no. 10;
49/48, 25/24, 35/32, 10/9, 245/216, 7/6, 175/144, 5/4, 35/27, 49/36, 25/18,
1225/864, 35/24, 49/32, 14/9, 25/16, 175/108, 5/3, 245/144, 7/4, 49/27, 175/96,
50/27, 35/18, 2

MARION(NO.15)-24; Marion's 7-limit Scale no. 15;
36/35, 15/14, 54/49, 8/7, 6/5, 60/49, 5/4, 9/7, 27/20, 48/35, 135/98, 10/7,
72/49, 3/2, 54/35, 8/5, 45/28, 80/49, 12/7, 432/245, 9/5, 90/49, 27/14, 2

MARION(NO.19)-25; Marion's 7-limit Scale no. 19;
21/20, 15/14, 35/32, 9/8, 189/160, 6/5, 135/112, 5/4, 9/7, 21/16, 27/20, 7/5,
45/32, 10/7, 3/2, 54/35, 63/40, 45/28, 27/16, 7/4, 9/5, 15/8, 27/14, 63/32, 2

MARION(NO.26)-24; Marion's 7-limit Scale no. 26;
28/27, 16/15, 49/45, 28/25, 784/675, 7/6, 32/27, 56/45, 32/25, 98/75, 4/3, 7/5,
196/135, 112/75, 14/9, 8/5, 49/30, 224/135, 392/225, 16/9, 49/27, 28/15, 49/25, 2

MCCLAIN-18; McClain's 18-tone scale, c.f. page 143 of The Myth of Invariance;
135/128, 9/8, 75/64, 625/512, 5/4, 81/64, 675/512, 45/32, 375/256, 3/2, 25/16,
405/256, 27/16, 225/128, 15/8, 243/128, 125/64, 2

METAL/BAR-13; Metal bar scale, c.f. McLaren, Xenharmonicon 15, pp.31-33;
128.442, 191.007, 264.247, 378.214, 394.918, 520.840, 555.813, 642.342, 724.750,
759.727, 885.821, 1039.735, 1193.099

MEYER-19; Max Meyer, c.f. Doty, David, 1/1 August 1992 (7:4) p.1 and 10-14;
16/15, 10/9, 9/8, 8/7, 7/6, 6/5, 5/4,, 4/3, 7/5, 10/7, 3/2, 8/5, 5/3, 12/7, 7/4,
16/9, 9/5, 15/8, 2

MEYER-29; Max Meyer, c.f. Doty, David, 1/1 August 1992 (7:4) p.1 and 10-14;
525/512, 135/128, 35/32, 567/512, 9/8, 75/64, 315/256, 5/4, 81/64, 21/16,
675/512, 175/128, 45/32, 729/512, 375/256, 189/128, 3/2, 25/16, 405/256,
105/64, 27/16, 7/4, 225/128, 945/512, 15/8, 243/128, 125/64,, 63/32, 2

MIXOL/CHRO(NO.1)-24; Mixolydian chromatic tonos;
22/21, 11/10, 22/19, 44/37, 11/9, 11/8, 11/7, 44/27, 88/53, 22/13, 11/6, 2,
44/21, 11/5, 44/19, 88/37, 22/9, 11/4, 22/7, 88/27, 176/53, 44/13, 11/3, 4

MIXOL/DIAT-24; Mixolydian diatonic tonos;
22/21, 11/10, 11/9, 22/17, 11/8, 22/15, 11/7, 22/13, 44/25, 11/6 44/23, 2, 44/21,
11/5, 22/9, 44/17, 11/4, 44/15, 22/7, 44/13, 88/25, 11/3, 88/23, 4

MIXOL/ENH)-24; Mixolydian enharmonic tonos;
22/21, 11/10, 44/39, 8/7, 22/19, 4/3, 11/7, 8/5, 176/109, 44/27, 88/49, 2, 44/21,
11/5, 88/39, 16/7, 44/19, 8/3, 22/7, 16/5, 352/109, 88/27, 176/49, 4

MIXOL/PIS-15; The Diatonic PIS in the Mixolydian Tonos;
11/10, 11/9, 11/8, 11/7, 22/13, 11/6, 2, 44/21, 11/5, 22/9, 11/4, 22/7, 44/13,
11/3, 4

*NOVARO-23; 9-limit diamond with 21/20, 16/15, 15/8 and 40/21 added for
evenness;*
21/20, 16/15, 10/9, 9/8, 8/7, 7/6, 6/5, 5/4, 9/7, 4/3, 7/5, 10/7, 3/2, 14/9, 8/5,
5/3, 12/7, 7/4, 16/9, 9/5, 15/8, 40/21, 2

NOVARO-49; 1-15 diamond, c.f. Novaro: Sistema Natural base del Natural-Aproximado, 1927, p. 49;
16/15, 15/14, 14/13, 13/12, 12/11, 11/10, 10/9, 9/8, 8/7, 15/13,, 7/6, 13/11, 6/5, 11/9, 16/13, 5/4, 14/11, 9/7, 13/10, 4/3, 15/11, 11/8, 18/13, 7/5, 10/7, 13/9, 16/11, 22/15, 3/2, 20/13, 14/9, 11/7, 8/5, 13/8, 18/11, 5/3, 22/13, 12/7, 26/15, 7/4, 16/9, 9/5, 20/11, 11/6, 24/13, 13/7, 28/15, 15/8, 2

PARTCH-29; Partch 11-limit Diamond;
12/11, 11/10, 10/9, 9/8, 8/7, 7/6, 6/5, 11/9, 5/4, 14/11, 9/7, 4/3, 11/8, 7/5, 10/7, 16/11, 3/2, 14/9, 11/7, 8/5, 18/11, 5/3, 12/7, 7/4, 16/9, 9/5, 20/11, 11/6, 2

PARTCH-43; Harry Partch's 43-tone pure scale;
81/80, 33/32, 21/20, 16/15, 12/11 11/10, 10/9, 9/8, 8/7 7/6, 32/27, 6/5, 11/9, 5/4, 14/11, 9/7, 21/16, 4/3, 27/20, 11/8, 7/5, 10/7, 16/11, 40/27, 3/2, 32/21 14/9, 11/7, 8/5, 18/11, 5/3, 27/16, 12/7, 7/4, 16/9, 9/5, 20/11, 11/6, 15/8, 40/21, 64/33 160/81, 2

PENTADEKANY(NO.1)-15; 2)6 1.3.5.7.11.13 Pentadekany;
13/12, 55/48, 7/6, 5/4, 65/48 11/8, 35/24, 143/96, 77/48, 13/8, 5/3, 7/4 11/6, 91/48, 2

PENTADEKANY(NO.2)-15; 2)6 1.3.5.7.9.11 Pentadekany (1.3 tonic);
33/32, 9/8, 55/48, 7/6, 5/4 21/16, 11/8, 35/24, 3/2, 77/48, 5/3, 7/4, 11/6, 15/8, 2

PERRETT-14; Perrett's 14-tone system (subscale of tierce-tone);
21/20, 9/8, 7/6, 5/4, 21/16 4/3, 7/5, 3/2, 63/40, 5/3, 7/4, 15/8, 63/32, 2

PERRETT/TIERCE-19; Perrett Tierce-Tone;
21/20, 35/32, 9/8, 7/6, 6/5, 5/4, 21/16, 4/3, 7/5, 35/24, 3/2 63/40, 8/5, 5/3, 7/4, 9/5 15/8, 63/32, 2

PHRYG/PIS-15; The Diatonic PIS in the Phrygian Tonos;
9/8, 9/7, 18/13, 3/2, 18/11, 9/5, 2, 36/17, 9/4, 18/7, 36/13, 3/1, 36/11, 18/5, 4

PHRYGIAN/CHRO-24; Phrygian Chromatic Tonos;
18/17, 9/8, 6/5, 36/29, 9/7, 18/13, 3/2, 36/23, 8/5, 18/11, 9/5, 2, 36/17, 9/4, 12/5, 72/29, 18/7, 36/13, 3/1, 72/23, 16/5, 36/11, 18/5, 4

PHRYGIAN/DIAT-24; Phrygian Diatonic Tonos;
18/17, 9/8, 9/7, 4/3, 18/13, 36/25, 3/2, 18/11, 12/7, 9/5, 36/19, 2 , 36/17, 9/4, 18/7, 8/3, 36/13, 72/25, 3/1, 36/11, 24/7, 18/5, 72/19, 4

PIANO-19; Enhanced Piano Total Gamut, c.f. 1/1 vol. 8/2 January 1994;
135/128, 16/15, 9/8, 7/6, 6/5, 5/4, 4/3, 45/32, 64/45, 3/2, 14/9, 8/5 (?), 5/3, 27/16, 7/4, 16/9, 15/8, 63/32, 2

PIPEDUM-31; Scale with homophonic intervals 81/80 225/224 1029/1024;
8 505/8 192, 135/128, 2 205/2 048, 567/512, 9/8, 147/128, 1 215/1 024, 19 845/16 384, 315/256, 81/64, 1 323/1 024, 21/16, 178 605/131 072, 2 835/2 048, 45/32, 11 907/8 192, 189/128, 3/2, 25 515/16 384, 405/256, 6 615/4 096, 1 701/1 024, 27/16, 441/256, 3 645/2 048, 59 535/32 768, 945/512, 243/128, 3 969/2 048, 63/32, 2

PURE-19; 5-limit 19-tone scale;
25/24, 135/128, 16/15, 9/8, 75/64, 6/5, 5/4, 4/3, 27/20, 45/32, 3/2, 25/16, 8/5, 5/3, 27/16, 225/128, 9/5, 15/8, 2

PURE-22; 5-limit 22-tone scale;
25/24, 16/15, 27/25, 9/8, 75/64, 6/5, 5/4, 32/25, 125/96, 4/3, 25/18, 36/25, 3/2, 25/16, 8/5, 5/3, 125/72, 9/5, 15/8, 48/25, 125/64, 2

SCALATRON-19; Scalatron (tm) 19-tone scale, 1974;
25/24, 16/15, 9/8, 75/64, 6/5, 5/4, 125/96, 4/3, 45/32, 36/25, 3/2, 25/16, 8/5, 5/3, 225/128, 9/5, 15/8, 125/64, 2

SCHIDLOF-21; Schidlof;
81/80, 21/20, 15/14, 9/8, 7/6, 135/112, 100/81, 5/4, 4/3, 27/20, 7/5, 10/7, 3/2, 14/9, 45/28, 5/3, 7/4, 25/14, 50/27, 15/8, 2

SIMS-18; E. Sims' 18-tone mode;
25/24, 13/12, 9/8, 7/6, 29/24, 5/4, 21/16, 11/8, 23/16, 3/2, 25/16, 13/8, 27/16, 7/4, 29/16, 15/8, 31/16, 2

SIMS(NO.2)-20; Sims II;
33/32, 17/16, 35/32, 9/8, 37/32, 19/16, 39/32, 5/4, 21/16, 11/8, 23/16, 3/2, 25/16, 13/8, 27/16, 7/4, 29/16, 15/8, 31/16, 2

SIMS-24; See his article, Reflections on This and That, 1991 p.93-106;
33/32, 25/24, 17/16, 13/12, 35/32, 9/8, 37/32, 7/6, 19/16, 29/24, 39/32, 5/4, 21/16, 11/8, 23/16, 3/2, 25/16, 13/8, 27/16, 7/4, 29/16, 15/8, 31/16, 2

SMITH-19; Roger K. Smith, "Multitonic" scale, JI version;
21/20, 35/32, 9/8, 7/6, 6/5, 5/4, 21/16, 4/3, 7/5, 35/24, 3/2, 14/9, 8/5, 5/3, 7/4, 9/5, 28/15, 35/18, 2

SQUARE-13; Tonality square with generators 1, 3, 5, 7;
8/7, 7/6, 6/5, 5/4, 4/3, 7/5, 10/7, 3/2, 8/5, 5/3, 12/7, 7/4, 2

STEL/EIKO-70; Stellated Eikosany 3 out of 1 3 5 7 9 11;
385/384, 49/48, 45/44, 33/32, 25/24, 135/128, 77/72, 693/640, 35/32, 847/768, 495/448, 9/8, 1155/1 024, 55/48, 147/128, 297/256, 7/6, 75/64, 105/88, 77/64, 315/256, 99/80, 5/4, 121/96, 81/64, 245/192, 165/128, 21/16, 385/288, 693/512, 11/8, 539/384, 45/32, 363/256, 63/44, 275/192, 231/160, 35/24, 165/112, 189/128, 3/2, 385/256, 55/36, 49/32, 99/64, 25/16, 63/40, 605/384, 35/22, 77/48, 45/28, 105/64, 5/3, 27/16, 55/32, 7/4, 99/56, 385/216, 315/176, 231/128, 175/96, 11/6, 15/8, 121/64, 77/40, 495/256, 35/18, 55/28, 63/32, 2

STEL/HEX(NO.1)-14; Stellated two out of 1 3 5 7 hexany, also dekatesserany, mandala, tetradekany;
21/20, 15/14, 35/32, 9/8, 5/4, 21/16, 35/24, 3/2, 49/32, 25/16 105/64, 7/4, 15/8, 2

STEL/HEX(NO.3)-14; Stellated Tetrachordal Hexany based on Archytas's Enharmonic;
28/27, 16/15, 784/729, 448/405, 256/225, 35/27, 4/3, 48/35, 112/81, 64/45, 1792/1215, 224/135, 16/9, 2

STEL/HEX(NO.4)-14; Stellated Tetrachordal Hexany based on the 1/1 35/36 16/15 4/3 tetrachord;
36/35, 1296/1225, 16/15, 192/175, 256/225, 9/7, 4/3, 48/35, 112/81, 64/45, 256/175, 288/175, 16/9, 2

STONE-16; Tom Stone's Guitar Scale;
17/16, 9/8, 19/16, 5/4, 21/16, 11/8, 23/16, 3/2, 25/16, 13/8, 27/16 7/4, 29/16, 15/8, 31/16, 2

TETRAPHONIC-31; 31-tone Tetphonic Cycle, conjunctive form on 5/4, 6/5, 7/6 and 8/7;
50/49, 25/24, 50/47, 25/23, 10/9, 25/22, 50/43, 25/21, 50/41, 5/4 60/47, 30/23, 4/3, 15/11, 60/43, 10/7, 60/41, 3/2, 49/32, 147/94, 147/92, 49/30, 147/88, 147/86, 7/4, 84/47, 42/23, 28/15, 21/11, 84/43, 2

TONOS(DIA/PIS/15)-15; Diatonic PIS in the new Tonos-15;
11/10, 11/9, 11/8, 22/15, 22/13, 11/6, 2, 44/21, 11/5, 22/9, 11/4, 44/15, 44/13, 11/3, 4

TONOS(DIA/PIS/17)-15; Diatonic PIS in the new Tonos-17;
12/11, 6/5, 4/3, 24/17, 8/5, 24/13, 2, 48/23, 24/11, 12/5, 8/3, 48/17, 3/1, 24/7, 4

TONOS(DIA/PIS/19)-15; Diatonic PIS in the new Tonos-19;
14/13, 7/6, 14/11, 28/19, 14/9, 7/4, 2, 56/27, 28/13, 7/3, 28/11, 56/19, 28/9, 7/2, 4

TONOS(DIA/PIS/21)-15; Diatonic PIS in the new Tonos-21;
8/7, 16/13, 4/3, 32/21, 32/19, 16/9, 2, 32/15, 16/7, 32/13, 8/3, 64/21, 64/19, 32/9, 4

TONOS(DIA/PIS/23)-15; Diatonic PIS in the new Tonos-23;
9/8, 9/7, 18/13, 36/23, 12/7, 9/5 2/1, 36/17, 9/4, 18/7, 36/13, 72/23, 24/7, 18/5, 4

TONOS(DIA/PIS/25)-15; Diatonic PIS in the new Tonos-25;
9/8, 9/7, 18/13, 36/25, 18/11, 9/5, 2, 36/17, 9/4, 18/7, 36/13, 72/25, 36/11, 18/5, 4

TONOS(DIA/PIS/27)-15; Diatonic PIS in the new Tonos-27;
10/9, 5/4, 10/7, 40/27, 5/3, 40/21, 2, 40/19, 20/9, 5/2, 20/7, 80/27, 10/3, 80/21, 4

TONOS(DIA/PIS/29)-15; Diatonic PIS in the new Tonos-29;
11/10, 11/9, 11/8, 44/29, 22/13, 11/6, 2, 44/21, 11/5, 22/9, 11/4, 88/29, 44/13, 11/3, 4

TONOS(DIA/PIS/31)-15; Diatonic PIS in the new Tonos-31;
23/22, 23/20, 23/18, 46/31, 23/14, 23/13, 23/12, 2, 23/11, 23/10, 23/9, 92/31, 23/7, 46/13, 4

TONOS(DIA/PIS/31/NO.2)-15; Diatonic PIS in the new Tonos-31B;
12/11, 6/5, 4/3, 48/31, 12/7, 24/13, 2, 48/23, 24/11, 12/5, 8/3, 96/31, 24/7, 48/13, 4

TONOS(DIA/PIS/33)-15; Diatonic PIS in the new Tonos-33;
12/11, 6/5, 4/3, 16/11, 8/5, 16/9, 2, 48/23, 24/11, 12/5, 8/3, 32/11, 16/5, 32/9, 4

TRI(NO.1)-19; 3:5:7 Tritriadic 19-tone Matrix;
50/49, 36/35, 7/6, 25/21, 6/5, 60/49, 49/36, 25/18, 7/5, 10/7, 36/25 72/49, 49/30, 5/3, 42/25, 12/7, 35/18, 49/25, 2

TRI(NO.2)-19; 3:5:9 Tritriadic 19-tone Matrix;
27/25, 10/9, 9/8, 6/5, 100/81, 5/4, 4/3, 27/20, 25/18, 36/25, 40/27, 3/2, 8/5, 81/50, 5/3, 16/9, 9/5, 50/27, 2

TRI(NO.3)-19; 4:5:6 Tritriadic 19-tone Matrix;
25/24, 16/15, 10/9, 9/8, 6/5, 5/4, 32/25, 4/3, 25/18, 36/25, 3/2, 25/16, 8/5, 5/3, 16/9, 9/5, 15/8, 48/25, 2

TRI(NO.4)-19; 4:5:9 Tritriadic 19-tone Matrix;
81/80, 10/9, 9/8, 100/81, 5/4, 81/64, 32/25, 25/18, 45/32, 64/45, 36/25, 25/16, 128/81, 8/5, 81/50, 16/9, 9/5, 160/81, 2

TRI(NO.5)-19; 5:7:9 Tritriadic 19-tone Matrix;
50/49, 49/45, 10/9, 81/70, 98/81, 100/81, 63/50, 9/7, 7/5, 10/7, 14/9, 100/63, 81/50, 81/49, 140/81, 9/5, 90/49, 49/25, 2

TRI(NO.6)-19; 6:7:8 Tritriadic 19-tone Matrix;
49/48, 9/8, 8/7, 7/6, 9/7, 64/49, 21/16, 4/3, 49/36, 72/49, 3/2, 32/21, 49/32, 14/9, 12/7, 7/4, 16/9, 96/49, 2

TRI(NO.7)-19; 6:7:9 Tritriadic 19-tone Matrix;
28/27, 54/49, 9/8, 8/7, 7/6, 98/81, 9/7, 4/3, 49/36, 72/49, 3/2, 14/9, 81/49 12/7, 7/4 16/9, 49/27, 27/14, 2

TRI(NO.8)-19; 7:9:11 Tritriadic 19-tone Matrix;
99/98, 126/121, 81/77, 98/81, 11/9, 121/98, 14/11, 9/7, 162/121, 121/81, 14/9, 11/7, 196/121, 18/11, 81/49, 154/81, 121/63, 196/99, 2

TRI(NO.9)-19; 4:5:7 Tritriadic 19-tone Matrix;
50/49, 35/32, 28/25, 8/7, 49/40, 5/4, 32/25, 64/49 7/5, 10/7, 49/32, 25/16, 8/5, 80/49, 7/4, 25/14, 64/35, 49/25, 2

TRIAPHONIC-17; 17-tone Triaphonic Cycle, conjunctive form on 4/3, 7/6 and 9/7;
28/27, 14/13, 28/25, 7/6, 28/23, 14/11, 4/3, 112/81, 56/39, 112/75, 14/9, 21/13 42/25, 7/4, 42/23, 21/11, 2

UR-PARTCH-39; Ur-Partch curved keyboard, published in Interval;
49/48, 33/32, 22/21, 16/15, 12/11, 10/9, 9/8, 8/7, 7/6, 6/5, 11/9, 5/4, 14/11, 9/7, 21/16, 4/3, 15/11, 11/8, 7/5, 10/7, 16/11, 22/15, 3/2, 32/21, 14/9, 11/7, 8/5, 18/11, 5/3, 12/7, 7/4, 16/9, 9/5, 11/6, 15/8, 21/11, 64/33, 96/49, 2

VOGEL-21; Vogel's 21-tone Archytas system, c.f. Divisions of the tetrachord;
28/27, 16/15, 9/8, 7/6, 32/27, 6/5, 896/729, 512/405, 4/3, 112/81, 64/45, 3/2, 14/9, 128/81, 8/5, 3 584/2 187, 2 048/1 215, 16/9, 448/243, 256/135, 2

ZOOMOO-31; Zoomoozophone tuning based on Partch's. Base freq. 392 Hz;
16/15, 12/11, 11/10, 10/9, 9/8, 8/7, 7/6, 6/5, 11/9, 5/4, 14/11, 9/7, 4/3, 11/8, 7/5, 10/7, 16/11, 3/2, 14/9, 11/7, 8/5, 18/11, 5/3, 12/7, 7/4, 16/9, 9/5, 20/11, 11/6, 15/8, 2

Appendix B

Historical organs with subsemitones, 1468–1721

This Appendix consists of two parts:
B.1 Chronological overview,
B.2 Bibliographical sources

Entries marked with the sign ● refer to organs which either are extant or have been reconstructed with semitones. Country codes: A = Austria, CH = Switzerland, D = Germany, DK = Denmark, E = Spain, F = France, GB = Great Britain, I = Italy, NL = Netherlands, PL = Poland, S = Sweden

Acknowledgement   The author[1] would like to express his gratitude to all those who contributed in the most substantial way and sharing ideas to my article and list of organs in this book (For the catalogue of these organs with many details, distributions of subsemitones and other mechanical systems, including the present state, and historical information, c.f. the paper I. Ortgies:Subsemitones in organ bult between 1468 and 1721: Introduction and commentary with an annotated catalog. GOArt Research Report, vol. 3(2003), S. Jullander (ed.), GOArt Publication no. 10, 196 pp., Göteborg 2003): S. Bicknell, London (GB); D. de Boer, Nuenen (NL); F.-H. Gress, Dresden (D); J. den Hertog, Gouda (NL); M. Hochgartz, Münster (D); M. Kirnbauer, Basel (CH); A. Leuthold, Zürich (CH); U. Pape, Berlin (D); P. Peeters, Göteborg (S); W. Rehn, Männedorf, (CH); J. Speerstra, Göteborg (S); V. Timmer, Leek,(NL); R. Wilhelm, Braunschweig (D); D. Wraight, Coelbe (D).

---

[1]Ibo Ortgies, Göteborg University, Dept. of Musicology, GOArt, P. O. Box 200, SE-405 30 Göteborg, Sweden
This research is an output of the project of the Göteborg Organ Art Center, Göteborg University, Sweden.

B.1   Chronological overview

Place (country)	building	time	builder
Cesena (I)	Cathedral	1468	A. Molighi
Lucca (I)	Cathedral St. Martino	1480	D.(di Maestro Lorenzo)
Padua (I)	Basilica St. Antonio	1489	A. Dilmani
unknown (D)	unknown	1499–1511?	2 unknown builders
Venice ? (I)	St. Sofia ?	bef. 1521	?
Bologna (I)•	St. Petronio	1528–1531	G. B. Facchetti
Bologna (I)	Cathedral	1532	G. B. Facchetti
Arrezzo (I)•	Cathedral	1534	L. da Cortona
Granada (E)	Cathedral	1542	M. Hernández
Cremona (I)	Cathedral	1544	G. B. Facchetti
Florence (I)	Cathedral	1546	B. di Argenta
Florence (I)	St. Maria Novella	before 1563	?
Mantua (I)•	St. Barbara	1565	G. Antegnati
Florence (I)	Cathedral	1567	?
Granada (E)	Cathedral	1568	F. Vázquez
Florence (I)	St. Trinita	1571	O. Zeffirini
Granada (E)	Cathedral	1578	D. L. and J. P. de Sanforte
Palermo (I)	Cathedral	1593	R. la Valle
Palermo (I)	St. Martino delle Scale	1594	R. la Valle
Bologna (I)•	St. Petronio	1596	B. Malamini
Ferrara (I)	Castello Estense	before 1598	?
Rome (I)•	St. Giovanni in Laterano	1598–1599	L. Blasi
Genua (I)	Cathedral before	1604	G.&G. Angelo Vitani of Pavia
Mantua (I)	St. Andrea	1604	G.&G. Angelo Vitani of Pavia
Verona (I)	Accad. Filarmonica	1604	B. Virchi
Milan (I)	St. Marco	1604–1611	C. Antegnati
Lucca (I)	Accad. di T. Raffaelli	before 1609	A. Lucchese
Dresden (D)•	Schloßkirche	1612	G. Fritzsche
Bückeburg (D)	Stadtkirche	1615	E. Compenius
Schöningen (D)	Schloßkapelle	1616	G. Fritzsche
Sondershausen (D)	Dreifaltigkeitskirche	1616–1617	G. Fritzsche
Wolfenbüttel (D)	Hauptkirche BMV	1620–1624	G. Fritzsche
Bayreuth (D)	Stadtkirche	1618–1619	G. Fritzsche
Braunschweig (D)	St. Katharinenkirche	1621–1623	G. Fritzsche
Stockholm (S)	Tyska Kyrkan	1625–1628	G. Herman and P. Eisenmenger

Braunschweig (D)	St. Ulrici	1626	G. Fritzsche
Sønderborg (DK)•	Slotskapell	1626	Bartolomäus [Zencker]
Rome (I)	St. Trinita . dei Pellegrini	1627	P. di Michelangelo D. da Mombaroccio
Regensburg (D)	Stadtmuseum	1627	St. Cuntz
Hamburg (D)	St. Maria Magdalena	1629	G. Fritzsche
Braunschweig (D)	St. Martini	1629–1630	G.Fritzsche
Hamburg (D)	St. Katharinen	1631–1633	G. Fritzsche
Paris (F)	St. Nicolas des Champs	1632–1636	C. Carliers
's-Hertogenbosch (NL)	St. Jan	1634–1636	G. & G. van Hagerbeer
Hamburg (D)	St. Jakobi	1635–1636	G. Fritzsche
Amersfoort (NL)	St. Joris	1636	G.&G. van Hagerbeer
Caltanisetta (I)	Chiesa Madre	1638	A. R. la Valleson
Sciacca (I)	St. Margherita	1639	G. Sutera & V. Monteleone
The Hague (NL) /s'Gravenhage	Hofkapel	1641	G. & G. van Hagerbeer
Wien (A)	Franziskanerkirche	1642	J. Wückherl
Plauen (D)	St. Johannis	1642(1651?)	J. Schedlich
Alkmaar (NL)	St. Laurenskerk	1643–45/46	G.& G. van Hagerbeer
Skara (S)•	Länsmuseum	1643–1651	N. Manderscheidt
Bamberg (D)•	St. Gangolf	1644	N. Manderscheidt
Ostheim v.d.Rhön (D)	Orgelbaumuseum	1646	N. Manderscheidt
Finale Emila (I)	Chiesa del Rosario	1647	A. Colonna
Schöningen (D)	St. Vincenz	1647	J. Weigel
Lübeck (D)	St Aegiden	1645–1648	F. Stellwagen
Annabg.-Buchholz (D)	St. Annen	1652	J. Schedlich
Luzern (CH)	St. Maria in der Au	1653	S. Manderscheidt
Fribourg (CH)•	St. Nicolas	1654–1657	S. Manderscheidt
Weimar (D)	Schloßkirche	1657–1658	L. Compenius
Rome (I)	St. Agostino	1658	G. Catarinozzi & G. Testa
Alcalá (E)	St. Diego	1659	J. de Hechabarría
Breslau/Wroclaw (PL)	St. Vinzenz	ca. 1660	J. C. B. Waldhauser
Trapani (I)	Monastery of St. Andrea	1665	A.&G. de Simone, F. Romano
Fribourg (CH)•	Hôpital des Bourgeois	1675	S. Manderscheidt
Gävle (S)	Heliga Trefaldighets Kyrka	ca.1670	S. Manderscheidt
Durham (GB)	Cathedral	1684–1685	B. Smith

London (GB)	Temple Church	1683–1687	B. Smith
Tolosa (E)	St. María	1693-1694	J. de Hechabarría
Frankfurt/O. (D)	St. Marienkirche	1691–1695	M. Schurig
Stockholm (S)	Storkyrkan	1690–98	G. Woitzig
Uppsala (S)●	Cathedral	1710	I. Risberg
St. Urban (CH)●	Klosterkirche	1716–1721	J.& V. F. Bossart

B.2   Bibliographical sources

● van Biezen, J.: Het Nederlandse orgel in de Renaissance en de Barok, in het bijzonder de school van Jan van Covelens. Deel 1. (Utrecht) 1995. (= Gierveld, G. J. (ed.): Muziekhistorische Monografieèn 14.)

● Burgemeister, L.: Der Orgelbau in Schlesien. bearb. v. Hermann Josef Busch, D. Großmann und R. Walter. Frankfurt 1973.

● Clutton, C.—Niland, A.: The British Organ. London 1982.

● Dähnert, U.: Historische Orgeln in Sachsen. Ein Orgelinventar. Frankfurt 1980. (= 70. GdO-Veröffentlichung.)

● Dupont, W.: Geschichte der musikalischen Temperatur. (Diss.). Nördlingen 1935. (Reprint: Lauffen/Neckar 1986.)

● Erici, E.: Orgelinventarium (ed. by Axel Unnerbäck). Stockholm 1988.

● Fock, G.: Arp Schnitger und seine Schule. Kassel 1974.

● Freeman, A.: Father Smith otherwise Bernhard Schmidt, being an Account of a Seventeenth Century Organ Maker. London 1926. (Facs. ed. by John Rowntree. Oxford 1977.)

● Gress, F.-H.: Die Gottfried-Fritzsche-Orgel der Dresdner Hofkapelle. Untersuchungen zur Rekonstruktion ihres Klangbildes. In: Alfred Reichling (ed.): Acta Organologica. Bd. 23. Kassel, 1993. (=146. Veröffentlichung der Gesellschaft der Orgelfreunde).

● Gurlitt, W.: Zwei archivalische Beiträge zur Geschichte des Orgelbaus in Braunschweig aus den Jahren 1626 und 1631. In: Braunschweigisches Magazin, 1913.

● Jeppesen, K.: Die italienische Orgelmusik am Anfang des Cinquecento. Bd. 1, Kopenhagen, 1943, 1960.

● Ruiz Jiménez, J.: Organería en la Diócesis de Granada 1492–1625. Granada 1995.

● Kjersgaard, M.—Beerståhl, N. F.: Bjurumsorgeln. In: Västergötlands Fornminnesförenings Tidskrift 1973–1974, pp. 213–258.

● Kjersgaard, M.: Dokumentation af et Manderscheidt-positiv. In: Dansk Orgelaarbog 1985–1986. København 1987, pp. 132–175.

● Klotz, H.: Über die Orgelkunst der Gotik der Renaissance und des Barock. Kassel 1975.

● Lade, G.: Orgeln in Wien. Wien 1990.

● Lindley, M.: Stimmung und Temperatur. (in Zaminer Frieder (ed.): Geschichte der Musiktheorie, Band 6. Darmstadt 1987.)

● Lunelli, R.: Der Orgelbau in Italien in seinen Meisterwerken vom 14. Jhd. bis zur Gegenwart. Mainz 1956. (German translation by C. Elis and P. Smets.)

- Meister, W. T.: Die Orgelstimmung in Italien und Süddeutschland vom 14. bis zum Ende des 18. Jahrhunderts. Lauffen am Neckar, 1991.
- van Nieuwkoop, H. (ed.): Het Historische Orgel in Nederland 1479–1725. Amsterdam 1997.
- Padgham, Ch.: The well-tempered Organ. Oxford 1986.
- Panetta, V. J., Jr. (ed./transl.): Treatise on Harpsichord Tuning by Jean Denis. Cambridge 1987.
- Pape, U.: Die Orgeln der Stadt Braunschweig. Wolfenbüttel 1966. (= Norddeutsche Orgeln, Heft 2.)
- Pape, U.: Die Orgeln der Stadt Wolfenbüttel. Berlin 1973.
- Praetorius, M.: Syntagmatis Musici Tomus Secundus. De Organographia. Wolfenbüttel 1619. Facsimile ed. by W. Gurlitt. Kassel 1958, 1980. (= Documenta Musicologica Band 1, XIV).
- Ratte, F. J.: Die Temperatur der Clavierinstrumente. Kassel 1991.
- Reuter, R.: Orgeln in Spanien. Kassel 1986.
- Schindler, J.-P.: Der Nürnberger Orgelbau des 17.Jahrhunderts. Michaelstein/Blankenburg 1991. = Thom, Eitelfriedrich (ed.): Sonderbeitrag 10. des Instituts für Aufführungspraxis der Musik des 18. Jahrhunderts.
- Schindler, J.-P.: Die Nürnberger Stadtorgelmacher und ihre Instrumente. Nürnberg 1995. (=Verlag des Germanischen Nationalmuseums. Ausstellungskatalog).
- Schlick, A.: Spiegel der Orgelmacher und Organisten. Mainz 1511. Facsimile: Transcription and notes by Elizabeth Berry Barber. Buren 1980.
- Schneider, T.: Die Orgelbauerfamilie Compenius. (Diss.). Berlin, 1937. (= Archiv für Musikforschung 2(1937).)
- Seydoux, F.: Die abenteurliche Odyssee eines bedeutsamen Instruments oder Das Freiburger Pedalpositiv von Sebald Manderscheidt aus dem Jahre 1667. In: Pio Pellizzari (ed.): Musicus Perfectus. Studio in onore di Luigi Ferdinando Tagliavini "prattico & specolativo" nella ricorrenza del LXV° compleanno. Bologna 1995. pp. 49–106.
- Seydoux, F.: L'Orgue de Chœer de la Cathédrale de St.Nicolas. In: Cathédrale St.Nicola Fribourg. Inauguration de L'Orgue de Chœer restauré. 1999. pp. 14–34.
- Seydoux, F.: L'Orgue de Chœer de la Cathédrale de St.Nicolas, Fribourg. In: La Tribune de l'Orgue 1/51. March 1999. pp. 4–12.
- Smets, P. (ed.): Orgeldispositionen. Eine Handschrift aus dem XVIII. Jahrhundert, im Besitz der Sächsischen Landesbibliothek, Dresden. Kassel 1931.
- Stembridge, Ch.: The Cimbalo Cromatico and other Italian Keyboard Instruments with Nineteen or More Divisions to the Octave. In: Performance Practice Review 6. 1993. pp. 33–59.
- Wraight, D.—Stembridge, Ch.: Italian Split-keyed Instruments with Fewer than Nineteen divisions to the Octave. In: Performance Practice Review 7. 1994. pp. 150-181. [In this article the part dealing with "Italian Organs with Split keys" has been written by Ch. Stembridge.]
- Stembridge, Ch.: Italian organ music to Frescobaldi. In: Nicholas Thistlethwaite and Geoffrey Webber: The Cambridge Companion to the Organ. Cam-

bridge 1998. pp. 148–163.
- Tagliavini, L. F.: Notes on Tuning Methods in Fifteenth-Century Italy. In: Fenner Douglass et. al. (eds.): Charles Brenton Fisk. Organ Builder. Volume One. (Essays in his Honor). Easthampton (Mass.) 1986. pp. 191–199.
- L'orgue espagnol de l'église Saint-Laurent à Lausanne. In: La Tribune de l'Orgue, N° 4, December 1991, pp.27–28.
- Unnerbäck, A.: The Organ in the German Church - The organ in Övertorneå. In: Ericsson, L. W. (ed.): Övertorneåprojektet. Luleå 1997.
- Vente, M. A.: Die Brabanter Orgel. Amsterdam 1963.
- Vogel, H.: Zur Stimmung der Orgel in der Deutschen Kirche in Stockholm. In: Ericsson, L. W. (ed.): Övertorneåprojektet. Luleå (1997).
- Ericsson, L. W. (ed.): Övertorneåprojektet. Om dokumentationen av orgeln i Övertorneå och rekonstruktionen av 1684 års orgel i Tyska kyrkan. Luleå 1997.
- Rüdiger W.: Die kürzlich wiederentdeckten Orgeltabulaturen in Wolfenbüttel. In: Hans Davidsson and Sverker Jullander (ed.): Proceedings of the Göteborg International Organ Academy 1994. Göteborg 1995. (= Skrifter från Musikvetenskapliga avdelningen, Göteborgs universitet, no. 39, 1995).
- Williams, P.: The European Organ 1450–1850. London 1966, 1982.

Index